兽医微生物学精要
（双语版）

Essentials of Veterinary Microbiology
(Bilingual Edition)

宋厚辉　杨永春　主编

科学出版社
北京

内 容 简 介

本书为中英双语，内容包括细菌学和病毒学两大部分。兽医微生物学领域的发展日新月异，本书对微生物的分类命名、分子微生物学、流行病学和诊断方法等内容进行了更新。为了进一步贯彻落实乡村振兴战略，讲好"三农"故事，本书的相关章节中引入了抗菌药物耐药性、兽医公共卫生、绿色生态低碳健康养殖等课程思政内容。本书聚焦兽医学和畜牧学领域，以简洁明了的形式介绍了微生物形态学、遗传学、耐药性、免疫学、生物安全和疫病防控等内容。兽医微生物学的最新进展也在相关章节进行了体现。本书还提供了在线课程、彩图、视频和习题等数字资源，以方便读者阅读理解。

本书是一部新形态教材，可作为动物医学和动物科学专业学生以及国际留学生了解兽医微生物及其所致疾病的参考用书。

图书在版编目（CIP）数据

兽医微生物学精要：汉、英／宋厚辉，杨永春主编．—北京：科学出版社，2023.6

科学出版社"十四五"普通高等教育本科规划教材　浙江省普通高校"十三五"新形态教材　国家级一流本科课程配套教材

ISBN 978-7-03-074505-7

Ⅰ.①兽…　Ⅱ.①宋…　②杨…　Ⅲ.①兽医学－微生物学－教材－汉、英　Ⅳ.①S852.6

中国版本图书馆CIP数据核字（2022）第253341号

责任编辑：林梦阳／责任校对：宁辉彩
责任印制：赵　博／封面设计：蓝正设计

科 学 出 版 社 出版
北京东黄城根北街16号
邮政编码：100717
http://www.sciencep.com

北京市金木堂数码科技有限公司印刷
科学出版社发行　各地新华书店经销

＊

2023年6月第　一　版　开本：787×1092 1/16
2025年5月第三次印刷　印张：17 1/2
字数：420 000

定价：79.80 元

（如有印装质量问题，我社负责调换）

编写人员名单
Editorial Committee

前　言

　　岁月荏苒，从浙江农林大学动物科技学院和动物医学院筹建并成立以来，不知不觉担任"兽医微生物学"主讲教师已有十余载，流光容易把人抛，红了樱桃，绿了芭蕉。学院管理工作固然繁忙，但担任"兽医微生物学"主讲教师仍其乐无穷，这也鞭策我每次讲授务求精准。备课虽然难免有些紧张，需要尽可能体现兽医微生物领域的最新进展，同时不忘PPT美化再美化。教学相长，教学本身也是视觉的呈现过程，否则会被学生耻笑了去。每次阅读文献时，也不忘告诫自己：疏于专业的学院领导是多么愚昧而无知，笃学之，慎思之，明德之，尚行之。为天地立心，为生民立命，为往圣继绝学，为万世开太平，且行且珍惜。

　　"兽医微生物学"博大精深，每年医学和兽医学领域的进展层出不穷，在国际上已经发展为同一医学（One Medicine）。微生物的分类也不断变化，基因型和血清型也在逐年增多，若干年后个别成员可能已另立新科或新属。随着结构生物学的发展，微生物的结构已能清晰呈现，但新的科学问题又随之而生，比如非洲猪瘟目前尚无有效的疫苗可用。多数高等院校动物医学专业逐渐从4年制变更为5年制，尤其近年来随着伴侣动物医学的发展，一些病原微生物在小动物临床医学中的重要性也越来越凸显，但由于课时所限，终不能在课堂上尽显"兽医微生物学"之美，经常顾此失彼，挂一漏万。

　　在兽医教育学家陆承平教授的醍醐灌顶教导之下，在本课程教学团队诸位同仁的辛苦努力之下，浙江农林大学"兽医微生物学"课程先后获批浙江省一流本科课程和国家级一流本科课程，课程团队获得了浙江省教师教学比赛一等奖。虽然团队成员在教学方面均有所成长，但总觉得本教学团队还是缺少了点什么。若能将平时在课程上讲授的心得体会、最新进展、录制的课堂视频、线上课程等呈现在《兽医微生物学精要》（双语版）教材中，岂不更妙。于是乎，本书主编与科学出版社高教农林生物分社周万灏社长进行了积极有效的沟通。最终，本教材也得到了浙江省高等教育学会的支持，列为浙江省普通高校"十三五"新形态教材建设项目。遂联系中国农业大学、南京农业大学、华中农业大学、浙江大学、扬州大学、华南农业大学、贵州大学、安徽农业大学、河南农业大学、河南科技大学、青岛农业大学、聊城大学、龙岩学院等13所高校兽医微生物学教研室诸位同仁，启动了《兽医微生物学精要》（双语版）新形态教材的编写工作。考虑到浙江农林大学已有外国留学生考取动物医学专业，本书的语言定为中英双语。

　　本书借鉴了南京农业大学陆承平教授和刘永杰教授主编的《兽医微生物学》（第六版）编写体例，化繁为简，中英对照。编委们分工如下：宋厚辉负责绪论、第1章、第2章、第3章、第4章，以及全书所有章节图表的收集整理、视频拍摄和初稿校对工作；杨杨负责第5章；管迟瑜负责第6章；杨永春负责第7章、第8章；孙静负责第9章；程昌勇负责第10章、第13章；夏菁负责第11章、第12章；尹会方负责第14章、第17章；陈颖钰负责第15章、第28章；刘文华负责第16章；司振书负责第18章、第26章；韩军负责第19章；王永强负责第20章、第21章、第34章；王桂军负责第22章、第23章；钱琨负责第24章；温贵兰负责第25章；何雷负责第27章、第29章、第30章；曹伟胜负责第31章；王新卫负责第32章、第33章。林梦阳为本书责任编辑。管迟瑜、曾欢、宋厚辉、方维焕、蔡畅、刘萍、周兴东负责全书的英文校对工作。

陆承平和方维焕担任主审。两位先生为人清正、严谨治学；在校稿过程中逐字逐句、逐一求证、精雕细琢、精益求精；先生之风，山高水长，虽不能至，心向往之。

由于编者水平有限，几易其稿，成书仓促，缺点和问题在所难免，请广大师生批评指正，意见请反馈至浙江农林大学。也希望本书每隔3～5年修订一次，及时更新和勘误，以反映本学科的最新进展。

主编

宋厚辉

2023年6月

Preface

As time goes by, I have been lecturing Veterinary Microbiology for more than ten years since the establishment of the College of Animal Science and Technology, and the College of Veterinary Medicine of Zhejiang A&F University. *Time and tide will not wait for a man forlorn: with cherry red spring dies, when green banana sighs*. Although the daily management of the college is busy, it is still a lot of fun to be a teacher of Veterinary Microbiology, which also motivates me to be precise in every class. My time is always precious in the preparation of lessons since I would like to deliver the latest advances in veterinary microbiology together with the best designed PowerPoint slides. Teaching benefits students as well as teachers. Teaching itself is also a visual presentation process; otherwise it will be ridiculed by students. When reading the literatures, I always remind myself that a college leader who neglects his/her profession is ignorant, as *Analects of Confucius* mentioned study extensively, reflect carefully, discriminate clearly, practice earnestly. Chinese philosopher and politician Zhang Zai in Song dynasty said: *to ordain conscience for Heaven and Earth, to secure life and fortune for the people, to continue lost teachings for past sages, to establish peace for all future generations. Treasuring while you are behaving*.

Every year, progress in the fields of medicine and veterinary medicine emerges in an endless stream, and it has developed into One Medicine internationally. Veterinary Microbiology is broad and profound. The classification of microorganisms is constantly changing, and the genotypes and serotypes are increasing year by year. After a few years, individual members may have established new families or new genera. With the development of structural biology, the structure of microorganisms has been clearly presented. But new scientific problems have also arisen. For example, there is no effective vaccine for African swine fever currently. Most of the universities in China have gradually changed Veterinary Medicine from a 4-year degree to a 5-year degree. In recent years, with the development of companion animal medicine, the importance of some pathogenic microorganisms in small animal clinics has become more and more prominent. However, due to the limitation of class hours, the beauty of Veterinary Microbiology cannot be fully displayed, and one often misses the other. I always attended to one thing but lost the sight of others, just like the old saying, *for one thing cited, ten thousand may have been left out*.

Under the inspiring teaching of Professor Lu Chengping, a prominent veterinary educator, and the hard work of colleagues in the teaching team of this course, the course Veterinary Microbiology was given the first-class award by Zhejiang province and lectures won the first prize in the teaching competition as well. This course was approved as a national first-class undergraduate course. Although the team members kept growing during the teaching, I always feel that there is something still missing. It would be great to add the lecturers' experiences, latest advances, recorded videos, online courses, *etc.* to the book. After an active and effective communication with Zhou Wanhao, the director of the Higher Education Agriculture, Forestry, and Biology Division of Science Press, this book gained the support from Zhejiang Province Association for Higher Education. It was also listed in the 13th Five-Year Plan New Form Textbook Project for Colleges and Universities of Zhejiang province. Veterinary Microbiology professionals in 13 universities were contacted to start the editing of this book, including China Agricultural University, Nanjing Agricultural University, Huazhong Agricultural University, Zhejiang University, Yangzhou University, South China Agricultural University, Guizhou University, Anhui Agricultural University, Henan Agricultural University, Henan University of Science and Technology, Qingdao Agricultural University, Liaocheng University, Longyan University, *etc.* Considering that there are already foreign students admitted to the College of Veterinary Medicine at Zhejiang A&F University, the language of this book is set to be bilingual in both Chinese and English.

This book followed the structure of *Veterinary Microbiology* (Sixth Edition) edited by Professor Lu Chengping and Professor Liu Yongjie in Nanjing Agricultural University, with simplified content in both Chinese and English. Song Houhui is responsible for the collection and organization of pictures in all chapters, video recording and proofreading of the first draft and the writing of Introduction, Chapter 1, Chapter 2, Chapter 3, Chapter 4. Yang Yang wrote Chapter 5. Guan Chiyu wrote Chapter 6. Yang Yongchun wrote Chapter 7 and Chapter 8. Sun Jing wrote Chapter 9. Cheng Changyong wrote Chapter 10 and Chapter 13. Xia Jing wrote Chapter 11 and Chapter 12. Yin Huifang wrote Chapter 14 and Chapter 17. Chen Yingyu wrote Chapter 15 and Chapter 28. Liu Wenhua wrote Chapter 16. Si Zhenshu wrote Chapter 18 and Chapter 26. Han Jun wrote Chapter 19. Wang Yongqiang wrote Chapter 20, Chapter 21 and Chapter 34. Wang Guijun wrote Chapter 22 and Chapter 23. Qian Kun wrote Chapter 24. Wen Guilan wrote Chapter 25. He Lei wrote Chapter 27, Chapter 29 and Chapter 30. Cao Weisheng wrote Chapter 31. Wang Xinwei wrote Chapter 32 and Chapter 33. Managing editor is Lin Mengyang. Guan Chiyu, Zeng Huan, Song Houhui, Fang Weihuan, Cai Chang, Liu Ping and Zhou Xingdong did the English proofreading of the book.

Lu Chengping and Fang Weihuan are the Reviewers-in-Chief. In the process of proofreading, the two Reviewers-in-Chief are rigorous in their academic studies and upright. They sought evidence for each point and looked up for confirmation carefully. Their temperament is as elegant as lofty mountain and flowing river. I do not think I can reach their highness, but my heart yearns for it.

Due to the limited knowledge of editors, the book was revised a few times and rushed to be published. Shortcomings and problems are inevitable in the book. We wish readers can point out our mistakes and we will correct them. Please provide your precious feedback to Zhejiang A&F University. We also hope that this book could be revised every three to five years with timely updates and errata to reflect the latest progress in the discipline.

Editor-in-Chief

Song Houhui

July 2023

"中科云教育"平台数字化课程登录路径

- 电脑端

  注册、登录"中科云教育"平台（www.coursegate.cn），搜索"兽医微生物学精要"并报名学习，可查看各种数字资源。

- 手机端

  扫描课程码，报名学习，注册后登录，再次扫描本课程码，进入课程，可查看各种数字资源。

兽医微生物学精要课程码

授课视频

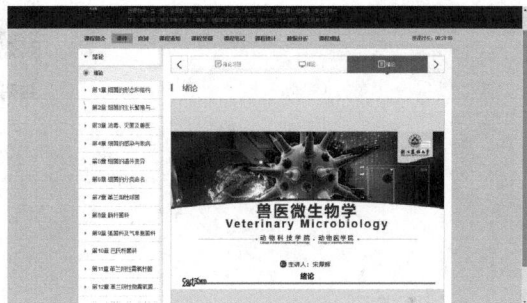

课件PPT

一、单项选择题（共8题）

1. 布尼亚病毒的核酸类型是怎样的？ What is the nucleic acid type of Bunyaviruses? (2分)

   ○ A.双链RNA double stranded RNA

   ◉ B.分节段的负链RNA segmented negative strand RNA

   ○ C.双链DNA double stranded DNA

   ○ D.单链正链RNA Single positive RNA

2. 以下关于禽流感病毒的叙述哪项是不正确的？ Which of the following statements about Avian influenza virus is not true? (2分)

   ◉ A.双链DNA DNA double stranded DNA

   ○ B.具有囊膜 have envelope

   ○ C.基因组为8个节段负链的RNA The genome of the virus contains eight segments of negative-sense RNA

   ○ D.具有血凝作用 It has haemagglutination

3. 禽流感病毒的核酸类型是怎样的？ What is the nucleic acid type of avian influenza virus? (2分)

   ○ A.双链DNA double stranded DNA

   ○ B.双链RNA double stranded RNA

   ○ C.分节段的负链RNA segmented negative strand RNA

   ○ D.单链正链RNA Single positive RNA

学习自测

《兽医微生物学精要（双语版）》数字教材入口

- 电脑端

  登录科学出版社数字教材平台（https://book.coursegate.cn/index），搜索"兽医微生物学精要（双语版）"，进入教材主页。

- 手机端

  扫描二维码进入教材主页。

扫码阅读

目　录

前言

绪论 ·· 1

第1节　兽医微生物学发展历史 ········ 1

第2节　兽医微生物学 ················· 3

第1章　细菌的形态和结构 ··········· 5

第1节　细菌的大小和形态 ············ 5

第2节　细菌的基本结构和特殊

结构 ························· 7

第3节　革兰氏染色 ·················· 8

第4节　细菌细胞壁的主要功能 ········ 9

第2章　细菌的生长繁殖与生态 ····· 13

第1节　细菌的代谢过程 ·············· 13

第2节　细菌的生长繁殖 ·············· 16

第3节　细菌的人工培养 ·············· 17

第4节　细菌群落生长的调控 ·········· 18

第5节　细菌的生态 ·················· 20

第3章　消毒、灭菌及兽医微生物

实验室生物安全 ··········· 23

第1节　物理消毒法 ·················· 23

第2节　化学消毒法 ·················· 26

第3节　抗生素和细菌素 ·············· 27

第4节　兽医微生物实验室生物

安全 ························· 27

第4章　细菌的感染与致病机理 ····· 30

第1节　细菌的致病性和毒力 ·········· 30

第2节　细菌的毒力因子及分泌

系统 ························· 33

第3节　机会致病菌 ·················· 36

第5章　细菌的遗传变异 ············ 38

第1节　细菌常见的变异现象 ·········· 38

第2节　细菌遗传的物质基础 ·········· 38

第3节　细菌变异的机制 ·············· 41

第6章　细菌的分类命名 ············ 44

第1节　细菌的分类地位 ·············· 44

第2节　细菌的命名 ·················· 45

第3节　细菌分类鉴定的标准 ·········· 46

第4节　细菌的鉴定程序 ·············· 46

第7章　革兰氏阳性球菌 ············ 48

第1节　葡萄球菌属 ·················· 48

第2节　链球菌属 ···················· 50

第8章　肠杆菌科 ··················· 53

第1节　埃希菌属 ···················· 53

第2节　沙门菌属 ···················· 56

第3节　耶尔森菌属 ·················· 57

第9章　弧菌科及气单胞菌科 ······· 59

第1节　弧菌科 ······················ 59

第2节　气单胞菌科 ·················· 61

第10章　巴氏杆菌科和黄杆菌科 ···· 63

第1节　巴氏杆菌属 ·················· 63

第2节　曼氏杆菌属 ·················· 65

第3节　放线杆菌属 ·················· 66

第4节　格拉菌属 ···················· 67

第5节　嗜血杆菌属 ·················· 70

第6节　禽杆菌属 ···················· 70

第7节　里氏杆菌属 ················ 72

第11章　革兰氏阴性需氧杆菌 ····· 73
　第1节　布氏杆菌属 ················ 73
　第2节　假单胞菌属 ················ 76
　第3节　伯克霍尔德菌属 ············ 77
　第4节　博德特菌属 ················ 78
　第5节　弗朗西斯菌属 ·············· 79
　第6节　不动杆菌属 ················ 80

第12章　革兰氏阴性微需氧菌和厌氧菌 ··· 83
　第1节　弯曲菌属 ·················· 83
　第2节　螺杆菌属 ·················· 84

第13章　革兰氏阳性无芽孢杆菌 ··· 86
　第1节　李氏杆菌属 ················ 86
　第2节　丹毒丝菌属 ················ 89

第14章　革兰氏阳性产芽孢杆菌 ··· 91
　第1节　芽孢杆菌属 ················ 91
　第2节　梭菌属 ···················· 94

第15章　分枝杆菌属及相似属 ····· 98
　第1节　分枝杆菌属 ················ 98
　第2节　放线菌属 ·················· 101
　第3节　嗜皮菌属 ·················· 103

第16章　螺旋体 ···················· 106
　第1节　螺旋体的一般特性 ·········· 106
　第2节　疏螺旋体属 ················ 108
　第3节　短螺旋体属 ················ 109
　第4节　钩端螺旋体属 ·············· 110

第17章　支原体 ···················· 113
　第1节　支原体的一般特性 ·········· 113
　第2节　猪的支原体 ················ 116
　第3节　禽的支原体 ················ 119
　第4节　牛羊的支原体 ·············· 121
　第5节　嗜血支原体 ················ 122

第18章　立克次体和衣原体 ········ 125
　第1节　立克次体 ·················· 125
　第2节　衣原体 ···················· 129

第19章　病毒的结构和分类 ········ 132
　第1节　病毒的结构特征 ············ 132
　第2节　病毒的化学组成 ············ 135
　第3节　病毒的分类 ················ 136

第20章　病毒的复制 ················ 139
　第1节　一步生长曲线 ·············· 139
　第2节　吸附、穿入与脱壳 ·········· 140
　第3节　生物合成 ·················· 142
　第4节　组装与释放 ················ 143

第21章　病毒的变异和演化 ········ 144
　第1节　突变 ······················ 144
　第2节　诱变 ······················ 146
　第3节　基因重组 ·················· 147
　第4节　病毒组分之间的相互作用 ···· 149
　第5节　遗传变异与病毒演化 ········ 149

第22章　病毒与细胞的相互作用 ···· 153
　第1节　病毒的细胞培养 ············ 153
　第2节　病毒与细胞的相互作用 ······ 155
　第3节　病毒引致的非杀细胞变化 ···· 158

第23章　病毒的致病机理 ·········· 159
　第1节　病毒的入侵、扩散和排放 ···· 159
　第2节　病毒的持续性感染 ·········· 161
　第3节　病毒感染对宿主组织和
　　　　　器官的损伤 ················ 162
　第4节　病毒感染对免疫系统的
　　　　　损伤 ······················ 163

第24章　病毒的检测 ················ 166
　第1节　病毒的分离和鉴定 ·········· 166
　第2节　病毒滴度的测定 ············ 169
　第3节　病毒颗粒的检测 ············ 170
　第4节　病毒感染的血清学检测 ······ 170
　第5节　病毒核酸的检测 ············ 174

第25章　双链DNA病毒 ·············· 176
　第1节　痘病毒科 ·················· 176
　第2节　非洲猪瘟病毒科 ············ 179

第3节　腺病毒科 ················183

第26章　疱疹病毒目 ·············187
　第1节　疱疹病毒科的主要特征 ·····187
　第2节　牛传染性鼻气管炎病毒 ·····189
　第3节　伪狂犬病毒 ············190
　第4节　禽传染性喉气管炎病毒 ······191
　第5节　马立克病病毒 ··········192
　第6节　鸭瘟病毒 ············193

第27章　单链DNA病毒 ···········195
　第1节　细小病毒科 ············195
　第2节　圆环病毒科 ············200
　第3节　细环病毒科 ············202

第28章　逆转录病毒 ·············204
　第1节　逆转录病毒科的
　　　　主要特征 ············204
　第2节　甲型逆转录病毒属 ········205
　第3节　乙型逆转录病毒属 ········207
　第4节　丙型逆转录病毒属 ········209
　第5节　丁型逆转录病毒属 ········210
　第6节　慢病毒属 ············211

第29章　双链RNA病毒 ···········215
　第1节　呼肠孤病毒科 ··········215

第2节　双RNA病毒科 ··········218

第30章　单负链病毒目 ···········221
　第1节　副黏病毒科 ············221
　第2节　弹状病毒科 ············225
　第3节　波纳病毒科 ············226

第31章　分节段的负链RNA病毒 ·····228
　第1节　正黏病毒科 ············228
　第2节　近布尼亚病毒科 ·········231
　第3节　白纤病毒科 ············232
　第4节　内罗病毒科 ············234
　第5节　汉坦病毒科 ············234

第32章　套式病毒目 ·············236
　第1节　冠状病毒科 ············236
　第2节　动脉炎病毒科 ··········245
　第3节　托巴套式病毒科 ·········247

第33章　微RNA病毒目 ···········248
　第1节　微RNA病毒科 ··········248
　第2节　嵌杯病毒科 ············253

第34章　其他正链RNA病毒 ········256
　第1节　戊肝病毒科 ············256
　第2节　披膜病毒科 ············256
　第3节　黄病毒科 ············258

Contents

Preface

Introduction ·································· 1

    Section 1   History of veterinary
             microbiology ····················· 1

    Section 2   Veterinary microbiology ······· 3

Chapter 1   Bacterial structure and
               morphology ················· 5

    Section 1   Bacterial size and
              morphology ····················· 5

    Section 2   Basic and special structures
              of bacteria ····················· 7

    Section 3   Gram staining ················ 8

    Section 4   The main functions of
              bacterial cell walls ··············· 9

Chapter 2   Bacterial growth and
               ecology ····················· 13

    Section 1   Bacterial metabolism ·········· 13

    Section 2   Growth of bacteria ·············· 16

    Section 3   Cultivation of bacteria ········· 17

    Section 4   Regulation of bacterial
              growth ························· 18

    Section 5   The ecology of bacteria ········ 20

Chapter 3   Disinfection, sterilization and
               biosecurity in the veterinary
               microbiology laboratory ········ 23

    Section 1   Physical disinfection ··········· 23

    Section 2   Chemical disinfection ········· 26

    Section 3   Antibiotics and
              bacteriocins ···················· 27

    Section 4   Biosafety of veterinary
              microbiology laboratory ······· 27

Chapter 4   Bacterial infection and
               pathogenic mechanisms ········ 30

    Section 1   Pathogenicity and virulence
              of bacteria ····················· 30

    Section 2   Virulence factors and secretion
              systems of bacteria ············· 33

    Section 3   Opportunistic pathogens ······ 36

Chapter 5   Genetic variations of bacteria ···· 38

    Section 1   Common phenomena of
              bacterial variations ············· 38

    Section 2   Material basis of bacterial
              inheritance ····················· 38

    Section 3   Mechanism of bacterial
              variations ······················ 41

Chapter 6   Taxonomy and nomenclature
               of bacteria ···················· 44

    Section 1   Taxonomy of bacteria ·········· 44

    Section 2   Nomenclature of bacteria ····· 45

    Section 3   Criteria for taxonomic
              identification of bacteria ······ 46

    Section 4   Procedures of bacterial
              identification ··················· 46

Chapter 7　Gram-positive cocci·······48
　　Section 1　*Staphylococcus*·······48
　　Section 2　*Streptococcus*·······50
Chapter 8　*Enterobacteriaceae*·······53
　　Section 1　*Escherichia*·······53
　　Section 2　*Salmonella*·······56
　　Section 3　*Yersinia*·······57
Chapter 9　*Vibrionaceae* and
　　　　　　Aeromonadaceae·······59
　　Section 1　*Vibrionaceae*·······59
　　Section 2　*Aeromonadaceae*·······61
Chapter 10　*Pasteurellaceae* and
　　　　　　Flavobacteriaceae·······63
　　Section 1　*Pasteurella*·······63
　　Section 2　*Mannheimia*·······65
　　Section 3　*Actinobacillus*·······66
　　Section 4　*Glaesserella*·······67
　　Section 5　*Haemophilus*·······70
　　Section 6　*Avibacterium*·······70
　　Section 7　*Riemerella*·······72
Chapter 11　Gram-negative aerobic
　　　　　　bacilli·······73
　　Section 1　*Brucella*·······73
　　Section 2　*Pseudomonas*·······76
　　Section 3　*Burkholderia*·······77
　　Section 4　*Bordetella*·······78
　　Section 5　*Francisella*·······79
　　Section 6　*Acinetobacter*·······80
Chapter 12　Gram-negative microaerobes
　　　　　　and anaerobes·······83
　　Section 1　*Campylobacter*·······83
　　Section 2　*Helicobacter*·······84
Chapter 13　Non-spore-forming Gram-
　　　　　　positive bacilli·······86
　　Section 1　*Listeria*·······86

　　Section 2　*Erysipelothrix*·······89
Chapter 14　Spore-forming Gram-positive
　　　　　　bacilli·······91
　　Section 1　*Bacillus*·······91
　　Section 2　*Clostridium*·······94
Chapter 15　*Mycobacteria* and similar
　　　　　　genera·······98
　　Section 1　*Mycobacterium*·······98
　　Section 2　*Actinomyces*·······101
　　Section 3　*Dermatophilus*·······103
Chapter 16　*Spirochaete*·······106
　　Section 1　General characteristics of
　　　　　　spirochaetes·······106
　　Section 2　*Borrelia*·······108
　　Section 3　*Brachyspira*·······109
　　Section 4　*Leptospira*·······110
Chapter 17　*Mycoplasma*·······113
　　Section 1　General characteristics of
　　　　　　Mycoplasma·······113
　　Section 2　*Mycoplasma* in pigs·······116
　　Section 3　*Mycoplasma* in poultries·······119
　　Section 4　*Mycoplasma* in bovine
　　　　　　and caprine·······121
　　Section 5　*Mycoplasma haemophilus*·······122
Chapter 18　*Rickettsia* and *Chlamydia*·······125
　　Section 1　*Rickettsia*·······125
　　Section 2　*Chlamydia*·······129
Chapter 19　Structure and taxonomy of
　　　　　　viruses·······132
　　Section 1　Structural characteristics
　　　　　　of viruses·······132
　　Section 2　Chemical composition of
　　　　　　viruses·······135
　　Section 3　Taxonomy of viruses·······136
Chapter 20　Virus replication·······139

Section 1　One-step growth curve ⋯⋯139

Section 2　Attachment, penetration and
　　　　　uncoating⋯⋯⋯⋯⋯⋯⋯⋯140

Section 3　Biosynthesis ⋯⋯⋯⋯⋯⋯142

Section 4　Assembly and release ⋯⋯143

**Chapter 21　Virus variation and
　　　　　evolution** ⋯⋯⋯⋯⋯⋯144

Section 1　Mutation⋯⋯⋯⋯⋯⋯⋯144

Section 2　Mutagenesis ⋯⋯⋯⋯⋯⋯146

Section 3　Recombination⋯⋯⋯⋯⋯147

Section 4　Interactions among the
　　　　　virus components ⋯⋯⋯149

Section 5　Genetic variation and viral
　　　　　evolution ⋯⋯⋯⋯⋯⋯149

**Chapter 22　Interaction between virus
　　　　　and cell** ⋯⋯⋯⋯⋯⋯153

Section 1　Cell culture of viruses⋯⋯153

Section 2　Virus-cell interactions⋯⋯155

Section 3　Virus-induced noncytocidal
　　　　　changes ⋯⋯⋯⋯⋯⋯⋯158

**Chapter 23　Pathogenic mechanisms of
　　　　　viruses** ⋯⋯⋯⋯⋯⋯⋯159

Section 1　Virus invasion, spreading
　　　　　and shedding ⋯⋯⋯⋯⋯159

Section 2　Persistent virus infection⋯161

Section 3　Damage to host tissues
　　　　　and organs by virus
　　　　　infection ⋯⋯⋯⋯⋯⋯162

Section 4　Damage to the immune
　　　　　system by virus infection⋯163

Chapter 24　Virus detection ⋯⋯⋯⋯166

Section 1　Virus isolation and
　　　　　identification ⋯⋯⋯⋯⋯166

Section 2　Determination of virus
　　　　　titration ⋯⋯⋯⋯⋯⋯⋯169

Section 3　Detection of virus
　　　　　particles ⋯⋯⋯⋯⋯⋯⋯170

Section 4　Serologic detection of
　　　　　viruses⋯⋯⋯⋯⋯⋯⋯⋯170

Section 5　Viral nucleic acid
　　　　　detection ⋯⋯⋯⋯⋯⋯⋯174

**Chapter 25　Double-stranded DNA
　　　　　viruses** ⋯⋯⋯⋯⋯⋯⋯176

Section 1　*Poxviridae*⋯⋯⋯⋯⋯⋯176

Section 2　*Asfarviridae*⋯⋯⋯⋯⋯179

Section 3　*Adenoviridae*⋯⋯⋯⋯⋯183

Chapter 26　*Herpesvirales* ⋯⋯⋯⋯187

Section 1　Main features of
　　　　　Herpesviridae⋯⋯⋯⋯187

Section 2　Infectious bovine
　　　　　rhinotracheitis virus⋯⋯189

Section 3　Pseudorabies virus⋯⋯⋯190

Section 4　Avian infectious
　　　　　laryngotracheitis virus ⋯191

Section 5　Marek's disease virus ⋯⋯192

Section 6　Duck plague virus⋯⋯⋯193

**Chapter 27　Single-stranded DNA
　　　　　viruses** ⋯⋯⋯⋯⋯⋯⋯195

Section 1　*Parvoviridae* ⋯⋯⋯⋯⋯195

Section 2　*Circoviridae* ⋯⋯⋯⋯⋯200

Section 3　*Anelloviridae*⋯⋯⋯⋯⋯202

Chapter 28　Retroviruses ⋯⋯⋯⋯⋯204

Section 1　Main characteristics of
　　　　　Retroviridae⋯⋯⋯⋯⋯204

Section 2　*Alpharetrovirus*⋯⋯⋯⋯205

Section 3　*Betaretrovirus*⋯⋯⋯⋯207

Section 4　*Gammaretrovirus*⋯⋯⋯209

Section 5　*Deltaretrovirus*⋯⋯⋯⋯210

Section 6　*Lentivirus*⋯⋯⋯⋯⋯⋯211

Chapter 29　Double-stranded RNA
　　　　　　　viruses ················215
　　Section 1　*Reoviridae*················215
　　Section 2　*Birnaviridae*················218
Chapter 30　Mononegavirales················221
　　Section 1　*Paramyxoviridae*················221
　　Section 2　*Rhabdoviridae*················225
　　Section 3　*Bornaviridae*················226
Chapter 31　Segmented negative-
　　　　　　　stranded RNA viruses················228
　　Section 1　*Orthomyxoviridae*················228
　　Section 2　*Peribunyaviridae*················231
　　Section 3　*Phenuiviridae*················232
　　Section 4　*Nairoviridae*················234

　　Section 5　*Hantaviridae*················234
Chapter 32　Nidovirales················236
　　Section 1　*Coronaviridae*················236
　　Section 2　*Arteriviridae*················245
　　Section 3　*Tobaniviridae*················247
Chapter 33　Picornavirales················248
　　Section 1　*Picornaviridae*················248
　　Section 2　*Caliciviridae*················253
Chapter 34　Other positive-stranded RNA
　　　　　　　viruses················256
　　Section 1　*Hepeviridae*················256
　　Section 2　*Togaviridae*················256
　　Section 3　*Flaviviridae*················258

绪　论
Introduction

内容提要　微生物学是研究微生物的分类、形态、生理、遗传变异、生态分布及其与人类关系的一门学科。兽医微生物学与人和动物健康、乡村振兴、生物安全和公共卫生等息息相关。

Introduction　Microbiology is a discipline that studies taxonomy of the microorganisms and their morphology, physiology, heredity and genetic variation, distribution in different ecosystems, and relationship with human beings. Veterinary microbiology is closely related to human and animal health, rural revitalization, biosecurity and public health.

第1节　兽医微生物学发展历史
Section 1　History of veterinary microbiology

我们的祖先对微生物早有认识和应用，在商代（公元前1600～前1046年）的甲骨文中就有关于酒的记载。兽医微生物学与医学微生物学的发展一直相伴而行，但兽医微生物学作为一门独立的科学，是19世纪以后的事情。

微生物学的发展可概括为三个阶段。

第一阶段：形态学时期。1683年，荷兰人列文虎克（1632～1723，**图0-1**）用自制的显微镜（**图0-2**）首次观察到微生物，从此以后，人们对微生物的形态、排列、大小等有了初步的认识，但由于自然发生论的存在，认识仅限于形态学方面。直到1861年，法国的微生物学家巴斯德（1822～1895，**图0-3**）通过盛有肉汤的鹅颈瓶（**图0-4**）和普通烧瓶实验否定了微生物自然发生论，解开了人们的精神枷锁，使人们对微生物的形态有了更多的认识，并能逐步解决生活和生产中出现的人

Our ancestors' understanding and application of microorganisms can be traced back to the oracle-bone inscriptions of "jiu" (wine) in the Shang Dynasty (1600-1046BC). Veterinary microbiology has undergone development in parallel with human microbiology since its very beginning. However, it was not until the 19th century that veterinary microbiology was recognized as a discipline. The development of microbiology can be summarized into three stages:

The first stage was on morphological aspects. Antonie van Leeuwenhoek (1632-1723, **Figure 0-1**), a Dutch microbiologist, observed microorganisms in 1683 for the first time with a self-made microscope (**Figure 0-2**). From then on, microbes were primarily recognized by their morphology, such as shape, arrangement and size because of the dominance of autogenesis theory. It was not until 1861 that Louis Pasteur (1822-1895, **Figure 0-3**), a French microbiologist, disproved the autogenetic theory of microorganisms by showing microbial growth in a conventional flask, but not in a swan-neck bottle (**Figure 0-4**), both being filled with broth. Since then, humans have unwound the spiritual shackles towards better understanding of the morphology of microorganisms

图 0-1　安东尼·列文虎克（1632～1723）肖像
Figure 0-1　Portrait of Antonie van Leeuwenhoek
(1632-1723)

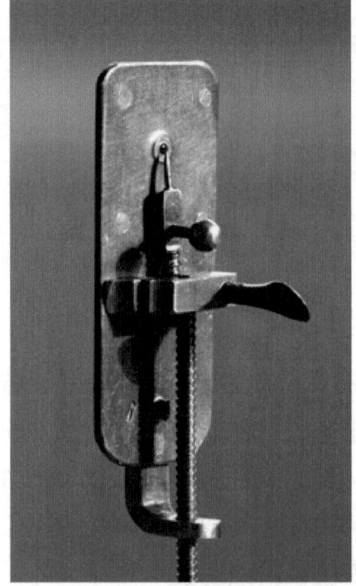

图 0-2　安东尼·列文虎克显微镜复制品
Figure 0-2　A replica of a microscope made by
Antonie van Leeuwenhoek

图 0-3　路易·巴斯德（1822～1895）肖像
Figure 0-3　Portrait of Louis Pasteur (1822-1895)

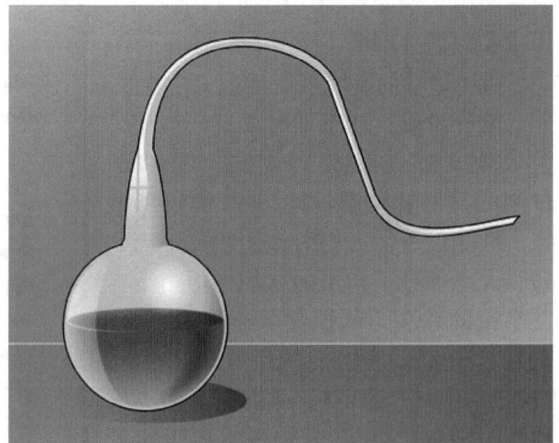

图 0-4　巴斯德使用的鹅颈瓶
Figure 0-4　Swan-neck bottle used by Pasteur

畜疾病问题。

第二阶段：生理学及免疫学的奠基时期。这个时期大约从1870年到1920年，微生物学已经发展成一门独立的学科，在理论上、技术上和生产上都取得了一定的成就，涌现了一批杰出的微生物学家，如因发明白喉的血清疗法而获得1901年首届诺贝尔生理学或医学奖的德国医学家埃米尔·阿道夫·冯·贝林；因发现蚊子是传播疟疾的媒介而获得1902年诺贝尔生理学或医学奖的英国微生物学家罗纳德·罗斯；在肺结核研究中作出杰出贡献而获得1905年诺贝尔生理学或医学奖的德国细菌学家罗伯特·科赫，以及人类历史上第一个动物病毒——口蹄疫病毒的发现者德国微生物学家弗里德里希·奥古斯特·约翰内斯·洛弗勒和病理学家保罗·弗罗施。

第三阶段：近代及现代微生物学时期。大约从1920年起至今，微生物学科的发展越来越深入，在理论研究、技术创新及实际应用方面都取得了重要进展。随着电子显微镜、单克隆抗体、组学、分子克隆等技术的发展和应用，兽医微生物学的发展也在不断呈现日新月异的变化。

and started to apply such knowledge to solve problems in human and animal diseases in their daily lives and husbandry.

The second stage focused on the foundation of physiology and immunology. During this period, from 1870 to 1920, microbiology developed into an independent discipline and made certain achievements in theory, technology and application. A number of outstanding microbiologists emerged, such as Emil Adolf von Behring, a German medical scientist who received the first Nobel Prize in Physiology or Medicine in 1901 for his work on serum therapy, especially its application against diphtheria; Sir Ronald Ross, a British doctor who received the Nobel Prize in Physiology or Medicine in 1902 for his work on malaria infestation in the mosquito gut; Robert Koch, a German physician and one of the founders of bacteriology, who received the Nobel Prize in Physiology or Medicine in 1905 for his discoveries on pulmonary tuberculosis; and Friedrich August Johannes Löffler, a German microbiologist who, along with pathologist Paul Frosch, succeeded in demonstrating the first animal virus that could cause foot and mouth disease in artiodactyls.

The third stage covers modern and contemporary microbiology. Since 1920, microbiology has achieved in-depth development with great progress in theoretical research, technological innovation and practical applications. With the development and application of electron microscopy, monoclonal antibodies, omics, molecular cloning and other technologies, the development of veterinary microbiology is constantly evolving.

第2节 兽医微生物学
Section 2 Veterinary microbiology

兽医微生物学是研究与动物疾病、人畜共患病和食品安全相关的重要微生物形态与结构、生物学特性、遗传变异、抗原性、毒力因子、致病机制、鉴别检测以及防控策略的学科。在保障动物健康、维护公共卫生安全、保护生态环境方面发挥重要作用。其研究范围不仅限于家畜、家禽、伴侣动物、实验动物、水生动物，以及野生动物等的微生物，还涉及生物安全、食品安全、实验室安全、国境安全等诸多领域；研究深度达到基因和分子

Veterinary microbiology studies the microorganisms that are closely related to animal diseases, zoonotic diseases and food safety regarding their morphology and structure, biological characteristics, heredity and genetic variations, antigenicity, virulence factors, pathogenesis, differential identification, and preventative control strategies. It plays important roles in protecting animals from infections, safeguarding public health and preserving ecosystems. The research scope of veterinary microbiology includes not only the microbes that infect livestock, poultry, companion animals, laboratory animals, aquatic species and wild animals, but also those related to biological safety, food safety, laboratory safety, border security and other fields. Pathogenetic studies

层面，涉及毒力因子与宿主互作及其机制。目前已明确记载的人兽共患微生物有800余种，其中60%的人类病原微生物来自动物，75%的新发传染病为人兽共患病，如炭疽、狂犬病、鼠疫、结核病等。因此，兽医微生物学与人和动物健康密切相关，是现代医学和生命科学的重要组成部分。

have extended into the individual genes encoding virulence molecules and the mechanisms governing their interaction with host factors. More than 800 zoonotic microorganisms have been documented so far. Of them, about 60% of human pathogens are derived from animals, and 75% of emerging infectious diseases are zoonotic, *e.g.*, anthrax, rabies, plague, tuberculosis, *etc*. Therefore, veterinary microbiology is closely related to human and animal health and exists as an important component of modern medicine and life sciences.

思考题　Questions

1. 为什么要学习兽医微生物学？

2. 兽医微生物学在乡村振兴中的地位和作用是什么？

1. Why do we study veterinary microbiology?

2. What roles does veterinary microbiology play in rural revitalization?

第1章 细菌的形态和结构
Chapter 1 / Bacterial structure and morphology

内容提要 细菌具有一定的大小、形态及排列方式。细菌直接用肉眼看不到，必须借助光学显微镜或电子显微镜才能观察，因此测定细菌大小的单位是微米和纳米。

Introduction Bacterial morphology refers to the size and shape of bacterial cells and their arrangement. Bacteria are invisible by naked eyes. Light or electron microscopes are used to magnify bacteria for visual observation or imaging. The units of bacterial size commonly used are micrometers and nanometers.

第1节 细菌的大小和形态
Section 1 Bacterial size and morphology

细菌主要由三种基本类型组成，即球状、杆状和螺旋状，分别对应：球菌、杆菌和螺形菌。例如：葡萄球菌为球菌；大肠杆菌为杆菌；霍乱弧菌为螺形菌（**图1-1**）。细菌的大小以生长在适宜条件

Bacterial cells can be spherical, rod-like or spiral in shape, and are referred to as cocci, rods or spirochaetes, respectively. For example, *Staphylococcus* is a coccus, and *Escherichia coli* is a rod, while *Vibrio cholerae* is a spirochaete (**Figure 1-1**). The size of bacteria is determined from the logarithmic growth phase of their young cultures

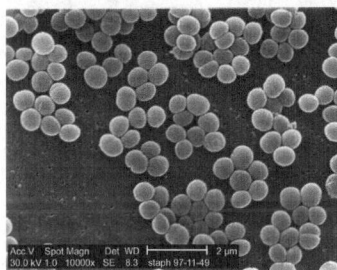

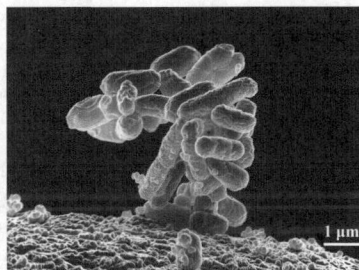

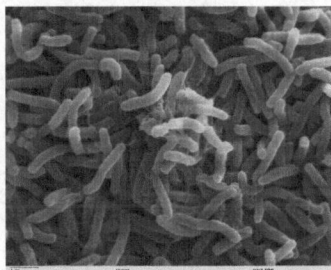

球菌：葡萄球菌（*Staphylococcus*） 杆菌：大肠杆菌（*Escherichia coli*） 螺形菌：霍乱弧菌（*Vibrio cholerae*）

图1-1 细菌的形态：球菌、杆菌和螺形菌
细菌呈球形、细长形或螺旋状，分别称为球菌、杆菌或螺形菌。图像来源见英文。

Figure 1-1 Bacterial morphology: coccus, rod and spirochaete
Bacteria can have spherical, slender or spiral shapes, and are referred to as cocci, rods or spirochaetes, respectively. Source: *Staphylococcus*: Janice Carr, the Centers for Disease Control and Prevention's Public Health Image Library (PHIL), with identification number #6486; *E. coli*: Eric Erbe, the Agricultural Research Service, the research agency of the United States Department of Agriculture, with the ID K11077-1; *Vibrio cholerae*: T. J. Kirn. http: //remf.dartmouth.edu/images/bacteriaSEM/source/1.html.

和培养基中指数期的幼龄培养物为标准；各种细菌的大小是相对稳定的，具有种的特征，可以作为初步鉴定的依据。

in appropriate media at optimal temperature. Bacteria size can be used for presumptive identification since different bacterial species are relatively stable and have characteristic morphology.

小 知 识
Extended Knowledge

蛭弧菌中的牛顿力学
Newton's Laws of Motion in *Bdellovibrio*

为什么微生物会呈现不同的形态呢？这与其在长期演化过程中所处的生存环境是分不开的，比如蛭弧菌。蛭弧菌的形状是螺杆状，外形如同螺丝钉，其特别之处在于尾巴处有一长长的鞭毛。蛭弧菌可以捕食它的同类；当环境中出现其他革兰氏阴性菌时，蛭弧菌头会与之发生物理碰撞，然后利用其身体的优势高速旋转。根据力学原理，当螺丝钉高速旋转时，将快速钻入物体内部。同理，高速旋转的蛭弧菌在几秒钟之内能破坏革兰氏阴性菌的细胞壁，并钻入与之接触的细菌体内进行繁殖。这就是会使用牛顿力学的蛭弧菌（**图1-2**）。

Why do microorganisms come in different shapes? Along bacterial evolution, their living environment has a huge impact on them. For example, *Bdellovibrio* is a spirochaete resembling a screw used by construction workers. It features a long flagellum at its tail. It practices cannibalism. When encountering other Gram-negative bacteria, *Bdellovibrio* collides with them and then rapidly spins its own body. Based on the principles of force, a rapid spinning spiral nail can quickly puncture through its target object. Similarly, *Bdellovibrio* damages the cell wall of Gram-negative bacteria within seconds, enters and proliferates there. This is an example of how *Bdellovibrio* has mastered Newton's Laws of Motion (**Figure 1-2**).

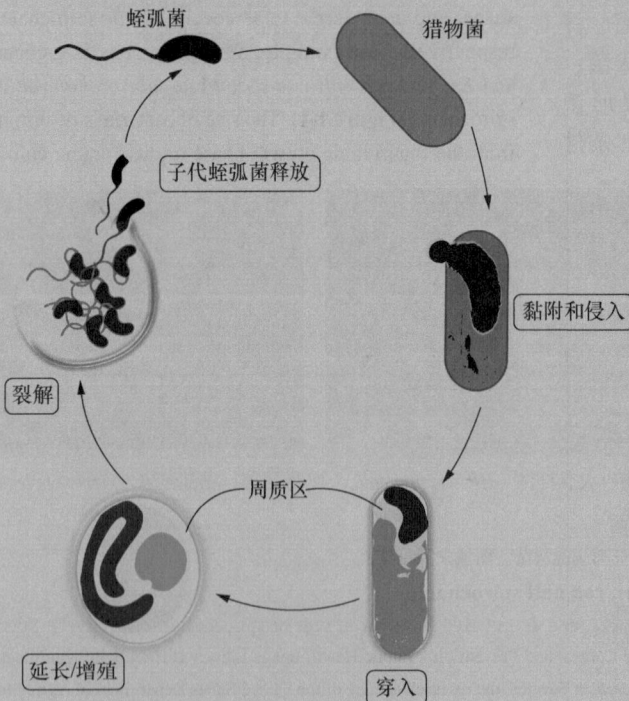

图1-2　"捕食"中的蛭弧菌
蛭弧菌吸附革兰氏阴性菌后，穿入猎物细菌周质区并进入内部增殖。4 h内可裂解菌体释放子代蛭弧菌。图像来源见英文。

Figure 1-2　*Bdellovibrio* in "prey"
Bdellovibrio attaches to a Gram-negative bacterium after contact and penetrates into the prey's periplasmic space. Once inside, elongation occurs and progeny cells are released within 4 h. Source: Madigan MT. 2011. Brock Biology of Microorganisms. New York: Pearson Education.

第2节　细菌的基本结构和特殊结构
Section 2　Basic and special structures of bacteria

细菌的结构由基本结构和特殊结构组成。基本结构就是构成细菌基本骨架的结构，从外到内依次是细胞壁、细胞膜和细胞质，其中，细胞质中有间体、核糖体和核酸以及各种内含物。细菌的特殊结构是指某些细菌的特有结构，例如，荚膜、鞭毛、菌毛和芽孢等（图1-3）。

The structures of bacteria consist of basic and special components. Basic components make up the basic skeletons of bacteria. From the outer to the inner layer, a bacterium is composed of the cell wall, cell membrane and cytoplasm. In the cytoplasm, there are mesosomes, ribosomes and nucleoid. Special components are present only in some bacteria, *e.g.*, capsules, flagella, pili and spores (**Figure 1-3**).

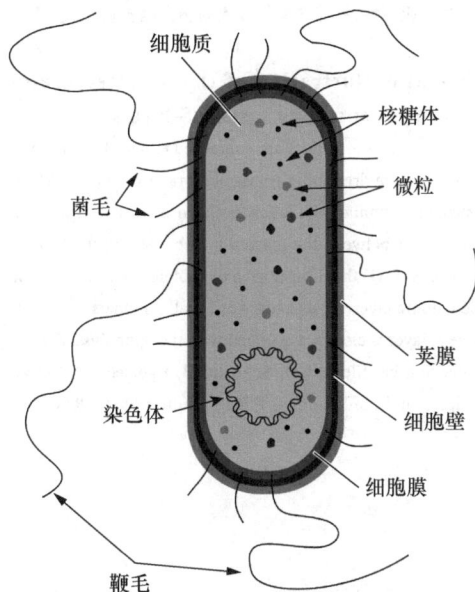

图1-3　细菌的结构
图像来源见英文。
Figure 1-3　Structures of a typical bacterial cell
Source: Quinn P J, Markey B K, Leonard F C, et al. 2015. Review of Veterinary Microbiology. 2nd ed. New York: John Wiley & Sons.

荚膜是指细菌在细胞壁的外面产生的一种包围整个菌体，主要起保护作用的黏液样物质。S层是指部分革兰氏阳性细菌的一种特殊的表层结构，由单一的蛋白质亚单位组成，规则排列，呈类晶格状（图1-4）。S层是一种简单的生物膜，类似荚膜的屏障作用。鞭毛是指突出于菌体表面的细长丝状蛋白结构，具有抗原性，为细菌的运动器官。菌毛是指在菌体上的一种比鞭毛数量多、形状较直、直径较细、长度较短的毛发状细丝，亦称纤毛或伞

Capsules are mucoid substances surrounding the cell wall of bacteria as a protective layer. Surface S-layer is a unique external structure of some Gram-positive bacteria composed of homogenous subunits of a certain protein regularly aligned to form lattices (**Figure 1-4**). S-layer is a simple form of a biofilm and serves as a barrier similar to capsules. Flagella are filamentous proteinaceous projections on the bacterial surface that are antigenic. They afford the bacteria motility. Another filamentous structure of the bacteria is pili. A single bacterium has more pili than its flagella/flagellum. Pili, also known as fimbriae, are straighter, more slender and shorter than flagella. Pili are cylinders

毛。菌毛是一种空心的蛋白质管，具抗原性，分为普通菌毛和性菌毛两类。鞭毛和菌毛的区别在于长短不同，长的是鞭毛，短的是菌毛。芽孢是细菌在菌体内形成一个内生的孢子，称为芽孢。细菌的芽孢内有核酸，在适宜的环境中，芽孢可以独立繁殖为一个新的细菌个体。

primarily composed of helically arranged oligomeric pilin proteins with antigenicity. There are two kinds of pili, sex pili and ordinary pili. The major difference between a flagellum and a pilus is their length. The longer one is flagellum, and the shorter one is a pilus. The endospores formed inside the bacteria are called spores. Each spore contains a set of nucleoids. Under proper environmental conditions, a spore can become vegetative.

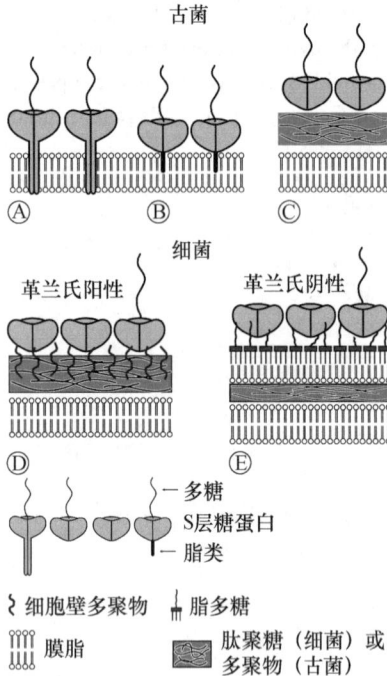

图1-4　细菌的S层结构

古菌S层糖蛋白晶格由具有柱状疏水跨膜结构域的蘑菇状亚基组成（A）或由脂修饰的糖蛋白亚基组成（B）；少数古菌在质膜和S层之间有一坚硬中间层（C）；在革兰氏阳性菌中，S层糖蛋白通过二级细胞壁聚合物与肽聚糖层结合（D）；在革兰氏阴性菌中，S层与外膜的脂多糖密切相关（E）。图像来源见英文。

Figure 1-4　Schematic illustration of the architecture of prokaryotic cell envelope containing surface S-layers

S-layers in archaea with glycoprotein lattices are composed either of mushroom-like subunits with pillar-like, hydrophobic trans-membrane domains (A), or lipid-modified glycoprotein subunits (B). A few archaea possess a rigid wall layer as an intermediate layer between the plasma membrane and the S-layer (C). In Gram-positive bacteria (D) the S-layer (glyco) proteins are bound to the rigid peptidoglycan-containing layer via secondary cell wall polymers. In Gram-negative bacteria (E) the S-layer is closely associated with the lipopolysaccharide of the outer membrane. Source: Sleytr UB, Schuster B, Egelseer EM, et al. 2014. S-layers: principles and applications. FEMS Microbiology Reviews, 38 (5): 823-864.

第3节　革兰氏染色
Section 3　Gram staining

一般来说，细菌需要染色后才能利用光学显微镜进行观察。最常见的染色方法是1884年丹麦医生革兰发明的革兰氏染色法（**图1-5**），至今仍在沿用。根据该方法，可以把细菌分为革兰氏阳性菌和阴性菌两大类。革兰氏染色法主要包括5个步骤。第1步：把细菌放到玻片上用火焰稍加热固定。第2步和第3步分别用结晶紫和碘酒进行着色，此时结晶紫和碘在细菌内部形成紫色复合物。第4步：用95%乙

In general, staining is required for observing bacteria under a light microscope. The most common method is Gram staining, invented by a Danish physician, Dr. Gram in 1884 (**Figure 1-5**) and has been used ever since. By this staining, bacteria can be grouped as Gram-positive or Gram-negative. There are 5 steps in Gram staining. The 1st step is to make a smear of the bacteria onto a glass slide and heat fixed. For step 2 and step 3, the slide is stained with crystal violet and then iodine, respectively. The crystal violet and iodine will form a purple complex within the bacterial cells. Step 4 is to decolorize with 95% alcohol. The cell wall of Gram-positive bacteria contains a thick layer of the peptidoglycan

醇脱色。由于阳性菌细胞壁含有网格状肽聚糖，当用95%乙醇脱色时，肽聚糖网格空隙缩小，菌体内部的结晶紫和碘复合物被限制在菌体内部，无法被乙醇脱色，细菌呈现紫色。革兰氏阴性菌肽聚糖少，当用95%乙醇脱色时，结晶紫和碘复合物从细胞壁间隙流出，此时的菌体为无色。第5步：用红色的沙黄染料对菌体进行复染。革兰氏阴性菌可被再次着色，因此在显微镜下观察时是红色的，如大肠杆菌和嗜血杆菌。由于革兰氏阳性菌在第4步因肽聚糖孔径变小且菌体内部已被结晶紫和碘复合物占据，沙黄无法对其进行再次着色。所以，革兰氏阳性菌在显微镜下观察时仍然是紫色的，如金黄色葡萄球菌、链球菌和李氏杆菌。

network. Upon treatment decolorized with 95% alcohol, the network shrinks and traps the purple complex of crystal violet and iodine inside the bacteria, that is, the bacterial cells are not decolorized and remain purple. On the other hand, Gram-negative bacteria contain only a much thinner layer of peptidoglycan. When decolorized by 95% alcohol, the purple complex slips through the cell wall. Thus, the Gram-negative bacterial cells become colorless at this point. Step 5, the final step, is to counterstain with the red colored safranin. The colorless Gram-negative bacteria after step 4 treatment could take up the safranin dye. Thus, the Gram-negative bacteria appear red under the microscope, such as *Escherichia coli* and *Haemophilus*. However, safranin can't enter the Gram-positive bacteria due to shrinking of the peptidoglycan network in step 4 and occupation of the purple complex inside the bacteria. That's why Gram-positive bacteria remain purple under the microscope, just like *Staphylococcus aureus*, *Streptococcus* and *Listeria*.

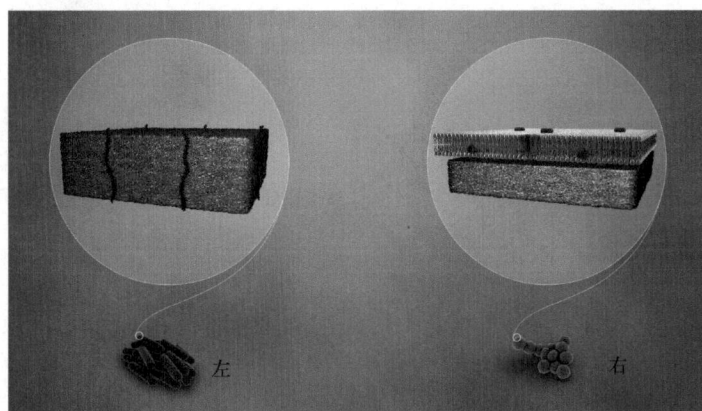

彩图

图1-5 细菌革兰氏染色示例
紫色的为具有较厚肽聚糖层的革兰氏阳性菌（左），粉红色的为革兰氏阴性菌（右）。

Figure 1-5　Diagram of bacterial Gram staining
Purple-stained Gram-positive bacteria (left) with a thick peptidoglycan layer and pink-stained Gram-negative bacteria (right).

第4节 细菌细胞壁的主要功能
Section 4　The main functions of bacterial cell walls

原生质球和原生质体：通过革兰氏染色，可以看出革兰氏阴性菌（**图1-6**）和阳性菌（**图1-7**）的细胞壁结构是不一样的。革兰氏阴性菌的细胞壁由外膜和质膜两层膜组成，中间是肽聚糖。而革兰氏阳

Spheroplast and protoplast: Gram staining indicates that there is difference in the cell wall structure between Gram-negative (**Figure 1-6**) and Gram-positive (**Figure 1-7**) bacteria. The cell wall of Gram-negative bacteria consists of an outer membrane, a plasma membrane and a thin layer of peptidoglycan between the two membranes. The wall of

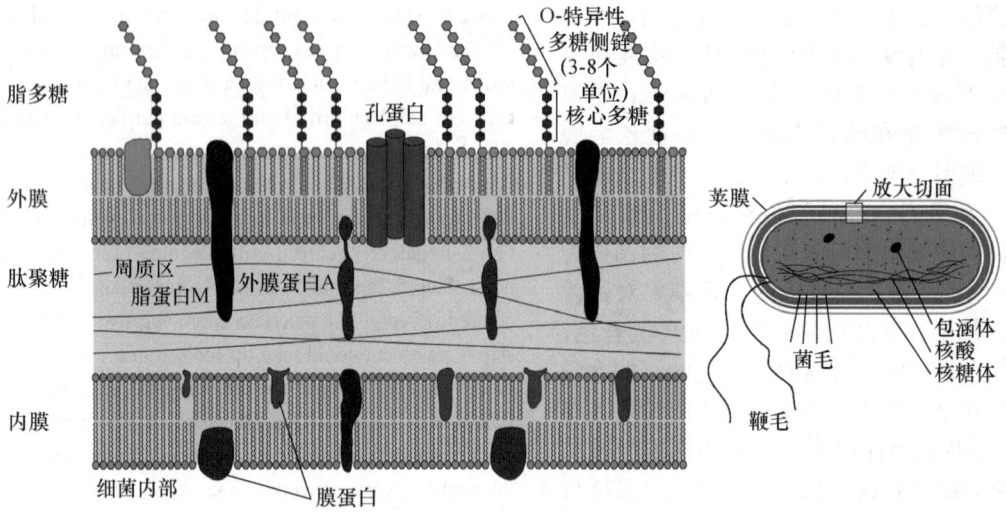

图1-6　革兰氏阴性菌细胞壁结构

Figure 1-6　Diagram of Gram-negative cell wall

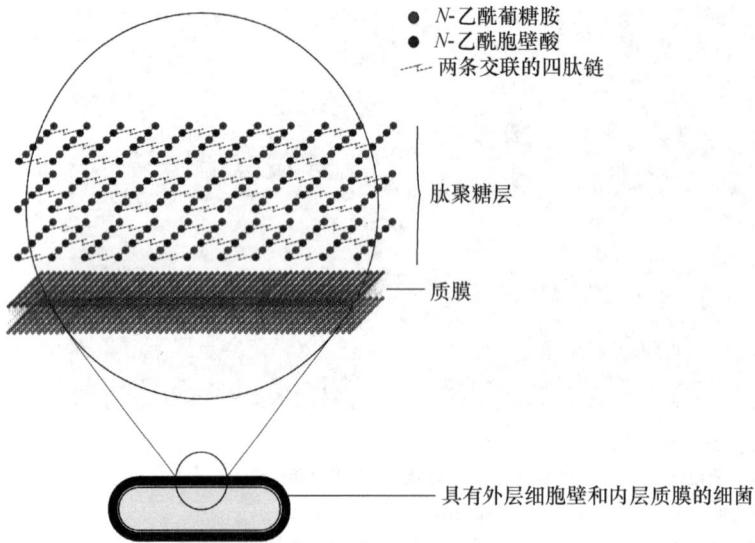

图1-7　革兰氏阳性菌细胞壁结构

Figure 1-7　Diagram of Gram-positive cell wall

性菌只由一层质膜组成，外面是若干层肽聚糖。肽聚糖为网格状结构，由寡肽、N-乙酰葡糖胺、N-乙酰胞壁酸聚合而成，因此称为肽聚糖。若细菌的肽聚糖结构被破坏，则细菌失去保护作用，如青霉素能封

Gram-positive bacteria only contain a plasma membrane and multiple thick layers of peptidoglycan. Peptidoglycan forms a mesh-like polymer composed of oligopeptides N-acetylglucosamines (NAG) and N-acetylmuramic acids (NAM). That's why it is called peptidoglycan. If the peptidoglycan structure is damaged, bacteria will lose their

闭肽聚糖合成关键酶的活性位点，导致 *N*-乙酰葡糖胺和 *N*-乙酰胞壁酸无法通过 *L*-丙氨酸和 *D*-丙氨酸之间的肽键相连（**图 1-8**）。失去肽聚糖的细菌像薄薄的气球一样，形成原生质球或原生质体。原生质球是部分除去细菌细胞壁内层的肽聚糖，形成仍有外膜包裹的菌体（革兰氏阴性菌），可为杆状或球状；原生质体是完全除去细菌的细胞壁的菌体（革兰氏阳性菌）。

armory. For example, penicillin can block the active site of peptidoglycan synthase, which prevents the connection between NAG and NAM via the peptide bonds between *D*-alanine and *L*-alanine (**Figure 1-8**). Without peptidoglycan, the cell wall becomes a fragile balloon dubbed spheroplast and protoplast. A spheroplast is a bacterial cell (Gram-negative) that partially lacks the peptidoglycan in the inner layer of the cell wall, but still has an outer membrane, and can be rod-shaped or spherical under the microscope. A protoplast is a bacterial cell (Gram-positive) with its complete cell wall lost.

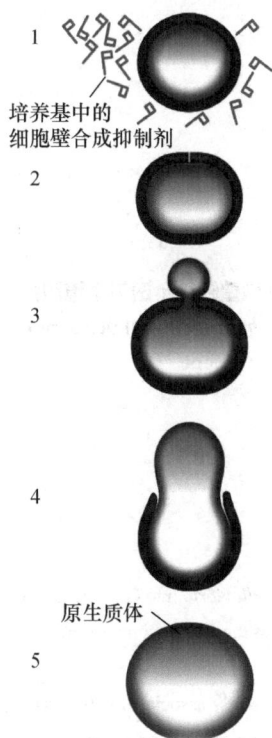

1　培养基中的细胞壁合成抑制剂

2

3

4

5　原生质体

图1-8　在细胞壁合成抑制剂（如青霉素、万古霉素）存在条件下的细菌细胞分裂示意图

1. 将青霉素（P）和分裂期的细菌一起加入培养基中；2. 细菌细胞开始生长，但不能合成新的细胞壁；3. 当细菌细胞继续生长时，被质膜覆盖的细胞质通过细胞壁的缝隙溢出；4. 细菌细胞壁的完整性进一步受到破坏。细胞继续增大，但无法将多余的细胞质"捏"成两个子细胞；5. 细胞壁完全脱落，形成球状体，相对于原始细胞极为脆弱。丧失细胞壁导致细菌失去对其形状的控制。即使最初的细菌是杆状的，最后也会形成球状的原生质球。

Figure 1-8　Diagram depicting the failure of bacterial cell division in the presence of a cell wall synthesis inhibitor (*e.g.*, penicillin, vancomycin)

1. Penicillin (P) is added to the growth medium with dividing bacteria. 2. The cell begins to grow but is unable to synthesize new cell wall to accommodate the expanding cell. 3. As the growth continues, cytoplasm covered by plasma membrane begins to squeeze out through the gap (s) in the cell wall. 4. Cell wall integrity is further violated. The cell continues to increase in size but is unable to "pinch off" the extra cytoplasmic material into two daughter cells. 5. The cell wall is shed entirely, forming a spheroplast, which is extremely vulnerable compared to the original cell. The loss of the cell wall also causes the cell to lose control over its shape, so even if the original bacterium is rod-shaped, the spheroplast is generally spherical.

L型细菌：原生质体和原生质球统称为L型细菌。L型细菌因最初在英国的李斯特预防医学研究所发现的这种细菌细胞壁缺陷的现象，取Lister英文首字母L命名。L型细菌一般呈多形性（**图1-9**），有球状和杆状，生长繁殖缓慢；在软琼脂平板上，能生长出"煎荷包蛋"样小菌落。

细胞壁的功能：细菌细胞壁通过"分隔"和"过滤"作用保护菌体。"保护"是指维持细菌外形和提高机械强度，保护细菌耐受低渗或高渗环境。"过滤"是指

L-form bacterium is an umbrella term for spheroplasts and protoplasts. Cell wall defective or L-form bacteria were first discovered in the Lister Institute of Preventive Medicine in Great Britain, so the scientists used the first letter from Lister to name this finding. L-form bacteria are polymorphic (**Figure 1-9**). They can be spherical or rod-shaped and grow slowly. On the soft agar plate, L-form bacteria grow into small colonies with a "fried eggs" appearance.

The function of cell walls: their main function is to offer protection of the bacterial cells by physical separation and filtration. The cell wall maintains the shape of the bacterium, enhances its mechanical strength and aids its

具有分子筛作用，阻挡有害物质进入菌体，维持菌体内外离子平衡。细菌细胞壁缺陷之后，细菌的形状和生长特性将发生改变。

tolerance to hypo- or hyper-osmotic environments. The cell wall also functions as molecule filters to block the entry of harmful substances and keeps the ionic balance across the bacterial cell. Therefore, a defective cell wall will alter the morphology and growth traits of bacteria.

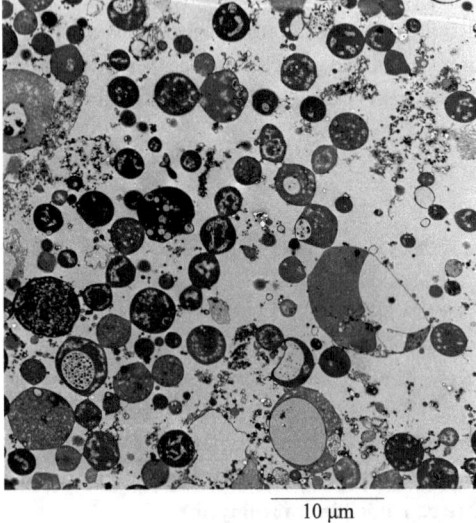

10 μm

图1-9　L型枯草芽孢杆菌（细胞壁缺陷）透射电镜照片
Figure 1-9　L-form *Bacillus subtilis* (cell wall deficiency) under transmission electron microscope

思考题　Questions

1. 蛭弧菌高速旋转并钻入革兰氏阴性菌是需要能量的，那么其能量从何而来？

1. Rapid spinning and drilling of *Bdellovibrio* into Gram-negative bacteria requires energy. Where does the energy come from?

2. 为何原生质球多见于革兰氏阴性菌而原生质体多见于革兰氏阳性菌呢？

2. Why are spheroplasts more commonly associated with Gram-negative bacteria, while protoplasts are more often associated with Gram-positive bacteria?

3. L型细菌在革兰氏染色中是什么颜色的？

3. What's the color of the L-form bacteria upon Gram staining?

4. L型细菌的形成条件是什么？

4. How are the L-form bacteria formed?

5. "绿水青山就是金山银山"。在绿色生态健康养殖过程中，哪些养殖环境变化可以导致环境中的细菌成为L型细菌？

5. "Lucid water and lush mountains are invaluable assets". In the green and ecology-friendly animal agriculture, what changes or husbandry practices can cause the bacteria in the environment to become L-form ones?

第2章　细菌的生长繁殖与生态
Chapter 2 / Bacterial growth and ecology

内容提要　细菌生长繁殖离不开菌体细胞内的代谢，涉及非常复杂的生物化学过程，按功能分为物质摄取、分解代谢和合成代谢三个过程。不同细菌生长繁殖的营养要求、酶系统、代谢产物等各不相同，形成多种多样的代谢类型，是细菌生化鉴定的重要基础。

Introduction　Bacterial growth is closely linked to the metabolism of bacterial cells that involves complex biochemical reactions. Bacterial metabolism can be functionally classified as three processes of uptake of nutrients, catabolism and anabolism. Different bacterial species vary in their nutritional requirements, enzymes and metabolites, thus forming different profiles of metabolic reactions that can serve as the basis for biochemical identification at the species level.

第1节　细菌的代谢过程
Section 1　Bacterial metabolism

细菌代谢是指菌体细胞内所有生物化学反应的总和，涉及细菌在适当环境（如人工培养基）中的生长繁殖和功能发挥相关的有机和无机物质转化。细菌的营养物质代谢有两方面作用，一是用于组成菌体细胞的各种成分，二是供给细菌代谢所需能量。此外，细菌代谢过程中还可产生具有医用价值的多种代谢产物。细菌代谢包括物质摄取、分解代谢和合成代谢三个过程。

物质摄取：按照营养物质的转运方式不同，可分为被动扩散和主动转运。被动扩散又有单纯扩散和易化扩散，前者指水、气体以及水溶性小分子物质可以直接自由地出入菌体细胞，不需能量（**图2-1**）；易化扩散是指大分子物质的选择性转运，如葡萄糖、氨基酸等需要借

Bacterial metabolism refers to all the biochemical reactions that occur within a bacterial cell to transform organic and inorganic substances into those required for bacterial growth and function in an appropriate chemical milieu (such as a bacterial culture medium). Metabolism functions in two ways for the bacteria: to provide various components that form the bacterial cells, and to supply energy to sustain metabolism itself. Besides, a host of metabolites generated during bacterial metabolism are of important medical values. Bacterial metabolism can be classified into three processes of uptake of nutrients, catabolism and anabolism.

Uptake of nutrients: this process can be divided into two major types, passive diffusion and active transport. Passive diffusion can be subdivided into simple diffusion and facilitated diffusion. The former applies to water, gas and water-soluble small molecules that can freely enter and exit bacterial cells with no requirement for energy (**Figure 2-1**). The latter pertains to larger molecules, such as glucose and amino acids, whose transport is selective

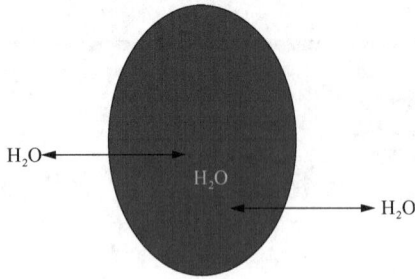

图2-1 单纯扩散

水、气体、水溶性小分子（乙醇、甘油），以及某些Na^+等小分子物质的单纯扩散，无须能量、无选择性、无特异性。

Figure 2-1 Simple diffusion

The simple diffusion of water, gas, water-soluble small molecules (ethanol, glycerol) and some small molecular substances, such as Na^+. This process requires no energy and has no selectivity or specificity.

助载体蛋白转运至细胞内。单纯扩散和易化扩散的共同点是物质从高浓度向低浓度扩散（**图2-2**）。主动转运是物质逆浓度梯度，从低浓度跨越细菌的细胞膜向高浓度转运，除载体蛋白外，还需要提供能量才能完成转运过程，如半乳糖以及钠钾泵介导的钠离子和钾离子转运（**图2-3**）。基团转位和特定金属离子的转运也属于主动转运，如葡萄糖和甘露醇的跨膜转运，除了需要载体蛋白和能量外，其在细胞膜上被磷酸化修饰，分别形成6-P-葡萄糖和6-P-甘露醇后进入菌体内，属于磷酸化基团转位（**图2-4**）。细菌生长必需的某些微量元素也通过载体蛋白转运，以铁离子转运为例，其进出细胞膜需要铁载体协助，铁载体运输的是化合价为3价的铁。

and requires carrier proteins. A common feature of simple diffusion and facilitated diffusion is that substances diffuse from high to low concentrations (**Figure 2-2**). Active transport is a process in which a substance is transported across the bacterial cell membrane from low to high concentrations with the help of both carrier proteins and energy. Examples include the transport of galactose and of Na^+ and K^+ ions via the sodium-potassium pump (**Figure 2-3**). Transfer of chemical groups and transport of metal ions are also considered as active transport. For instance, glucose and mannitol need to be phosphorylated on the bacterial cell membrane, a process that requires carrier proteins and energy, before they can enter the cell in the form of glucose-6-phosphate, and mannitol-6-phosphate. This process is referred to as phospho-group transfer (**Figure 2-4**). Some metal ions that serve as micronutrients indispensable to bacterial growth are also transported by carrier proteins. Taking iron transport as an example, a specific protein called siderophore is needed for the transport of ferric iron across the cell membrane.

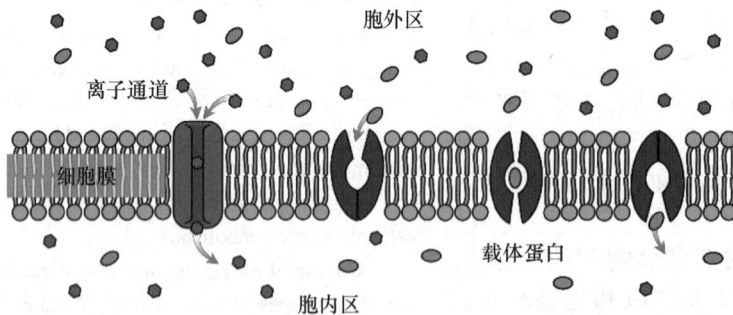

图2-2 易化扩散（离子通道和载体蛋白）
Figure 2-2 Facilitated diffusion in cell membrane, showing ion channels and carrier proteins

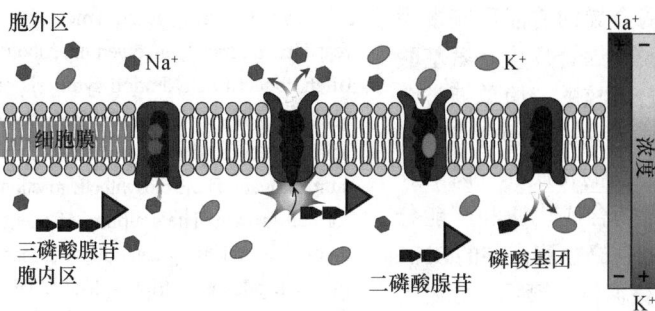

图2-3 主动转运

水解三磷酸腺苷与特定物质转运偶联——钠钾泵。

Figure 2-3 Active transport

The energy from the hydrolysis of ATP is directly coupled to the movement of a specific substance across a membrane— sodium-potassium pump.

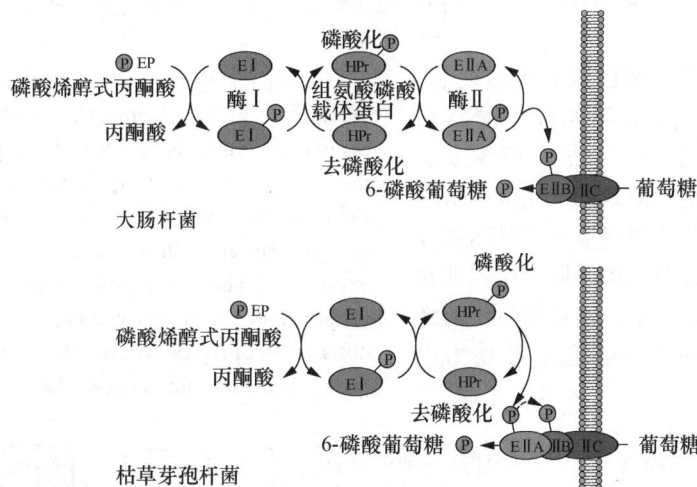

图2-4 基团转位——磷酸转移酶系统

葡萄糖、甘露醇等的转运，需要载体蛋白和能量，同时被磷酸化。

Figure 2-4 Group translocation—phosphotransferase system

The transport of glucose and mannitol requires carrier protein and energy. The molecules are phosphorylated after translocation.

分解代谢：指由酶介导并产生能量的化学反应，多为水解反应。细菌代谢所需的能量主要通过一系列氧化-还原反应获得，糖类是细菌获得能量的主要基质。有氧呼吸菌以氧分子作为电子受体，厌氧呼吸以硝酸根或硫酸根离子等无机分子作为电子受体，而发酵则以有机化合物为电子受体。

Catabolism: the enzyme-regulated chemical reactions that release energy are generally the ones involved in catabolism. Catabolic reactions are mostly hydrolytic. The bacterial cells gain energy mainly through a series of oxidation-reduction reactions using sugars as the major sources for energy production. In bacteria specialized in aerobic respiration, oxygen (O_2) serves as the final electron acceptor. In anaerobic respiration, inorganic substances, such as nitrate ions (NO_3^-) or sulfate ions (SO_4^{2-}), act as the final electron acceptors. In fermentation, organic compounds serve as the final electron acceptors.

合成代谢：由酶介导的需能反应多为合成代谢，常涉及脱水缩合反应。氨基酸合成蛋白质、核苷合成核酸、简单糖类合成多糖等均属于合成代谢。生物合成反应可以提供细菌生长所需的物质；三磷酸腺苷（ATP）分子使需能合成代谢和产能分解代谢偶联成为可能；聚合酶介导的DNA复制或RNA合成，核糖体、鞭毛、分泌系统等的组装均属于生物合成。

Anabolism: the enzyme-regulated energy-requiring reactions are mostly involved in anabolism. Anabolic processes often concern dehydration synthesis reactions. Examples of anabolic processes include the formation of proteins from amino acids, nucleic acids from nucleotides, and polysaccharides from simple sugars. These biosynthetic reactions generate the materials for cell growth. The coupling of energy-requiring anabolism and energy-releasing catabolism is made possible through the molecule adenosine triphosphate (ATP). Polymerase-mediated DNA replication and RNA synthesis as well as the assembly of ribosomes, pili and secretion systems all belong to biosynthesis.

第2节　细菌的生长繁殖
Section 2　Growth of bacteria

当细菌摄取了足够的营养物质之后，细菌的代谢活动增加，不断合成DNA和RNA，启动蛋白质等的合成，菌体以二分裂方式进行增殖（**图2-5**）。在液体培养基中，细菌的生长可用四个生长时期进行描述，分别是迟缓期、指数期、稳定期和衰亡期（**图2-6**）。其中，指数期生长的细菌细胞形态和染色最为典型，致病力也最强。

After taking up enough nutrients, bacterial cells exhibit a significant increase in metabolic activity, with continuous synthesis of DNA, RNA, proteins and other molecules, and replicate via binary fission (**Figure 2-5**). Four distinct phases of growth can be observed when the bacteria grow in liquid culture media: lag phase, exponential phase, stationary phase and decline phase (**Figure 2-6**).The exponential phase is the best timing for observation of typical bacterial morphology and staining , and the bacterial culture in this phase is also pathogenic.

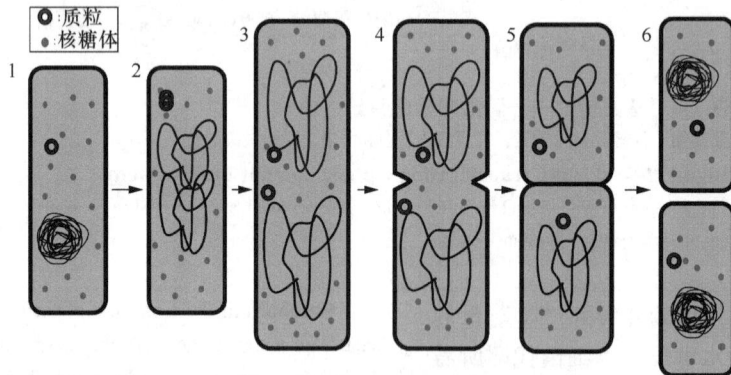

图 2-5　细菌以二分裂方式增殖
1. 分裂前的细菌有紧密缠绕的DNA；2. 细菌的DNA启动复制；3. 当菌体变大准备分裂时，DNA分开并被拉到两极；
4. 合成新的细胞壁并伴随细菌分离；5. 新细胞壁形成，细菌完成分裂；6. 新的子细胞含有紧密缠绕的DNA、核糖体和质粒。

Figure 2-5　Bacterial replication by binary fission
1. The bacterium has tightly coiled DNA before binary fission; 2. The DNA of the bacterium starts to replicate; 3. The DNA is pulled to the separate poles of the bacterium as it increases size to prepare for splitting; 4. The growth of a new cell wall begins with the separation of the bacterium; 5. The new cell wall fully develops, resulting in the complete split of the bacterium; 6. The new daughter cells have tightly coiled DNA, ribosomes, and plasmids.

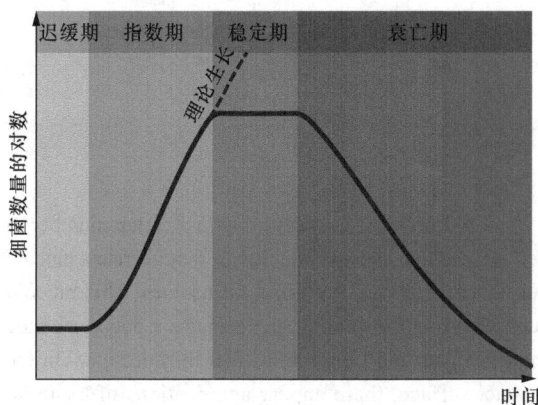

图2-6　细菌生长曲线
细菌在液体培养基中的生长可以分为四个时期：迟缓期、指数期、稳定期、衰亡期。

Figure 2-6 Bacterial growth curve
The growth of bacteria in batch culture can be modeled with four different phases: lag phase, log phase or exponential phase, stationary phase, and decline phase.

第3节　细菌的人工培养
Section 3　Cultivation of bacteria

虽然细菌在自然界环境中无处不在，但如果我们打算从动物身体或环境中分离某一种特定的细菌并非易事，需要明确什么样的培养基适合该菌，而且所选择的培养基只能培养该菌。

培养基：即人工配制的基质，含有细菌生长繁殖必需的营养物质。培养基的种类按照营养组成的差异可分为基础培养基和营养培养基；按照状态差异可分为固体培养基、半固体培养基和液体培养基；按照功能的差异可分为鉴别培养基、选择性培养基和厌氧培养基。对于某种特定细菌的培养，既可能是鉴别培养基也可能是选择性培养基，比如麦康凯培养基，既是鉴别培养基又是选择性培养基。麦康凯培养基含有胆盐，可以破坏革兰氏阳性菌细胞壁脂磷壁酸，革兰氏阳性菌在麦康凯培养基上不生长，因此可用来培养大肠杆菌等革兰氏阴性菌。虽然革兰氏阴性菌和阳性菌可在三糖铁培养基中生长，但不同的细菌在三糖铁培养基中呈现的颜色不同，因此可以相互鉴别，故称鉴别培养基。厌氧培养基是专为分离消化道微生物以及其他在厌氧环境中生长的细菌。不同细菌在不同的固体培养基表面呈现的菌落形状、大小和颜色不同，可相互区别。

Although bacteria are ubiquitous in natural environments, it might be hard to obtain a certain specific species of interest from animals or the environment. It is necessary to know the right culture media suitable for the target species, and the media used should be selective for the species of interest.

Culture media are man-made media that contain necessary nutrients for bacterial growth and replication. Culture media can be classified as basic media and enriched media based on the composition of nutrients, as solid media, semi-solid media and liquid broth by physical traits, or as differential media, selective media and anaerobic media by functionality. Either differential or selective media can be used for the cultivation of certain specific species. For example, the MacConkey medium is both differential and selective for culturing Gram-negative bacteria such as *E. coli*, because it contains bile salts that damage the lipoteichoic acids in the cell wall and inhibit the growth of the Gram-positive species. Both Gram-positive and negative bacteria can grow in the triple sugar iron (TSI) agar. However, the bacterial colonies are different in color between the two. That's why TSI agar is a differential medium. The anaerobic media are designed to culture microorganisms from the digestive tract and bacteria that require anaerobic conditions for growth. Different bacterial species exhibit different shapes, sizes and colors of the colonies in different solid media, which can be used for presumptive differential purposes.

第4节 细菌群落生长的调控
Section 4   Regulation of bacterial growth

生物被膜：细菌在培养基或接触物表面等特定环境中达到一定数量时分泌一些黏液样的物质，形成膜样结构将其自身缠绕其中，这一膜样结构即为生物被膜，包含多糖、纤维蛋白和脂蛋白等。细菌在固体接触物表面可以形成大量菌体聚集的膜样物（**图2-7**）。生物被膜具有结构和代谢复杂性，是细菌适应自然环境的一种生命现象，也是细菌应对不良环境的一种具有保护性的生长模式。包被有生物被膜的细菌称为被膜菌，细菌一旦形成生物被膜，对抗生素等药物和宿主免疫系统具有抵抗力；由于生物被膜中多糖的存在，被膜菌容易黏附到医疗器械表面，且难以清除，可引发医源性感染。

Biofilm: when bacteria grow to a certain number in culture media or specific conditions, they secrete a mucoid substance and wrap it around themselves. This mucoid substance is called biofilm and contains a polysaccharide matrix, fibrin and lipoproteins that they secrete. On the contact surface, there appear aggregations of bacterial cells with an envelope around themselves (**Figure 2-7**). The formation of biofilm has structural and metabolic complexity and indicates bacterial adaptation to the environments where they live. It also provides protection during their growth under harsh conditions. The bacteria coated with biofilm are called biofilm bacteria. Biofilm contributes to resistance to antibiotics and host immune systems. Polysaccharide-rich biofilms make it easy for bacteria to adhere to the surface of medical supplies, which is hard to clean off, thus causing iatrogenic infections.

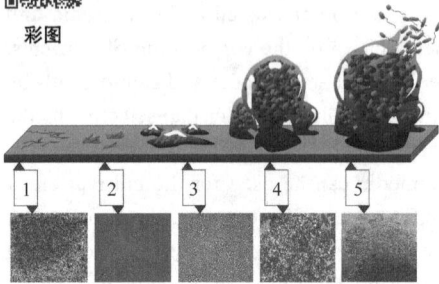

彩图

图2-7　生物被膜形成的五个阶段
1. 黏附；2. 不可逆黏附；3. 成熟Ⅰ；4. 成熟Ⅱ；5. 分散。图中每个阶段与正在形成生物被膜的铜绿假单胞菌显微照片相对应。所有显微照片显示比例相同。图像来源见英文。

Figure 2-7   Five stages of biofilm development
1. initial attachment; 2. irreversible attachment; 3. maturation Ⅰ; 4. maturation Ⅱ; and 5.dispersion. Each development stage in the diagram is paired with a photomicrograph of a developing *P. aeruginosa* biofilm. All photomicrographs are shown on the same scale. Source: Monroe D. 2007. Looking for chinks in the armor of bacterial biofilms. PLoS Bio, 5 (11): e307.

群体感应：当大量的细菌聚集在一起时，除了产生生物被膜之外，细菌之间还可以使用其特有的"语言"进行对话，这就是群体感应，又称密度感应。群体感应是指细菌通过其自发产生和释放的一些特定的信号分子，感知其浓度的变化，调节微生物群体活动的行为（**图2-8**）。群体感应可以产生生物被膜、菌体发光或者产生毒素等现象。细菌与细菌之间进行交流的

Quorum sensing: besides biofilm formation, when a large number of bacterial cells cluster, they communicate among themselves with a unique "language". This is called quorum sensing, also known as density sensing. The bacterial cells constitutively release special signal molecules that can be sensed by others within the population to regulate their activity in response to density changes, *e.g.* formation of biofilm, emission of fluorescence and production of toxins (**Figure 2-8**). The signal molecules are small and are known as autoinducers

革兰氏阴性菌群体感应

低密度　　　　　高密度　　　　生物被膜

细胞内部情况

2. 黏附并激活
调节蛋白

1. 通过细胞
膜扩散

3. 结合
DNA

4. 产生新
的AHL

5. 重复上述过程

细胞

酰基高丝氨酸内酯自诱导物

激活蛋白

染色体

AHL合酶

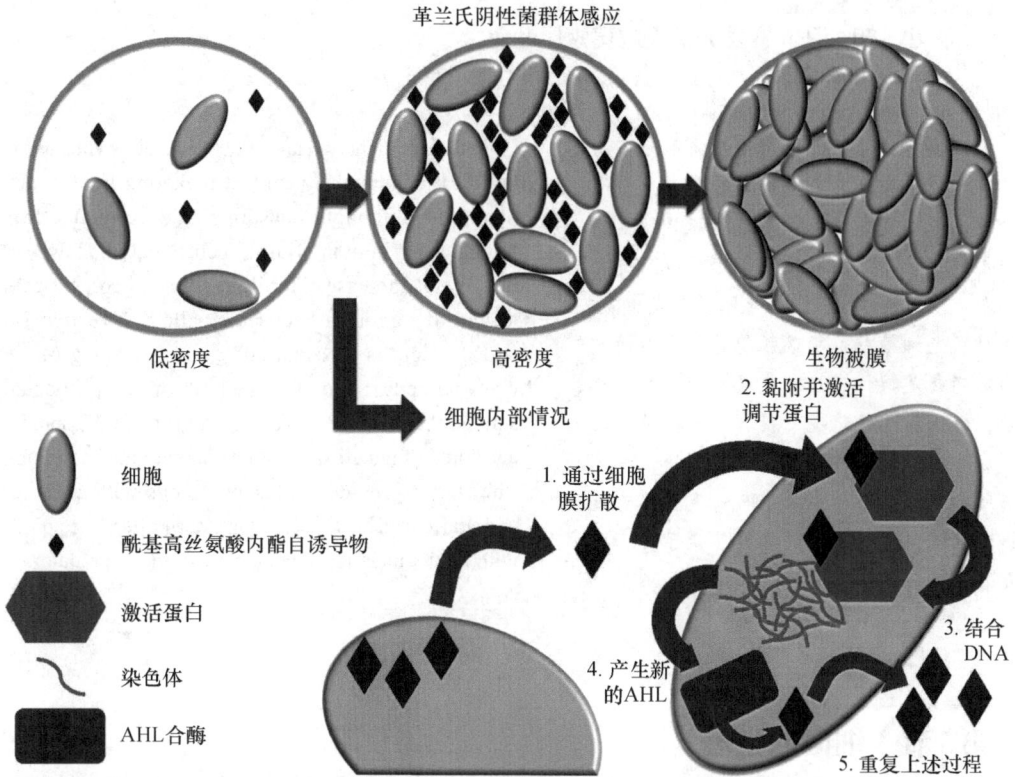

图2-8　群体感应示意图

1. 细菌将化学物质（自诱导物）分泌到环境中以感知周围的菌群数量。每种细菌都有其独特的自诱导物，对于革兰氏阴性菌而言，最常见的是酰基高丝氨酸内酯（AHL）。这种细胞与细胞间的密切接触和交流对生物被膜的形成至关重要。群体感应通过产生更多的细胞形成一定的密度。AHL可通过细胞膜从一个细胞扩散到邻近的另一细胞；2. AHL与激活蛋白黏附后再与细胞的染色体结合；3. 最后，通过AHL合酶与DNA结合产生更多的AHL，并分泌到胞外；4、5. 该过程不断重复，并使环境中的细菌数量持续增长，从而产生新的AHL和生物被膜。生物被膜中的多糖可增加被膜菌的致病性并阻止抗生素的渗透。

Figure 2-8　Quorum sensing diagram

1. Cells excrete chemicals (autoinducers) into their environment to measure the population around it. Every type of bacterium has its own type of autoinducer, and for Gram-negative bacteria acyl homoserine lactone (AHL) is the most common one. This close contact cell-to-cell communication is essential for biofilms to form. Quorum sensing ensures this density growth by producing more cells. An AHL will diffuse from one cell to another via the cellular membrane; 2. From there, it attaches onto an activator protein and then binds to the chromosomes; 3. Lastly, produce more AHL by AHL synthase binding to DNA and are secreted outside the cell; 4, 5. The process repeats and makes the population to create new AHL and the biofilm. Polysaccharides increase the pathogenicity of the biofilm and prevent the penetration of antibiotics.

"语言"为信息素，其本质是一种小分子物质。除AI-2（自诱导物）共有外，在革兰氏阴性菌和阳性菌的信息素有差别。革兰氏阴性菌信息素有高丝氨酸内酯类、大环内酯类、芳香族类、γ-丁内酯、喹诺酮类、羟基酮类等；革兰氏阳性菌信息素为寡肽类物质。

or pheromones. They differ between Gram-positive and Gram-negative bacteria, except for autoinducer 2 (AI-2), which is common to both. The pheromones of Gram-negative bacteria include homoserine lactones, macrolides, aromatics, γ-butyrolactones, quinolones, hydroxyketones, *etc.*, while those of Gram-positive bacteria are mostly oligopeptides.

小 知 识 / Extended Knowledge — 发光的夏威夷短尾鱿鱼 / Luminous Hawaiian bobtail squid

在漆黑的夜晚，夏威夷短尾鱿鱼在海水中游弋，在海面上不停地闪烁着"鬼火"一般的光芒。很多人以为夏威夷短尾鱿鱼有一个发光器官，但其实夏威夷短尾鱿鱼并非自己发光，而是依靠一种发光细菌而发出光芒，这种发光细菌就是费氏弧菌，费氏弧菌发光的本领就是群体感应。夏威夷短尾鱿鱼出生后，会从周围的环境中获取费氏弧菌并与之形成共生关系。鱿鱼为费氏弧菌提供栖息的场所和稳定的营养来源，而费氏弧菌为鱿鱼提供照明，在夜晚可以躲避潜伏于其下的捕食者。

In dark nights, a Hawaii bobtail squid swims in the ocean. It continuously emits a flickering light. Many people have thought that this might be related to a luminous organ in the Hawaii bobtail squid. However, this is not the case. The light actually comes from a kind of luminous bacterium called *Vibro fischeri* and the light is a product of quorum sensing of the bacterial cells. Soon after the birth of Hawaii bobtail squids, they acquire *V. fischeri* from the environment and form a mutualistic relationship with the bacterium. The squids provide the bacterial cells with a shelter and nutrients, and *V. fischeri* in return offers the squids with light which helps them away from the predators.

第5节 细菌的生态
Section 5 The ecology of bacteria

人和动物消化道、口腔、泌尿生殖道以及体表存在众多的微生物种群，共同维持机体的微生态平衡和健康状态。微生物种群之间的相互关系包括：

互生：指两种或两种以上微生物共同生存时，可互相受益。互生双方可以在自然界中单独存在，互生时又可使对方受益，如宿主体内的正常菌群。

共生：两种或多种微生物共同生活在一起，彼此互利，以致分离后单独不能很好地生活。例如，由蓝细菌（进行光合作用，为真菌提供养料）和真菌（产生有机酸分解矿物质供蓝细菌）组成的地衣。

拮抗：两种或两种以上微生物共同生长时，使双方或一方受害的现象。双方受害叫竞争，一方受害叫偏生，如放线菌和真菌（产生抗生素）对细菌的抑制现象。

There are many microbial floras in the digestive tract, oral cavity, urogenital tract and body surface of humans and animals. Maintenance of the microecological balance between the microbes and hosts or among the microbial population contributes to the health of the hosts. The relationship between or among microbial floras includes:

Mutualism/synergism: when two or more types of microorganisms live together, they can benefit from each other. Each one of the mutualistic partners can exist alone in nature and benefit each other when they become neighbors, *e.g.* the normal flora in the host.

Symbiosis/commensalism: two or more types of microorganisms live together and benefit each other, but they cannot live well alone, *e.g.* lichens are symbiotically composed of cyanobacteria which feed fungi with its photosynthesis, and fungi which produce organic acids to decompose minerals for cyanobacteria.

Antagonism: it is a phenomenon that two or more types of microorganisms grow together, causing damage to one side or both. If both sides suffer, it is called competition. If only one side is negatively affected, it is called amensalism. *e.g.* the inhibition of certain bacteria by actinomyce and fungi (producing antibiotics).

寄生：一种小型微生物生活在另一大型微生物的体内或体表，从中获取营养生长繁殖并使后者蒙受伤害或被杀死的现象，如噬菌体与细菌的关系。

正常菌群与其宿主之间的关系为共生关系，具体表现为：营养（如瘤胃微生物将纤维素类饲料降解为可被吸收的营养物质）、免疫（如肠道菌群维持宿主的正常免疫反应）和生物拮抗（如肠道正常菌群对外侵微生物的拮抗作用）。

益生菌：某些细菌或真菌有利于宿主肠道微生物区系的平衡，能抑制对宿主有害的微生物的生长，这些微生物被称为益生菌（如酸奶中的乳酸杆菌、双歧杆菌等）。

益生元：一些寡糖类物质（如菊粉、纤维素）添加给动物，不被宿主消化吸收，但是能选择性刺激宿主消化道内有益微生物的生长，对宿主产生有益作用，这类成分可称益生元。

菌群失调：如果宿主患病、外科手术、环境改变、滥用抗菌药物等，宿主机体某个部位正常菌群中的微生物种类和数量或栖居处发生改变，引起菌群失调。

悉生生物学是一门研究已知微生物状态动物体的科学。悉生动物是携带微生物悉数被掌握的动物，是医学和兽医学领域重要的模式生物。悉生动物主要分成三类：无菌动物、已知菌动物和无特定病原体动物。

无菌动物：不携带任何微生物的动物。

已知菌动物：在无菌动物基础上，有目的地接种，使其携带某些已知微生物的动物。

无特定病原体动物：指不存在某些特定的具有致病性或潜在致病性微生物的动物。

除了上述提到的三种悉生动物之外，实验动物还有以下两种：

清洁动物：来源于剖腹产，饲养于半屏障系统，体内外不含主要病原体的动物。

普通动物：开放条件下饲养的动物。

Parasitism: it is a phenomenon that a small microorganism lives in the body or on the body surface of another larger microorganism, from which it obtains nutrients, grows and causes damage or death to the latter, *e.g.* the relationship between phages and bacteria.

The relationship between the normal flora and its host is symbiotic, manifested as nutritional supplementation (*e.g.* rumen microorganisms assisting in the decomposition of fibrous fodder into absorbable nutrients), protective immunity (*e.g.* intestinal flora that help establish host normal immune responses), and biological antagonism (*e.g.* the antagonistic effects of normal non-pathogenic flora on pathogenic microorganisms).

Probiotics: some bacteria or fungi are beneficial to maintaining homeostasis of intestinal microflora and inhibit the growth of harmful microorganisms in the host. These microbes are called probiotics, *e.g. Lactobacillus, Bifidobacterium* in yogurt.

Prebiotics: some oligosaccharides (such as inulin and cellulose) added to animal feeds are not digested and absorbed by the host. However, they can selectively stimulate the growth of beneficial microorganisms in the digestive tract and have beneficial effects on the host. Such ingredients are called prebiotics.

Dysbacteria: the changes of bacterial flora are often triggered by illness, surgery, environmental changes, abuse of antibacterial drugs, *etc*. The type and quantity of bacterial flora at a certain part of the host body or environment may shift, causing dysbacteria.

Gnotobiotics or gnotobiology is a study exploring animals with a known microbiological state which are referred to as gnotobiotes. Such animals are important model organisms in the field of medicine and veterinary medicine. There are three types of gnotobiotes as follows: germ-free animals, gnotobiotic animals and specific pathogen-free animals.

Germ-free animals: the animals that do not carry any microorganisms.

Gnotobiotic animals: the animals that are derived from germ-free animals by inoculating defined species of microorganisms for specific purposes.

Specific pathogen-free animals: the animals that do not carry some specified pathogens or microorganisms of potential pathogenicity.

In addition to the three types of animals mentioned above, there are two other experimental animals as below:

Clean animals: the animals that are derived from caesarean section, raised in a semi-barrier system and do not carry major pathogens *in vivo* and on the body surface.

Conventional animals: the animals raised under open conditions.

思考题　Questions

1. 葡萄糖是如何通过细菌的细胞膜进行转运的?

2. 生物被膜与群体感应之间有何联系?

3. 从细菌与宿主和环境之间的生态关系角度阐述人与自然和谐共生的"同一世界·同一健康"理念。

1. How is glucose transported across bacterial cell membranes?

2. What is the relationship between biofilm and quorum sensing?

3. Please elaborate the concept of "One World, One Health" — harmonious coexistence between humans and nature from the view of ecological relationship between bacteria and hosts or environments.

第3章 消毒、灭菌及兽医微生物实验室生物安全

Chapter 3 / Disinfection, sterilization and biosecurity in the veterinary microbiology laboratory

内容提要 "东死鼠，西死鼠，人见死鼠如见虎；鼠死不几日，人死如垿堵。"瘟疫，自古以来都令人恐惧，那么如何预防和面对呢？这就需要了解消毒、灭菌及生物安全的重要性。

Introduction "The dead bodies of mice could be just as frightening as tigers to the people around. The death of mice would be followed by death of humans with piles of corpses." Plague has been a terrifying word in history. How do we prevent and deal with it when it happens? It is essential to understand the importance of disinfection, sterilization and biosecurity.

第1节 物理消毒法
Section 1 Physical disinfection

基本概念

消毒指杀灭固体物表面或液体中的病原微生物的方法。消毒只要求达到消除传染性的目的，而对非致病性微生物及其芽孢、孢子并不严格要求全部杀死。

灭菌指杀灭物体中所有病原微生物和非致病性微生物及其芽孢、霉菌孢子的方法。

防腐指阻止或抑制微生物生长繁殖的方法，微生物不一定死亡。

无菌指没有活的微生物的状态。

无菌操作指防止或杜绝任何微生物进入动物机体或其他物体的操作。

消毒灭菌的方法主要有物理消毒法和

Basic concept

Disinfection is the process of killing pathogenic microorganisms on any solid objects or in liquids. The goal is to eliminate the infectivity. It does not necessarily inactivate non-pathogenic microorganisms and their spores.

Sterilization is the process of killing all microorganisms, bacterial and fungal spores.

Antisepsis is a method to prevent or inhibit the growth of microorganisms.

Asepsis is a state in which there are no living microorganisms.

Aseptic techniques are the methods aimed at preventing any microorganisms from entering animal bodies or objects.

The methods of disinfection and sterilization may

化学消毒法两大类，其中物理消毒法包括：热力灭菌法、辐射灭菌法、超声波灭菌法、滤过除菌法等。

热力灭菌法是指采用高温的方法灭活微生物，包括干热灭菌法和湿热灭菌法两种。干热灭菌法是指直接用火焰灼烧或者热空气灭菌，适用于玻璃、瓷器、金属器械等。湿热灭菌法是指利用湿热的空气进行灭菌，适用于手术器械和各种液体培养基和液体食品等（图3-1）。湿热灭菌的经典代表是巴氏消毒法（图3-2），此方法是由法国著名微生物学家巴斯德发明的。巴氏消毒法包括三个类别，分别是：63～65 ℃维持 30 min；71～72 ℃ 维持 15 s；132 ℃维持 1～2 s，分别代表：低温、高温和超高温巴氏消毒。巴氏消毒的优点是可以杀灭液态食品中的病原菌或特定微生物，而又不致严重损害其营养成分和风味，主要用于葡萄酒、啤酒和牛奶的消毒。

be physical or chemical. Physical sterilization includes thermal sterilization, radiation sterilization, ultrasound sterilization, filtration, *etc.*

Thermal sterilization refers to the use of high-temperatures to inactivate microorganisms, including dry heating sterilization and moist heating sterilization. **Dry heating** includes burning fire or hot air, which is applicable to glass, china, and metal instruments. **Moist heating** uses moist and hot air, which is applicable to surgical instruments, liquid growth media and foods (**Figure 3-1**). A classic example of sterilization by moist heating is **pasteurization (Figure 3-2)**. It is named after the well-known French microbiologist, Louis Pasteur, who invented the method. There are three heat treatment options in pasteurization including 63-65 ℃ for 30 min, 71-72 ℃ for 15 s, or 132 ℃ for 1-2 s, referred to as low temperature, high temperature and ultra-high temperature pasteurization, respectively. The advantages of pasteurization include killing pathogenic bacteria or specific microorganisms in the liquid foods with maximally preserving the nutrients and flavors therein. It is widely used in the sterilization of wine, beer and milk.

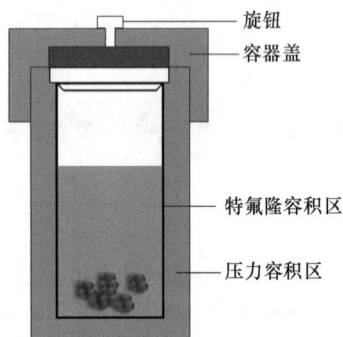

图3-1　高压灭菌容器示意图
图像来源见英文。

Figure 3-1　Schematic of typical hydrothermal autoclave equipment
Source: Yang G, Park SJ. 2019. Conventional and microwave hydrothermal synthesis and application of functional materials: a review. Materials (Basel), 12 (7): 1177.

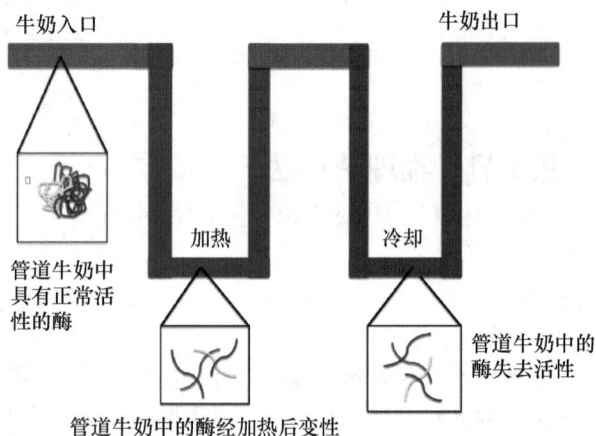

图3-2　巴氏消毒工艺概述
牛奶从左边进入管道，牛奶中含有具有功能活性的酶和蛋白质。当牛奶流经加热管道时酶和蛋白质将受热变性。因细胞功能受损，病原体生长停止。最后牛奶流经冷却管道，可防止发生美拉德反应和焦糖化。巴氏消毒还将管道中的细胞通过加热加压的方式崩解。

Figure 3-2　The general overview of the pasteurization process
The milk enters the pipe on the left, and the functional enzymes and proteins become denatured as the milk flows through the heating pipe. This helps to stop pathogen growth by damaging the functions of cells. As the milk flows through the cooling pipe, it prevents the occurrence of Maillard reaction and caramelization. The pasteurization process is also able to make the cells burst due to pressure and heating.

辐射灭菌法包括可见光、日光、紫外线、电离辐射、红外线、微波灭菌等。可见光灭菌效果最弱，但若将结晶紫、亚甲蓝、汞溴红、伊红、沙黄等染料加到培养基或涂在外伤表面，能增强可见光的杀菌作用，即**光感作用**。光感作用的原理是染料遇到可见光产生超氧离子，破坏细菌蛋白质中的色氨酸、酪氨酸、组氨酸、蛋氨酸和半胱氨酸，氧化核酸中的鸟嘌呤等。直射日光有强烈的杀菌作用。紫外线消毒法是一种常见的灭菌方法，实验室常用的杀菌灯紫外线波长为254 nm，可以通过形成胸腺嘧啶二聚体直接损伤细菌细胞的DNA（图3-3）。细菌受到死量的紫外线照射后，若3 h内再以可见光照射，则部分细菌又能恢复活力，即**光复活现象**。电离辐射包括放射性同位素射线、X射线等。红外线辐射灭菌指的是采用0.77～1000 μm的电磁波进行灭菌，其中以1～10 μm波长热效应最强。微波即超高频电磁波，灭菌常用的频率为915～2450 MHz。

Radiation sterilization includes the use of visible light, sunlight, ultraviolet light, ionizing radiation, infrared light and microwaves. Among them, visible light is the weakest in sterilizing activity. When dyes such as crystal violet, methylene blue, merbromin, eosin and safranin are added to the growth media or topically applied to wounds, the sterilizing activity of visible light is enhanced, a phenomenon called **photosensitization**. When exposed to visible light, these dyes produce superoxides that are destructive to tryptophan, tyrosine, histidine, methionine, cysteine in bacterial proteins and guanine in nucleic acid, etc. Ultraviolet light is commonly used for sterilization. The wavelength used in the laboratory is 254 nm, which can directly damage DNA by the formation of thymine dimers (**Figure 3-3**). **Photoreactivation** refers to the phenomenon that some bacteria pre-exposed to a lethal dose of ultraviolet light can restore their activity if subjected to visible light within 3 h. Ionizing radiation includes radioisotope rays, X-rays, *etc*. Infrared radiation refers to sterilization with electromagnetic waves of 0.77-1000 μm in wavelength, the ones with a wavelength from 1 μm to 10 μm having the strongest thermal effect. Microwave is an ultra-high frequency electromagnetic wave. The frequency for sterilization is 915-2450 MHz.

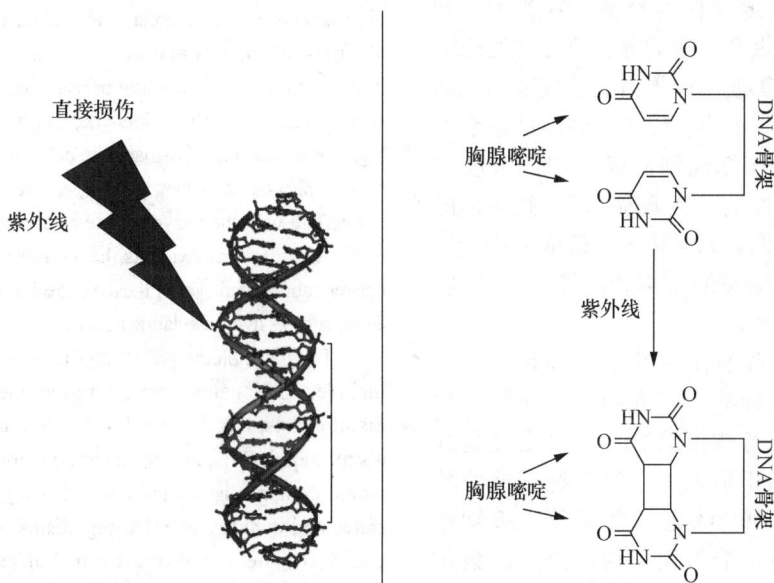

图3-3 紫外线直接损伤DNA
紫外线光子直接被DNA吸收（左），形成胸腺嘧啶-胸腺嘧啶-环丁烷二聚体（右）
Figure 3-3 Direct DNA damage by UV light
The UV-photon is directly absorbed by DNA (left). One of the possible reactions from the excited state is the formation of a thymine-thymine cyclobutane dimer (right).

超声波灭菌法是指使用20 000～200 000 Hz的声波，在细菌细胞表面产生空泡和10^5 kPa的瞬时高压从而产生空化或腔化作用，使细菌裂解。超声波破碎法可用于提取细菌细胞中的蛋白组分，在实际操作过程中可根据含有细菌的溶液体积选择合适的超声探头。

滤过除菌法也称过滤除菌，是指通过物理阻留作用使液体或空气通过特定孔径的滤膜从而除去其中的细菌等微生物的方法，如实验室常用的营养液、血清等除菌。

Ultrasound sterilization uses ultrasound ranging from 20 000 to 200 000 Hz in frequency for sterilization. It forms vacuoles on the surface of bacteria and generates an instantaneous pressure as high as 10^5 kPa. The consequential cavitation effect tears apart bacteria. This method is also used by scientists in the lab to extract intracellular proteins from bacterial cells. In practice, different probes can be chosen based on the volume of the bacterial suspension.

Filtration is a sterilization method that is achieved through the physical retention of bacteria by forcing liquid or air through filter membranes with a specific pore size. Culture broths and serum samples are commonly sterilized this way in the laboratory.

第2节 化学消毒法
Section 2　Chemical disinfection

化学消毒法是指使用化学消毒剂或防腐剂进行消毒。消毒剂是指用于杀灭病原微生物的化学药品，防腐剂是指用于抑制病原微生物生长繁殖的化学药品。目前市售的化学消毒剂种类繁多，有酚类、醇类、醛类、重金属盐类等。消毒剂的杀菌机制包括：①使菌体蛋白变性或凝固，如高浓度酚类、醇类、醛类、高浓度重金属盐等；②损伤细菌细胞膜，如低浓度酚类、表面活性剂、醇类等；③干扰细菌的酶系统和代谢，如氧化剂、低浓度重金属盐等；④改变核酸结构和功能，如染料、烷化剂等。

影响消毒剂作用的主要因素包括：①消毒剂的性质、浓度和作用时间，如消毒用的乙醇浓度为70%，浓度过高或过低都不利于消毒。②微生物所处的状态和数量，如70%乙醇能有效杀灭细菌繁殖体，但不能杀死芽孢。微生物数量越多，消毒时间越长。③环境温度，如温度每升高10 ℃，重金属盐杀菌作用增加2～5倍。④酸碱度，如戊二醛需要加入$NaHCO_3$才能发挥杀菌作用。⑤有机物，如环境中的蛋白质等有机物可以降

Chemical disinfection refers to the use of chemical disinfectants or preservatives for disinfection. Disinfectants are chemicals used to kill pathogenic microorganisms. Preservatives are chemicals used to inhibit the growth and reproduction of pathogenic microorganisms. A variety of disinfectants are commercially available, including phenols, alcohols, aldehydes and heavy metals. Mechanistically, disinfectants act to ①denature or coagulate bacterial proteins, such as phenol, alcohol, aldehyde, heavy metal salts, *etc.* at high concentrations; ②damage the cell membrane of bacteria, such as phenol, surfactant, alcohol, *etc.* at low concentrations; ③interfere with the enzyme systems and metabolic activities of bacteria, such as oxidants, heavy metal salts, *etc.* at low concentrations; and ④alter the structure and function of nucleic acids, such as dyes, alkylating agents, *etc.*

The main factors affecting the action of disinfectants include: ①the nature, concentration and contact time of disinfectants, *e.g.* the alcohol concentration for effective disinfection is 70%, and concentrations too high or too low would compromise disinfection efficacy. ②physiological states and quantity of microorganisms, *e.g.* 70% alcohol could kill the vegetative form, but cannot inactivate spores. The higher the bacterial load is, the longer the time required to kill the bacteria. ③environmental temperature, The effect of heavy metal salt sterilization is increased by 2-5 times for every 10℃ increase in temperature. ④pH, *e.g.* $NaHCO_3$ is necessary to modulate pH of the glutaraldehyde solution for effective sterilization.

低消毒效果。⑥药物的相互拮抗，如阴离子肥皂和阳离子苯扎溴铵混合使用时，将削弱消毒效果。

⑤organic substances, *e.g.* proteins in the environment can reduce the disinfection effect. ⑥mutual antagonism of disinfectants or chemicals, *e.g.* anionic soap and cationic benzalkonium bromide can weaken the disinfection effect when used together.

第3节 抗生素和细菌素
Section 3 Antibiotics and bacteriocins

抗生素是指某些微生物在代谢过程中产生的一类可以抑制或者杀死另外一些微生物的物质，如青霉素是青霉菌产生的一种次级代谢产物，这种代谢产物不会杀死青霉菌，但可以干扰革兰氏阳性菌细胞壁的合成而起到抗菌作用。

细菌素是指由细菌产生的一种具有抗菌作用的蛋白质或多肽，只作用于与它同种不同菌株的细菌及亲缘关系比较近的细菌，如乳酸链球菌素是乳酸链球菌产生的代谢产物，对单核细胞增生李氏杆菌等革兰氏阳性菌具有抑菌作用。

植物杀菌素是指黄连、金银花、马齿苋、板蓝根等植物中含有的杀菌物质。

Antibiotics are substances produced by certain microorganisms during metabolic processes and they can inhibit or kill other microorganisms. Penicillin is a secondary metabolic product produced by *Penicillium* fungi. It cannot kill *Penicillium* but can inactivate Gram-positive bacteria by disrupting their cell wall synthesis.

Bacteriocins are antibacterial proteins or peptides produced by certain bacteria. They only act on other strains of the same or closely related species. Nisin, a metabolic product of *Lactococcus lactis*, has inhibitory activity against some Gram-positive bacteria such as *Listeria monocytogenes*.

Herbs like *Coptis chinensis*, *Lonicera japonica* (honeysuckle), *Portulaca oleracea* and *Radix isatidis* also contain bactericidal substances referred to as **phytocidins**.

第4节 兽医微生物实验室生物安全
Section 4 Biosafety of veterinary microbiology laboratory

兽医微生物实验室生物安全主要是指病原微生物传入和传出以及对人、环境和动物的安全性。生物安全级别对应于病原微生物的危险级别，不同风险级别的微生物的实验室操作要求不同。生物安全级别的最低级是1级，最高级是4级。国际通用的生物危害标识是类似于4个圆圈的叠加形状图，颜色从黄色、橙色和红色之间渐变，分别代表1级到4级的递进（**图3-4**）。1级实验室又称P1实验室，对人和动物的风险最低，可从事已知对实验室工作人员和动物无明显致病性，对环境危害程度小，特性清楚的微生物，如大肠杆

Biosafety of the veterinary microbiology laboratory mainly refers to safety concerns related to pathogenic microorganisms, entering or be released from laboratories, posing risks to humans, environments and animals. Biosafety levels correspond to the risk levels of the pathogens used in the laboratory. Laboratories working with pathogens of different risk levels should be built and operated accordingly. It is universally accepted that biosafety level 1 (BSL-1) is of the lowest risk, while level 4 (BSL-4) is of the highest risk. The symbol for biosafety is composed of 4 overlapping circles with a color gradient from yellow to orange to red, standing for BSL-1 to BSL-4 (**Figure 3-4**). BSL-1, also known as P1, stands for the lowest risk to human beings and animals. It applies to laboratory settings where

菌。目前高校和科研院所常见的微生物实验室级别是1级和2级实验室。而P3和P4实验室活动以及实验室建设需要经过严格审批，所从事的病原微生物对人或动物可能造成严重危害，比如口蹄疫、结核杆菌的有关操作需要在P3实验室进行。P4实验室是目前人类所拥有生物安全等级最高的实验室。

personnel work with microbes that are well known to be of low risk, such as *E. coli*. BSL-1 and BSL-2 are the most common levels worldwide in universities and research institutes. BSL-3 and BSL-4 require strict scrutiny as they work with pathogenic microbes that pose significant threats to humans and/or animals, *e.g.* foot and mouth disease virus and *Mycobacterium tuberculosis* are handled only in a BSL-3 lab. The highest level defined so far is BSL-4.

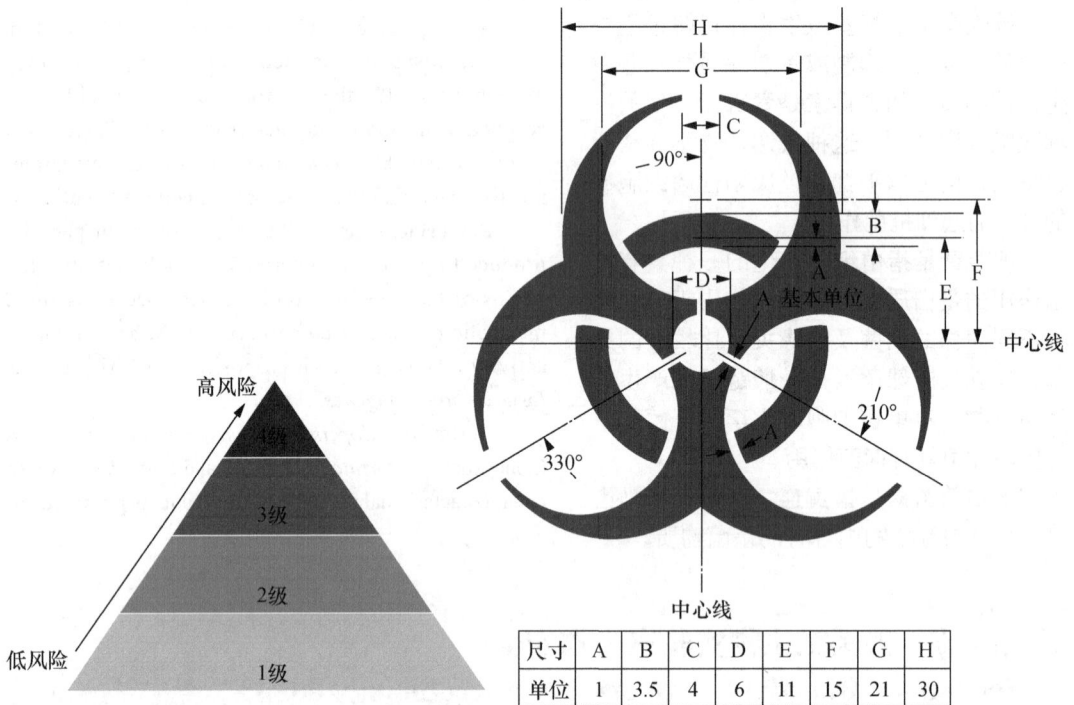

尺寸	A	B	C	D	E	F	G	H
单位	1	3.5	4	6	11	15	21	30

图3-4　生物安全级别1～4（左）和生物危害标识尺寸（右）示意图
美国疾病控制和预防中心将疾病按照生物危害级别进行分类，1级为最低风险，4级为最高风险。相应的实验室和设施称为BSL（生物安全级别）1～4或简称为P1～P4（病原体或防护级别）。生物危害标识最早于1966年由陶氏化学公司的一名负责容器产品研发的环境健康工程师查尔斯·鲍德温发明。

Figure 3-4　The diagram of BSL (biosafety level) 1-4 (left) and the biohazard symbol with dimensions (right)
The United States Centers for Disease Control and Prevention (CDC) categorize various diseases at different levels of biohazard, level 1 being minimum risk and level 4 being extreme risk. Laboratories and other facilities are categorized as BSL (biosafety level) 1-4 or as P1 to P4 in short (pathogen or protection level). The biohazard symbol was developed in 1966 by Charles Baldwin, an environmental health engineer working on containment products at Dow Chemical Company.

思考题 **Questions**

1. 非洲猪瘟病毒的分离、培养和鉴定应在什么级别的实验室进行，为什么？

1. What level of laboratory should be used for isolation, culture and identification of African swine fever virus and why?

2. 为了维护国家安全，防范和应对生物安全风险，保障人民生命健康，保护生物资源和生态环境，促进生物技术健康发展，推动构建人类命运共同体，实现人与自然和谐共生，《中华人民共和国生物安全法》已由第十三届全国人民代表大会常务委员会第二十二次会议通过，自2021年4月15日起施行。该法体现了什么理念？

2. In order to maintain national security, prevent and cope with biosecurity risks, ensure people's lives and health, protect biological resources and ecological environment, promote the development of biotechnology, facilitate the establishment of a community of shared future for mankind, and realize the harmonious coexistence of human and nature, the *Biosafety Law of the People's Republic of China* has been promulgated by the 22nd session of the Standing Committee of the 13th National People's Congress. It is approved and will come into effect on April 15, 2021. What concept does the law embody?

第4章 细菌的感染与致病机理
Chapter 4 / Bacterial infection and pathogenic mechanisms

内容提要 感染是微生物入侵宿主并在其体内特定组织细胞内繁殖；当感染对机体造成损害，宿主表现症状，即为感染性疾病；若病原可以通过感染动物直接或间接传给未感染的健康动物，引发疾病传播，即称之为传染病。但是，习惯上传染病的英文也用"infectious disease"表示。不同细菌可以引起多种不同的感染性疾病，临床症状严重程度从不明显到明显不等；病原感染而不表现症状称为无症状感染。

Introduction Infection is the invasion of the host by microorganisms with subsequent multiplication in cells of specific organs or tissues. Infectious disease occurs when the cells in the body are damaged as a result of the infection, and signs and symptoms of an illness appear. Contagious disease (or communicable disease) occurs if the causal pathogens could be passed from the infected to the uninfected animals by direct contact or indirect exposure. However, the term infectious disease is customarily used as an equivalent to the term contagious disease in the literature. Different pathogenic bacteria can cause a multitude of different infections, ranging in severity from inapparent to fulminating in clinical signs. Infections that do not cause clinical signs are called asymptomatic infections.

第1节 细菌的致病性和毒力
Section 1 Pathogenicity and virulence of bacteria

致病性是指一定种类的病原菌在一定条件下，能在宿主体内引起感染的能力，是细菌种的特征之一。

毒力是指病原菌致病力的强弱程度，是量的概念。同种细菌的不同菌株，其毒力不一样。因此，毒力是菌株的特征。

经典的科赫法则：如何才能确定某种特定病原菌的致病性？比如，我们如何才能确定某动物的伤口感染一定是由金黄色葡萄球菌引起的？100年前，德国内科医

Pathogenicity is the ability of a bacterial pathogen to cause infection under a certain condition, considered as one of the important traits of bacteria at the species level.

Virulence is a quantitative concept of bacterial pathogenicity indicating the degree of severity in causing infections. Within the same species, virulence can vary among different strains, so it is a character at the strain level.

Classical Koch's Postulates: how do we confirm the pathogenicity of a specific bacterium? For example, how do we know that the wound infection in an animal is caused by *Staphylococcus aureus*? A century ago, Robert

生——诺贝尔生理学或医学奖获得者罗伯特·赫尔曼·科赫（**图4-1**）同样也问了类似的问题，并提出了可以确定病原菌致病性的科赫法则（**图4-2**）。科赫法则的内容主要包括四点：第一，特殊的病原菌应在同一疾病中查见，在健康者中不存在；第二，此病原菌能被分离培养而得到纯种；第三，此纯培养物接种易感动物，能导致同样病症；第四，自实验感染的动物体内能重新获得该病原菌的纯培养。

Hermann Koch (**Figure 4-1**), a German physician and Nobel Prize laureate in Physiology and Medicine, wondered the same thing and developed Koch's Postulates (**Figure 4-2**). The postulates include four rules: firstly, the pathogen shall be present in all cases of the same disease and absent in healthy individuals; secondly, a pure culture of the pathogen could be isolated; thirdly, the same disease must be reproduced when a susceptible host is inoculated with the pure culture of the pathogen; fourthly, a pure culture of the same pathogen must be recovered from the inoculated hosts.

图4-1　罗伯特·赫尔曼·科赫（1843～1910）

德国内科医生，发现了细菌学领域的科赫法则，分离了炭疽、结核和霍乱的病原。

Figure 4-1　Robert Hermann Koch (1843-1910)

A German physician, known for discovering bacteriology Koch's Postulates of germ theory, and isolating anthrax, tuberculosis and cholera.

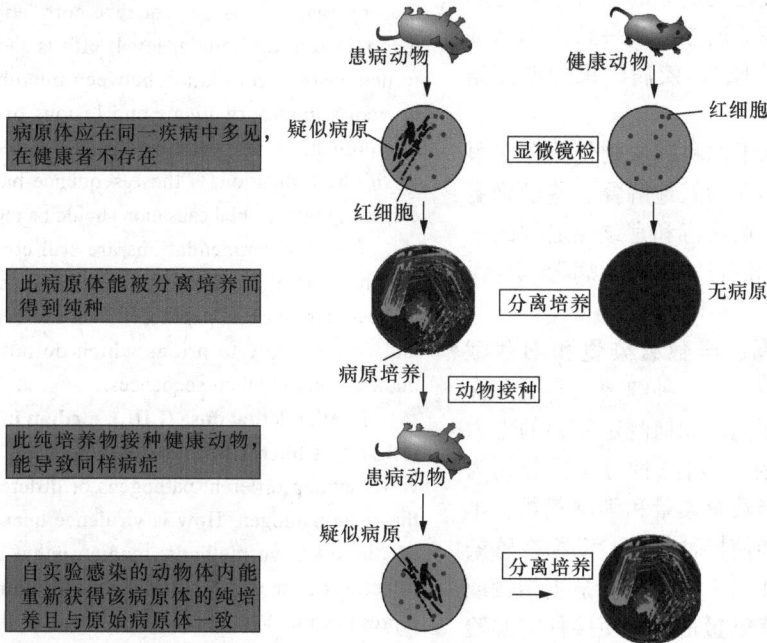

图4-2　科赫法则——旨在建立致病微生物和疾病之间的因果关系

Figure 4-2　Koch's Postulates designed to establish a causal relationship between a causative microbe and a disease

分子水平科赫法则：科学家发现经典科赫法则具有一定的局限性，比如有些病原菌感染宿主不引起临床发病，是健康带菌；或者迄今仍无法在体外人工培养；或无可用的易感动物等。因此，无法按照经典科赫法则进行验证。因此，美国科学家弗雷德里克斯和雷蒙于1996年提出了分子水平的科赫法则。①某种传染病的大多数病例中应存在病原的核酸序列。病原核酸主要存在于病变器官或组织中，无病变的器官中则无。②健康的宿主或组织中与病原体相关的核酸序列拷贝数应较少或没有。③随着疾病消失，病原体相关核酸序列的拷贝数应减少或检测不到。随着疾病复发，情况相反。④当核酸阳性早于临床症状，或者核酸拷贝数与疾病或病理的严重程度正相关时，核酸与疾病的关联为因果关系。⑤从已知核酸序列推断出的病原菌性质应与该菌的生物学特性一致。⑥应在细胞水平寻找组织-序列相关性：通过特异性原位杂交证明病原的核酸序列与病变区域或疑似病原所在区域一致。⑦该基于核酸序列的病原菌-疾病因果关系证据可重复验证。

以上分子水平的科赫法则仍然存在争议，因为它并不能很好地解释一些已确定的疾病，如乳头瘤病毒和宫颈癌之间的关系；也不适用朊病毒，因为朊病毒本身无核酸序列。

半数致死量、半数感染量和细菌载量：不同种细菌或同一种细菌的不同菌株致病性或毒力不同，如何测定细菌的毒力呢？测定细菌的毒力有3种方法，分别为半数致死量、半数感染量和细菌载量。**半数致死量**是指能使接种的实验动物在感染后一定时限内死亡一半所需的微生物量或毒素量。**半数感染量**是指能使接种的实验动物、鸡胚或细胞在一定时限内感染一半所需的微生物量或毒素量。无论是半数致死量还是半数感染量，对实验动物都有一

Koch's Postulates at the molecular level: however, there are limitations to Koch's Postulates. For example, some pathogens do not cause clinical signs in healthy individuals who act as carriers of such pathogens. Some pathogens are not culturable *in vitro* or there are no susceptible animal species available for artificial inoculation. In these cases, Koch's Postulates cannot be applied. American scientists Fredricks and Relman (1996) have suggested the following rules: ① a nucleic acid sequence belonging to a putative pathogen should be present in most cases of an infectious disease. Microbial nucleic acids should be found preferably in those organs or anatomic sites known to be diseased, and not in those organs that lack pathology. ②fewer, or no, copies of pathogen-associated nucleic acid sequences should occur in hosts or tissues without disease. ③with the resolution of disease, the copy number of pathogen-associated nucleic acid sequences should decrease or become undetectable. With clinical relapse, the opposite should occur. ④when sequence-positive detection predates clinical symptoms, or a sequence copy number correlates with the severity of disease or pathology, the sequence-disease association is more likely to be a causal relationship. ⑤the nature of the microorganism inferred from the available sequence should be consistent with the known biological characteristics of the organisms. ⑥tissue-sequence correlation should be determined at the cellular level: efforts should be made to demonstrate correlation between microbial sequence or visible microorganisms and lesions or areas where microorganisms are presumed to be located by specific *in situ* hybridization. ⑦these sequence-based forms of evidence for microbial causation should be reproducible.

These recommendations are still controversial in that they do not account well for some established causal associations, such as papillomavirus and cervical cancer, and do not apply to prions which do not contain any nucleic acids and their sequences.

Median lethal dose (LD_{50}), median infectious dose (ID_{50}) and bacterial load: pathogenicity or virulence varies among different pathogens or different isolates of the same pathogen. How is virulence quantified? There are three known methods: median lethal dose, median infectious dose and bacterial load. **Median lethal dose** represents the dose needed to kill half of the challenged animals within a period, and **median infectious dose** represents the dose needed to infect half of the challenged animals, embryonated eggs or cells within a period. To get a more accurate estimate, median infectious dose or lethal

定的数量要求。为此，科学家采用另外一种测量细菌毒力的方法——细菌载量。**细菌载量**是指感染动物不同组织脏器的细菌数量。

dose must be determined in a relatively large population. Due to this limitation, **bacterial load** is introduced, which is the number of bacteria recovered from different tissues and organs in an infected animal.

第2节 细菌的毒力因子及分泌系统
Section 2 Virulence factors and secretion systems of bacteria

为什么不同的细菌毒力是不一样的呢？这就需要了解细菌的毒力因子构成。细菌的毒力因子是指构成细菌对宿主致病性的物质，主要包括与细菌侵袭力有关的因子、毒素等。毒力因子可以使病原菌具备在机体内定植、突破机体的防御屏障、内化进入细胞，进而在体内繁殖和扩散，这种能力即感染力。细菌的菌毛、鞭毛等表面结构以及透明质酸酶、链激酶等细菌酶类皆为毒力因子。而病原菌正是具备了与感染力相关的毒力因子，才能建立感染。

细菌毒素是毒力因子的一部分，包括两种，一种是可以直接分泌到细胞外的毒素，称**外毒素**（图4-3）；另一种是与细胞壁相连不能分泌的毒素，称**内毒素**（图4-4）。但也有少数外毒素存在于菌体细胞的胞周间隙，只有当菌体细胞裂解后才能释放到细胞外。

外毒素是一种可溶性蛋白质，不耐热，毒性强，对宿主不致热，抗原性强，能刺激机体产生中和抗体。外毒素经0.4%甲醛溶液处理后，其毒性消失，但仍保留原来的抗原性，称类毒素。类毒素注射入动物体内，可产生抗体，称抗毒素。抗毒素能和毒素结合，可用于产生外毒素细菌感染的紧急治疗和预防，如破伤风梭菌感染。

内毒素是锚定在革兰氏阴性菌外膜上的脂多糖（图4-4）。与蛋白类外毒素相比，多糖耐热，毒性弱。内毒素由O抗原、核心多糖和类脂A组成，其中类脂A是内毒素的毒性部分，也是内毒素锚定在

Why do different bacteria exhibit varying levels of virulence? It is necessary to understand the components of virulence factors. Virulence factors refer to those that contribute to the pathogenicity of the bacteria to the hosts, and mainly include invasiveness-associated factors and bacterial toxins. Virulence factors enable the bacteria to colonize the host cell surface, bypass the host defense mechanisms and get internalized into the cells where they replicate and spread in the host, which can be generalized as infectivity. Structures on the bacterial surface such as pili and flagella, and bacterial enzymes such as hyaluronidase and streptokinase are virulence factors. Only bacteria that possess such factors related to infectivity can establish infection in the hosts.

Bacterial toxins are virulence factors and can be categorized into two types: **exotoxins** (**Figure 4-3**), which are directly released into the surroundings from bacterial cells, and **endotoxins** (**Figure 4-4**), which are attached to the cell wall and usually not released. There are a few exotoxins that are normally present in the periplasmic space and are released only when the bacterial cells are broken.

Exotoxins are soluble proteins in nature, heat sensitive, strongly toxic, non-febrile, and strongly antigenic to stimulate the body to produce neutralizing antibodies. Exotoxins can be inactivated by treatment with 0.4% formaldehyde. The products are called toxoids. Specific antibodies could be generated in the animals inoculated with toxoids. These antibodies are referred to as antitoxins. They can bind to and neutralize the exotoxins for emergency treatment or for prevention against exotoxin-producing bacteria, *e.g. Clostridium tetani*.

Endotoxins are lipopolysaccharides anchored to the outer membrane in the cell wall of Gram-negative bacteria (**Figure 4-4**). They are polysaccharides in nature. In contrast to protein-based exotoxins, polysaccharides are weakly toxic and heat resistant. The endotoxin molecule is composed of an O antigen, a core polysaccharide and

细菌（如肉毒梭菌）

激活的外毒素

加热或化学灭活

类毒素

外毒素

外毒素受体

抗体

体细胞

外毒素与受体互作

抗原抗体互作

宿主细胞死亡

宿主细胞存活

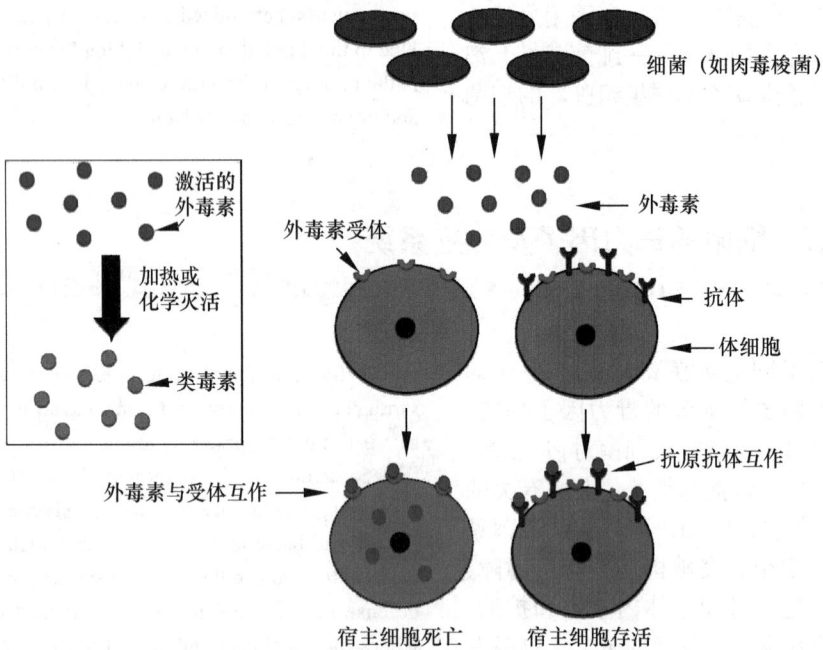

图4-3　细菌细胞分泌的外毒素

以肉毒梭菌外毒素为例，该毒素对体细胞有毒性。体细胞表面有抗体与外毒素结合，防止外毒素入侵。外毒素与抗体结合形成抗原-抗体复合物后被免疫系统锁定清除。否则，外毒素将与细胞表面的受体结合，抑制宿主蛋白质合成导致细胞死亡。加热或使用化学物质可对外毒素失活。失活的外毒素称为类毒素，对体细胞无害。

Figure 4-3　Exotoxins are secreted by bacterial cells

Clostridium botulinum exotoxins, for example, are toxic to somatic cells, which have antibodies on the cell surface to target exotoxins and bind to them, preventing the invasion of exotoxins. The binding of the exotoxin and antibody forms an antigen-antibody interaction, and the exotoxins are targeted for destruction by the immune system. If this interaction does not happen, the exotoxins bind to the exotoxin receptors that are on the cell surface and cause death of the host cell by inhibiting protein synthesis. Heat or chemicals can result in the deactivation of exotoxins. The deactivated exotoxins are called toxoids and they are not harmful to somatic cells.

O抗原

外核

PO_4^-
PO_4^-
PO_4^-
PO_4^-
PO_4^-

内核

类脂A

图4-4　脂多糖结构

脂多糖（LPS）又称内毒素，是一种由脂质和多糖组成的大分子，由O抗原、核心多糖（外核和内核通过共价键连接而成）和类脂A组成。脂多糖存在于革兰氏阴性细菌外膜中。

Figure 4-4　Structure of a lipopolysaccharide

Lipopolysaccharide (LPS), also known as an endotoxin, is a large molecule consisting of a lipid and a polysaccharide composed of O antigen, core polysaccharide (outer core and inner core joined by a covalent bond), and lipid A. It is found in the outer membrane of Gram-negative bacteria.

细胞膜上的特殊结构。内毒素在革兰氏阴性菌中普遍存在，虽然毒性弱，但中毒之后仍然能引起发热、呕吐和腹泻等症状，重则血液循环中白细胞骤减、弥散性血管内凝血等，导致休克，甚至死亡。海洋中有一种在外形上类似螃蟹的节肢动物，这种海洋生物诞生于4亿年前，虽然名为马蹄蟹，但是并非螃蟹，其学名为鲎。1964年，学者莱文和班发现鲎的血液遇内毒素呈凝胶状，检测内毒素的鲎试剂就是基于这一原理，但使用的不是血液，而是可被内毒素凝集的鲎变形细胞裂解物。

Ⅲ型分泌系统：革兰氏阴性菌能分泌诸多毒力因子。这些毒力因子是如何分泌到细菌细胞外的呢？目前在革兰氏阴性菌中已经发现9种分泌系统（Ⅰ～Ⅵ和Ⅷ～Ⅹ），结构和功能上研究比较透彻的是Ⅲ型分泌系统，能通过其针筒样结构将多种效应蛋白（毒力因子）直接分泌进入宿主细胞内。该系统存在于肠致病性大肠杆菌、沙门菌、耶尔森菌、铜绿假单胞菌等多种病原菌中；其结构复杂精细，由数十种蛋白参与形成基底复合体、针管状部件和针尖（转位子）（图4-5）。革兰氏阳性菌的分泌系统与革兰氏阴性菌不同，多以通用的分泌系统执行蛋白分泌功能，目前只发现在分枝杆菌中存在Ⅶ型分泌系统。

细菌毒力因子表达的调控：细菌毒力因子的表达受严谨调控，比如可感染人和动物的单核细胞增生李氏杆菌一旦离开人和动物体，暴露在肉、蛋、奶、瓜果蔬菜、冰箱、土壤、水体等食品或环境中时，其PrfA调控的毒力因子是不表达或低表达的，而受SigB调控的与环境抗性相关因子高表达。说明细菌能够感知外界环境的变化，如温度、渗透压、pH和氧含量等调节其生存所需要的因子。近年来发现，细菌的非编码小RNA和双组分信号转导系统也能调控毒力因子表达。非编码小RNA能导致mRNA降解或阻碍其翻

inner lipid A. Among them, lipid A causes toxic effects and serves as a structure embedding the lipopolysaccharide onto the cell membrane. Although endotoxins are weakly toxic, they could cause fever, vomiting, diarrhea and other symptoms in cases of food poisoning caused by Gram-negative bacteria. Shock or even death could occur when the intoxicated hosts are accompanied by a marked reduction of leukocytes in the blood, disseminated intravascular coagulation, *etc*. Horseshoe crabs, *Limulus polyphemus*, though referred to as crabs, are not crustaceans but arthropods. They live in the ocean and have been on the earth for about 400 million years. Two scientists, Levin and Bang (1964), reported their finding that limulus blood became coagulated upon contact with the endotoxins. The limulus test, now widely used for endotoxin detection, is based on this principle using the endotoxin-coagulable limulus amebocyte lysate instead of blood.

Type Ⅲ secretion system: Gram-negative bacteria secrete many virulence factors. How are these factors released outside bacterial cells? There are nine types of secretion systems (types Ⅰ to Ⅵ, and types Ⅷ to Ⅹ) identified so far in Gram-negative bacteria. Of them, the type Ⅲ secretion system (T3SS) is best characterized in terms of its structure and functions. T3SS is present in several major pathogens such as enteropathogenic *Escherichia coli*, *Salmonella*, *Yersinia*, *Pseudomonas* spp., *etc*., and is featured by its capability to deliver effector proteins as virulence factors directly into the host cells via its "syringe-like" apparatus. The system is an elegant assembly of more than twenty proteins forming the base complex, the filamentary needle tubing component, and the needle tip (translocon) **(Figure 4-5)**. The Gram-positive bacteria are quite different from the Gram-negative bacteria in that they mostly utilize the general secretion pathways for protein secretion, with only one specialized secretion system (type Ⅶ system or T7SS) found in *Mycobacterium* spp.

Regulation of expression of bacterial virulence factors: expression of virulence factors in bacteria is strictly regulated. For example, *Listeria monocytogenes* is a bacterium causing infection in humans and animals. Its virulence factors regulated by the PrfA protein will not be expressed or expressed at a low level, and the factors involved in environmental resistance could be elevated under SigB regulation, once the bacterium leaves the host and stays in foods or environments such as meat products, milk, eggs, vegetables, fruits, soil or water. This strongly suggests that bacteria can sense changes in temperature, osmotic pressure, pH and oxygen content in their surroundings. Studies in

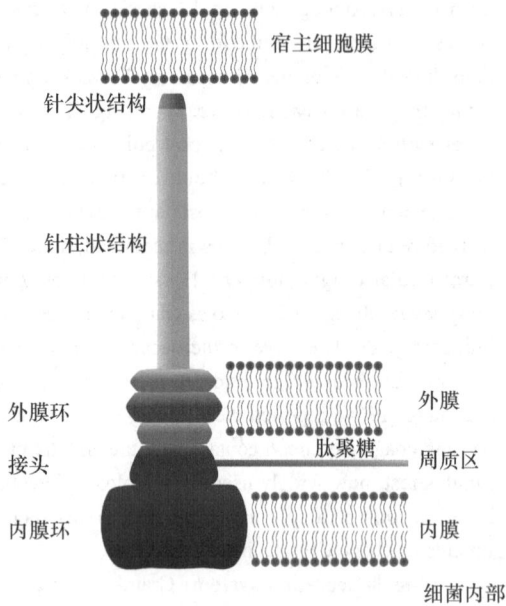

宿主细胞膜
针尖状结构
针柱状结构
外膜环　外膜
接头　肽聚糖　周质区
内膜环　内膜
细菌内部

图4-5　Ⅲ型分泌系统（T3SS）针状复合体示意图
T3SS 的标志是针状结构，当排出 ATP 酶时称为注射体。细菌蛋白质可通过针状结构直接从细菌细胞质分泌到宿主细胞质中。一个细菌细胞膜可存在几百个针状复合体，这是所有病原菌 T3SS 的普遍特征。

Figure 4-5　A schematic of the type Ⅲ secretion-system needle-complex
The hallmark of T3SS is the needle-like structure also called as injectisome when the ATPase is excluded. Bacterial proteins that need to be secreted pass from the bacterial cytoplasm directly into the host cytoplasm through the needle. A single bacterium can have several hundred needle complexes spread across its membrane. It has been proposed that the needle complex is a universal feature of all T3SSs of pathogenic bacteria.

译；双组分信号转导系统可像传感器一样感知细菌所处外界环境的变化，从而调控基因的表达。

recent years have discovered that expression of virulence factors is also regulated by non-coding microRNAs and two-component signal transduction systems. Noncoding microRNAs can degrade mRNA or interfere with translation. Two-component signal transduction systems are sensors of the bacteria to detect changes in environmental conditions, thus regulating the expression of target genes.

第3节　机会致病菌
Section 3　Opportunistic pathogens

除非实验条件下，完全无菌的人和动物在自然界中是不存在的。因此，人和动物体内含有诸多微生物，比如肠道微生物菌群、呼吸道微生物菌群等。但是这些微生物在健康宿主体内通常不致病，但当宿主的免疫屏障被打破或免疫功能异常时，其中一些潜在致病细菌会增殖到足够量，并穿越上皮屏障，入侵机体，造成感染并致病，称为机会致病菌。因此，机会致病菌也具有感染性。

On the earth, not a single sterile person or animal can be found, unless under experimental conditions, because bacteria establish intimate symbiotic relationship with humans and animals immediately after their birth. There are varieties of microbes in human and animal bodies, *e.g.* gut microbiota, microflora of the respiratory tract, *etc*. They do not usually cause infection unless there is a breach in the host defense barrier or when the individuals are immunocompromised. Some of the bacteria in the microflora that have pathogenic potential might seize the opportunity to outgrow, break the mucosal barrier and lead to systemic infections and clinical signs, hence the term opportunistic pathogens. Therefore, opportunistic pathogens are also infectious.

思考题　Questions

1. LD$_{50}$用于检测细菌毒力在临床实践中并不多见，原因是什么？

2. 在革兰氏阴性菌中，毒力因子是如何被分泌到细菌胞外的？

3. "爱国、敬业、诚信、友善"是公民基本道德规范。人长期处于悲伤、愤怒、酗酒、吸二手烟等极端情绪、不良习惯或生活环境中，为何容易被微生物感染？

1. The LD$_{50}$ testing is rarely used to detect bacterial virulence in clinical practice. Why?

2. In Gram-negative bacteria, how are virulence factors secreted out of the bacterial cells?

3. "Patriotism, dedication, integrity and friendship" are the basic moral norms of citizens. Why are people more vulnerable to infection by microorganisms when they are in extreme emotions such as sadness or anger, live in poor environments including second-hand smoking or have bad habits like alcohol addiction for a long time?

第5章 细菌的遗传变异

Chapter 5 / Genetic variations of bacteria

内容提要 据地质学家发现，38亿年前地球上已出现细菌。从地球出现原始生命开始迄今，细菌已成为目前地球上数量最多的一种生物。那么，细菌得以生存和演化的基础是什么？

Introduction According to the findings by geologists, bacteria have been living on the earth for 3.8 billion years. Since the emergence of living creatures, bacteria have become the most abundant organisms on the earth. What are their secret weapons for survival and evolution?

第1节 细菌常见的变异现象
Section 1　Common phenomena of bacterial variations

细菌的**遗传**是指亲代细菌与子代细菌的相似性，它使细菌的性状保持相对稳定，是细菌生存的基础。细菌的变异是指亲代与子代以及子代细菌之间的差异性，反映细菌的演化。

遗传变异使得细菌在培养基和显微镜下呈现多种形态及功能上的差异，这种变异称为表型变异，包括：形态和结构变异、菌落形态变异、毒力变异、耐药性变异、代谢变异、抗原性变异等。

Bacterial **Heredity** refers to the similarity between parental bacteria and their descendant generations, which keeps their traits relatively stable and serves as the basis for their existence. In contrast, variation refers to the difference between parental bacteria and their descendant generations or even among their descendants, reflecting the evolutionary potential of bacteria.

Because of genetic variations, bacteria may appear morphologically different in culture media or under microscope, or even functionally divergent. Such variations are called phenotypic variations, including changes in morphology and structure, appearance of the colony, virulence, drug resistance, metabolism, antigenicity, and so forth.

第2节 细菌遗传的物质基础
Section 2　Material basis of bacterial inheritance

细菌的表型变异是由基因型决定的。细菌体内的遗传物质除了基因组之外，还有一种特殊的遗传物质——游离于染色

Phenotypic variations of bacteria are determined by the genotype. In addition to the genome, there are also special genetic materials in bacteria called plasmids.

体之外的环状双螺旋DNA分子，即质粒（**图5-1**）。质粒按转移特性分为：接合性质粒和非接合性质粒。接合性质粒可以通过细菌的性菌毛在供体菌和受体菌之间进行DNA转移（**图5-2**）。按照编码的生物学性状，质粒可分为：致育质粒、毒力质粒和耐药质粒，分别编码性菌毛、毒力因子和耐药蛋白。

A plasmid is a circular double helix DNA molecule outside of the chromosome (**Figure 5-1**). Plasmids can be conjugative or non-conjugative by their transferring characteristics. DNA carried by conjugative plasmids can be transferred between the donor and recipient bacteria via sex pili (**Figure 5-2**). Based on the biological traits encoded in their genes, plasmids can be divided into fertility plasmids, virulence plasmids, and drug resistance plasmids, which encode sex pili, virulence factors, and proteins responsible for drug resistance, respectively.

图5-1 细菌及其染色体DNA和质粒

细菌为大椭圆形。在细菌内部，小到中等大小的圆圈表示质粒，长而细的闭合反复交叉线表示染色体DNA。

Figure 5-1 A bacterium with its chromosomal DNA and plasmids

The bacterium is drawn as a large oval. Within the bacterium, the small to medium size circles illustrate the plasmids, while the long thin closed line that intersects itself repeatedly illustrates the chromosomal DNA.

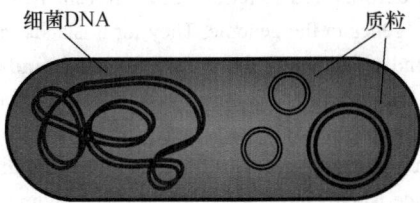

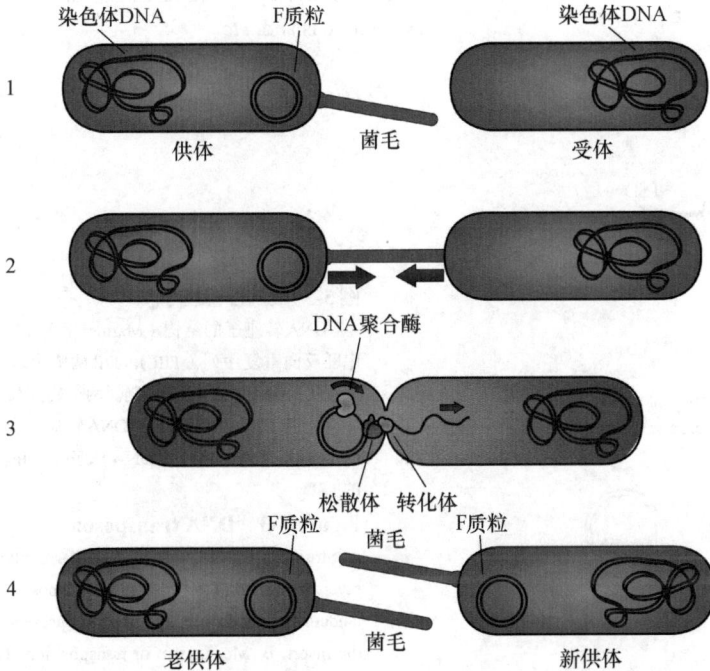

图5-2 细菌接合

1. 供体菌产生菌毛；2. 菌毛附着于受体菌细胞使两个细胞结合在一起；3. 质粒移动并产生缺刻，然后单链DNA转移入受体细胞；4. 质粒在两个细菌细胞内同时环化，合成第二链并产生菌毛；两个细菌细胞均成为供体。

Figure 5-2 Bacterial conjugation

1. Donor cell produces pilus; 2. Pilus attaches to the recipient cell, and brings the two cells together; 3. The mobile plasmid is nicked and a single strand of DNA is then transferred to the recipient cell; 4. Both cells recircularize their plasmids, synthesize the second strand, and reproduce pili; both cells are now donors.

在细菌基因组或质粒DNA上，有一些可移动的DNA片段。这些DNA片段可以在细菌基因组不同位置或基因组之间或质粒之间或基因组与质粒之间转移，称为转位因子，包括：插入序列、转座子（**图5-3**）、Mu噬菌体、基因盒-整合子系统等。转位因子必须具备特定的序列特征，比如末端反向重复序列，同时依赖特异性转位酶等。

病原菌的毒力基因在基因组中成群或成簇排列，分子结构与功能有别于细菌基因组其他基因，位于细菌基因组之内，就像"岛屿"一样，称为**毒力岛或致病性岛**，比如大肠杆菌、李氏杆菌、沙门菌的基因组中有很多毒力岛，多由基因水平转移获得。除了毒力岛之外，致病菌基因组中存在大量结构特征与毒力岛类似的基因结构，如分泌岛、抗性岛、代谢岛、共生岛等。

Certain DNA fragments in bacterial genomes or plasmids are able to switch their locations either between two loci on the same genome or between two genomes or between two plasmids or between a genome and a plasmid. Such DNA fragments are called transposable elements which include insertion sequences, transposons (**Figure 5-3**), Mu phage, gene cassette-integron system, *etc*. Transposable elements are not random but unique in sequence patterns, *e.g.* terminal inverted repeats. Transposable elements also require specific transposases.

The genes encoding virulence factors align in groups or clusters in the bacterial genome. Their molecular structures and functions are different from those of other genes in the genome. They form islands in the genome and are thus referred to as **virulence islands** or **pathogenicity islands**. They are found in bacteria such as *E. coli*, *Listeria*, *Salmonella*, *etc*. Besides, a great number of genomic structures bearing features similar to the virulence islands have been found in the genomes of pathogenic bacteria and named by various functions, such as secretory island, resistant island, metabolic island, symbiotic island, *etc*.

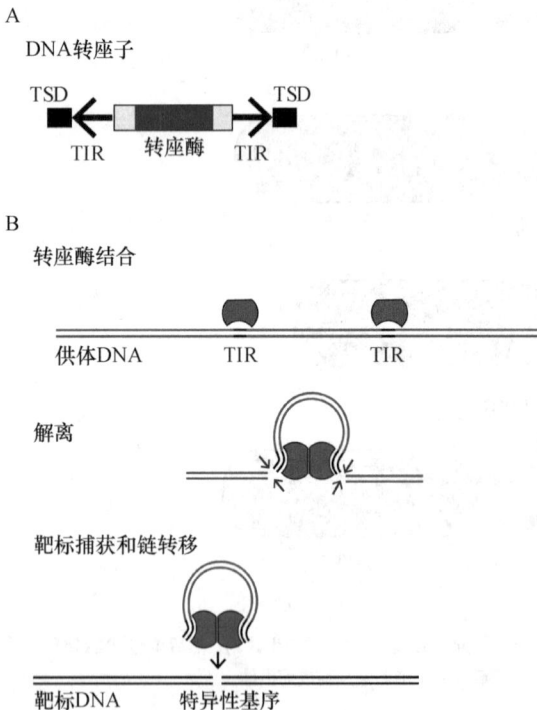

图5-3 DNA转座子

A. DNA转座子的结构（Mariner型）。转座酶基因两侧为串联反向重复序列（TIR），TIR两侧为两个短的串联位点重复（TSD）；B. 转座机制。两个转座酶识别并结合TIR序列，使DNA双链断裂。DNA转座酶复合物将携带的DNA插入基因组中特定DNA基序处，并在整合时产生短的TSD。

Figure 5-3 DNA transposon

A. Structure of DNA transposons (Mariner type). Two tandem inverted repeats (TIRs) flank the transposase gene. Two short tandem site duplications (TSDs) are present on both sides of the insert. B. Mechanism of transposition. Two transposases recognize and bind to the TIR sequences, join together and promote DNA double-strand cleavage. The DNA-transposase complex then inserts its DNA cargo at specific DNA motifs elsewhere in the genome, creating short TSDs upon integration.

第3节 细菌变异的机制
Section 3 Mechanism of bacterial variations

作为地球上数量最多的生物，细菌在几十亿年漫长的演化过程中离不开基因突变。基因突变是指生物细胞遗传物质DNA分子结构发生了稳定的可遗传的变化。它是生物进化和微生物演化的重要因素之一。如果染色体的一大段发生了变化，称为染色体畸变，包括：缺失、重复、插入、异位、倒位等。如果DNA链中一个或少数几个核苷酸发生了改变，称为点突变，包括：置换、转换、颠换、移码等。

供体细菌间接或直接地将部分遗传物质单向传递给受体菌，从而导致受体细菌发生基因重组的现象，称**基因转移**，主要包括：转化、转导、接合、原生质体融合和溶原性转换等方式。

转化是指供体菌游离的DNA片段直接通过热应激或者瞬时电穿孔等物理方法进入处于感受态的受体菌的过程，使受体菌获得新的性状。

转导是由噬菌体介导，把供体菌的DNA小片段携带到受体菌中，使受体菌获得供体菌的部分遗传性状的过程，包括普遍性转导和特异性转导（**图5-4**）。

普遍性转导是指噬菌体感染细菌后，在成熟和装配过程中，供体菌染色体上的任何DNA片段都有可能被装入噬菌体内，所以称普遍性转导。含有细菌DNA片段的噬菌体称为转导性噬菌体。若转导性噬菌体能再次将供体DNA片段整合到受体菌染色体DNA中，并能稳定复制，称为完全性转导；若转导性噬菌体不能将供体菌DNA片段重组到受体菌的染色体上，则称为流产转导。

特异性转导与普遍性转导类似，都是由噬菌体介导，但区别在于前者由温和型噬菌体介导，其DNA整合到细菌染色体上

As the most abundant living creatures on the earth that have undergone billions of years of evolution, gene mutations often occur in bacteria. Gene mutations refer to stable genetic changes of DNA molecules of a biological cell, which is one of the important factors in the evolution of living organisms including microorganisms. A mutation affecting a large portion of the chromosome is termed as chromosomal aberration, including deletion, duplication, insertion, translocation and inversion. A mutation affecting only one or very few nucleotides in the DNA sequence is termed as a point mutation. It can be in the form of replacement, transition, transversion and frameshift.

The phenomenon where a recipient bacterium accepts some genetic material directly or indirectly from a donor bacterium with the material incorporated into its genome is called as **transgenesis**. It can be achieved through transformation, transduction, conjugation, protoplast fusion and lysogenic conversion.

Transformation occurs when free DNA fragments from the donor bacterium are introduced into the recipient bacterium in a state of competence directly through physical methods such as heat shock or instantaneous electroporation. The recipient bacterium would have a new phenotype from the exogenous genetic material.

Transduction is mediated by phages which bring small DNA fragments of the donor bacterium into the recipient bacterium and allow expression of new traits in the recipient bacterium derived from the donor. It includes general transduction and specialized transduction (**Figure 5-4**).

In **generalized transduction**, any DNA fragment in the chromosomes of the donor bacterium has an equal chance to be incorporated into phages during maturation and assembly. The phage containing the fragments of bacterial DNA is called as transducing phage. A complete transduction is achieved if the newly integrated DNA fragments in the recipient's chromosome derived from the donor are stably replicated; in a case of integration failure, it is called as abortive transduction.

In **specialized transduction**, although mediated by phages as well, prophages are formed without causing bacterial lysis after the temperate phages integrate their DNA into the bacterial genome. The bacterium containing

普遍性转导   特异性转导

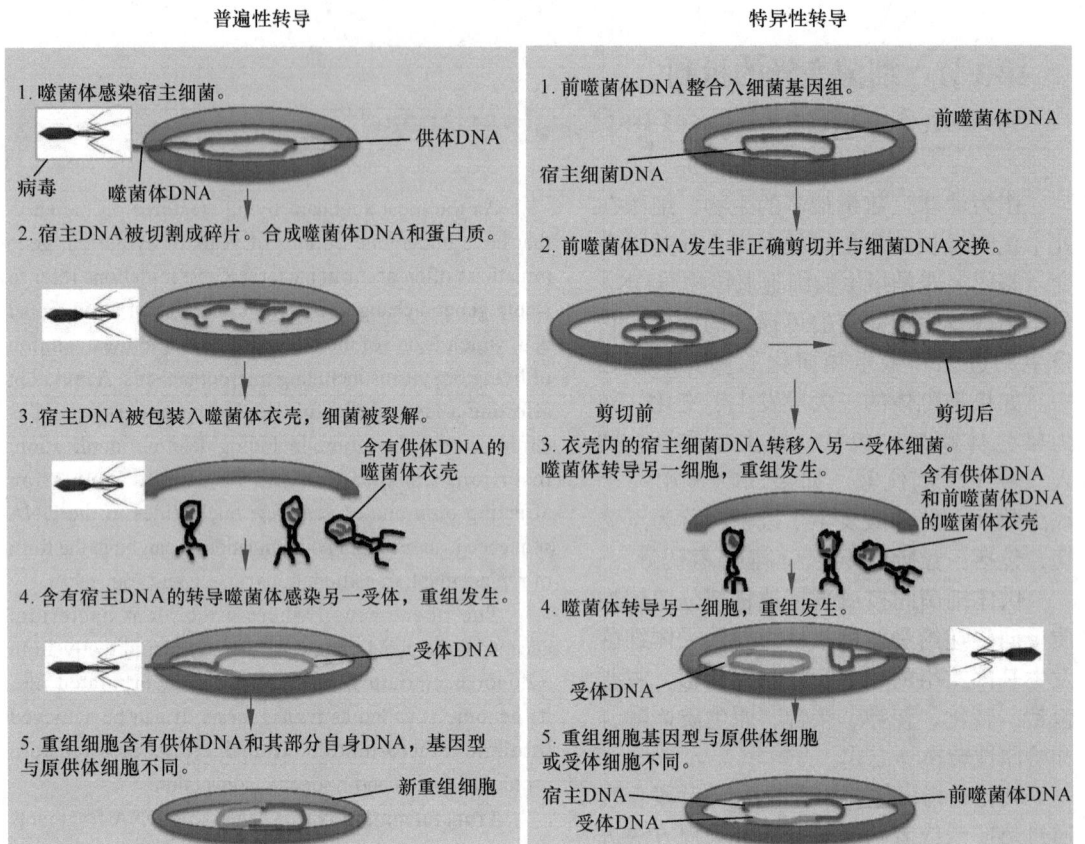

图5-4　普遍性转导（通过噬菌体将细菌任一基因转移入受体菌）和特异性转导（将细菌局部基因转移入受体菌）

普遍性转导容易发生且随机，而特异性转导依赖于基因在染色体上的位置和前噬菌体的不正确切割。

Figure 5-4　**Illustration of the difference between general transduction, which is the process of transferring any bacterial gene to a second bacterium through phage, and specialized transduction, which is the process of moving restricted bacterial genes to a recipient bacterium**

彩图

While generalized transduction can occur randomly and more easily, specialized transduction depends on the location of the genes on the chromosome and the incorrect excision of the prophage.

形成前噬菌体，细菌并没有被裂解，称为溶原菌。溶原菌受环境因素影响后可进入裂解周期，在前噬菌体从细菌染色体上切离时发生偏差，携带了细菌染色体部分片段，脱落后经复制组装成转导噬菌体。这种噬菌体再次感染时，只能将前噬菌体两旁的基因转移到受体菌内，故称特异性转导或局限性转导。

接合是指两个完整的细菌细胞通过性菌毛直接接触，由供体细菌将致育质粒

prophages is known as lysogeny. When the lysogenic cell is exposed to certain environmental stimuli, it may begin the lytic cycle. Imprecise excision could occur when the prophage DNA fragments are released from the host chromosome. Thus, the prophages may carry some chromosomal segments of the host cells and become transducing phages after cycles of replication and assembly. When the transducing phage reinfects a new host, it can only introduce the DNA fragments on the side of the prophage genome into the recipient bacterium. That's why it is called as specialized transduction, also known as restricted transduction.

DNA转移给受体细菌的过程。

原生质体融合是指选择具有优良性状的两个菌株细胞作为亲本，用人工的方法除去细胞壁，成为原生质体，使原生质体发生融合，形成带有双亲本菌株优良性状、遗传稳定的融合子。最常见的为酵母原生质体融合。

溶原转换是指温和噬菌体感染其宿主后，噬菌体基因与细菌基因组整合。噬菌体成为前噬菌体，而细菌则成为溶原菌，从而获得由噬菌体编码的某些性状。

Conjugation refers to the process in which fertility plasmid is transferred from the donor to the recipient bacterium via sex pili.

Protoplast fusion refers to the formation of a genetically stable fusant of two protoplasts, made by co-fusing two bacterial cells with excellent parental characteristics after artificial removal of their cell wall. The yeast fusant is the most common product of protoplast fusion.

Lysogenic conversion refers to the integration of phage genes into the bacterial genome upon infection of the host bacterium by a temperate phage. Then the phage becomes a prophage, while the bacteria become lysogens, thus obtaining some novel characters encoded by the phage genes.

思考题　Questions

1. 特异性转导与普遍性转导有何区别？

2. 从遗传和变异的角度考虑，耐药性存在哪些生物安全风险？如何应对微生物耐药？

1. What is the difference between specialized transduction and generalized transduction?

2. From the perspective of heredity and variation, what are the biosecurity risks of drug resistance? How to cope with the resistance of microbes to antimicrobials?

第6章 细菌的分类命名
Chapter 6 Taxonomy and nomenclature of bacteria

内容提要 "欲知平直，则必准绳；欲知方圆，则必规矩。"微生物与动植物一样也是有名称的，那么如何给细菌命名呢？这就需要了解细菌的分类与命名规则。

Introduction To evaluate the straightness of things, one must use a tightrope. To measure a rectangle or circle, one must use a ruler or compass. Every animal and plant has a name, and so does every microbe. How to name a bacterium? It is necessary to know the rules of bacterial taxonomy and nomenclature.

第1节 细菌的分类地位
Section 1 Taxonomy of bacteria

作为世界上数量最多的生物，细菌的分类与其他生物一样，遵循域、界、门、纲、目、科、属、种这一分类体系。目前国际上公认的三域学说包括：细菌域、古菌域和真核生物域。其中，古菌与细菌都属于原核生物，但其细胞壁组成不同，古菌成员属于极端嗜盐或高度耐热的极端微生物。细菌属于细菌域，分类比较简单，常见的分类阶元是属和种。其中，种是细菌分类的最基本单元。

在国际上，判断两株细菌属于同种还是同属的依据是16S rRNA保守区的序列差异。不同属的差异应大于5%；同源性大于98.65%的为同种。对于不同来源的同一种细菌的纯培养物，称为该菌的不同菌株。由于同一种细菌有很多菌株，因此菌株的命名相对比较随意，通常用地名、动物名称缩写加编号。把具有典型特征的某种细菌菌株进行系统鉴定后，作为标准菌

As the most abundant organisms in the world, just like other organisms, bacteria follow the taxonomic ranks of domain, kingdom, phylum, class, order, family, genus and species. Organisms can be classified into three domains: *Bacteria*, *Archaea* and *Eukarya*. *Bacteria* and *Archaea* are both prokaryotes, but the composition of the cell wall in *Archaea* is different from that in *Bacteria*. *Archaea* are extremophiles that are highly halophilic or thermophilic. Bacteria belong to the domain *Bacteria*, and their classification is relatively simple, with the most common taxonomic ranks being genus and species. Of them, species is the most basic unit in bacterial taxonomy.

Internationally, the degree of similarity of the sequences in the conserved regions of 16S ribosomal RNAs in the bacteria is used to determine whether two bacterial strains belong to the same species or genus. Dissimilarity of more than 5% suggests different genera. Similarity equal to or above 98.65% suggests the same species. The term "strain" refers to the pure culture of the same bacterial species from different sources. There could be various strains in the same bacterial species. The naming for each strain is relatively arbitrary and usually includes abbreviations of the location where it is isolated and the name of source animals, and a serial

株，也称参考菌株或模式菌株，保存于指定的菌种保藏机构，可根据需要索取，用于比较研究。

number. A strain of a particular bacterial species that is fully characterized could be defined as the type strain, also known as reference strain by microbiologists. Type strains are deposited in the assigned microbiological culture collection centers and available upon request for comparative research purposes.

第2节 细菌的命名
Section 2 Nomenclature of bacteria

细菌的命名采用属名＋种名的双名法。在英文书写时，种及以上阶元须斜体。当同一细菌的属名重复出现时，可缩写为属名的首字母大写。存在以下几种特殊情况，第一种情况：只确定属名，未确定种名的某一株细菌，书写时在属名的后面加上 sp. 小写正体，其中 sp. 为 species 的缩写。第二种情况：只确定属名，未确定种名的若干菌株，采用 spp. 小写正体，spp. 是 species 的复数形式。第三种情况：在属名、种名外还有亚种、变种、新种等在分类上属于"种"以下的名称，全部正体书写。

Binomial nomenclature is applied to bacteria. In scientific writing, species and higher rankings of the bacterial name should be italicized. The genus of a bacterium can be abbreviated to its capitalized initial if it repeatedly appears within the same article. There are some particular conditions one may encounter in scientific writing. When the species is not clearly defined and only the genus of the bacterium is known, the genus name is followed by a lowercased and non-italic "sp.", an abbreviation of the word species in a singular form. When referring to multiple strains of a bacterium with a defined genus but not species, lowercased and non-italic "spp.", an abbreviation of the word species in a plural form, is listed following the genus name. Occasionally, in bacterial nomenclature, taxonomic ranks below species such as subspecies, novel species and variant species are listed as well and they don't need to be italicized.

示例：Examples：

（1）属名重复出现，可缩写属名首字母。The genus name of a bacterium is abbreviated into its capitalized initial if it repeatedly appears from the second time in the body text.

E. coli（大肠杆菌）；*S. pullorum*（鸡白痢沙门菌）

（2）只确定属名，未确定种名的细菌。For bacteria with undefined species but known genus.

只确定属名，未确定种名的沙门菌 *Salmonella* sp.

只确定属名，未确定种名的若干株沙门菌 *Salmonella* spp.

（3）亚种。Subspecies (subsp.).

多杀性巴氏杆菌败血亚种 *Pasteurella multocida* subsp. septica

产气荚膜梭菌，新属新种 *Clostridium perfringens* gen. sp. nov.

关于以外国人的姓名命名的细菌，在翻译时要遵循相关规定。涉及外国人名译为汉语时去掉"氏"，名字中有字母发此音除外。例如，*Salmonella*，原来一直翻译为沙门氏菌，应为沙门菌（去掉氏）。

Chinese translation of bacterial species named after someone from other countries should follow the revelant rules. Transliteration of the person's name is practiced, with the word "shi" removed after the name, unless there were first few letters pronounced as "shi" in the name. The word "shi" literally means interrelated families or a clan in Chinese. For instance, *Salmonella* was translated into "Shamen shi" bacterium, and it is now corrected as "Shamen" bacterium in Chinese.

示例：Examples：

Brucella 译为布氏杆菌属（词头字母发音，保留氏）

Yersinia 译为耶尔森菌属（去掉氏）

Salmonella 译为沙门菌属（去掉氏）

Pasteurella 译为巴氏杆菌属（词头字母发音，保留氏）

Listeria 译为李氏杆菌属（词头字母发音，保留氏）

第3节 细菌分类鉴定的标准
Section 3　Criteria for taxonomic identification of bacteria

当临床上分离到一株细菌时，首先要确定该菌属于什么属、什么种。因此，细菌的鉴定要有一个标准。细菌的鉴定标准，可从传统方法和现代方法两个方面进行描述，传统方法描述细菌的表型特性，包括菌体形态、染色特性、培养物特征（如固体培养基上的菌落色泽、大小、形态等）、细胞壁结构、生化及抗原特征等；现代方法描述种系发生关系，即分析核酸序列特性，从分子水平确定分类地位，包括：①DNA（G＋C）含量：（G＋C）含量差异>5%、<10%时为不同种，（G＋C）含量差异>10%时为不同属；②核酸分子杂交：同源性>70%时为同种；③16S rRNA：根据相似性判定（测序结果与GenBank比对），大于98.65%时为同种；④全基因组测序。

When you find bacterial isolates from clinical samples, you surely want to know what genus or species they are. Criteria and relevant methods are, therefore, required for taxonomic identification. There are traditional and modern methodologies available to define the genus and species of clinical isolates. The traditional methodology describes the phenotypic characteristics of the bacteria, such as morphology, staining features, characteristics of the cultures (*e.g.* size, color and shape of colonies on solid agar media), cell wall structure, biochemical profiles, antigenicity, *etc.* The modern methodology analyzes the phylogenetic relationship of the bacteria by studying their nucleotide composition and identifies the bacteria at the molecular level. ①Guanine＋cytosine (G＋C) content of the genomic DNA: a difference over 5% but less than 10% in the (G＋C) content suggests different species, while a difference over 10% suggests different genera; ②Molecular hybridization of nucleic acids: homology over 70% in the nucleic acid sequence suggests the same species; ③16S rRNA: similarity over 98.65% between the sequence of the strain to be identified and that retrieved from GenBank of known species suggests the same species; ④Whole genome sequencing.

第4节 细菌的鉴定程序
Section 4　Procedures of bacterial identification

经典的细菌鉴定程序包括：病料采集、涂片、染色、镜检、分离培养、纯培养等步骤。得到纯培养物之后可以进行培养特性、生化特性、动物试验、血清学试验、分子生物学试验等进一步分析，其中基因测序可以对细菌基因型和进化关系进

Classical identification of a bacterium includes sample collection, slide preparation, staining, microscopic examination, primary isolation, pure culture acquisition, *etc.* Once the pure culture is obtained, analyses of its growth characteristics, biochemical traits, animal tests, serological and molecular tests can be performed. Gene sequencing can be used to accurately identify the genotype and phylogenetic

行精准鉴定。在上述步骤中，细菌的生化特性分析最为复杂，可借助全自动微生物生化鉴定仪，根据生化反应特性，结合其数据库中不同细菌的生化反应谱，可以判定革兰氏阳性菌和阴性菌所在的属和种；有些全自动微生物鉴定系统还可对其耐药性进行分析。

relationship of the bacteria. Among the steps above, biochemical analysis is the most complicated process, which can be facilitated by a fully automatic microbial biochemistry identification system. Such automated systems can determine the genus and species of Gram-positive and Gram-negative bacteria by comparing with the biochemical profiles of the bacteria in the database, and even the module for drug resistance testing could be built-in.

思考题　Questions

1. 请写出单核细胞增生李氏杆菌由域至种完整的分类命名。

2. 简述细菌的鉴定程序。

3. 简述何为标准菌株。

1. List the taxonomy of *Listeria monocytogenes* from domain to species.

2. Please describe in brief the procedures for bacteria identification.

3. What does a reference bacterial strain mean?

第7章 革兰氏阳性球菌
Chapter 7 / Gram-positive cocci

内容提要 有些细菌在显微镜下观察时如葡萄状成串或链球状成条，包括革兰氏阳性球菌中的葡萄球菌属和链球菌属成员。

Introduction Some bacteria appear as grape-like clusters or chain-like rods under the microscope, including members of the *Staphylococcus* and *Streptococcus* genera in Gram-positive cocci.

第1节 葡萄球菌属
Section 1 *Staphylococcus*

葡萄球菌属，顾名思义，其形态在显微镜下呈现葡萄状，因此称为葡萄球菌（**图7-1**）。该菌广泛存在于自然界，其中，致病性葡萄球菌可引起化脓性炎症、败血症、食物中毒。该菌主要特征是：无芽孢、无鞭毛，有的形成荚膜或黏液层；可用青霉素诱导成L型；革兰氏染色阳性；需氧兼性厌氧，触酶阳性，氧化酶阴性，无运动力；可在普通琼脂培养基、血琼脂上生长，在麦康凯培养基上不生长；最适生长温度为35～40 ℃，pH 7.0～7.5。

Staphylococcus, as the name implies, is a spherical-shaped bacterium that appears in clusters under the microscope, hence its name "*Staphylococcus*" (**Figure 7-1**). This bacterium is widely distributed in nature and some strains can cause pyogenic infections, sepsis, and food poisoning in humans. *Staphylococcus* is characterized by the absence of spores and flagella. Some strains can form a capsule or slime layer, can be induced to become L-forms by penicillin, and are Gram-positive. It is facultatively anaerobic, catalase-positive, oxidase-negative, and non-motile. *Staphylococcus* can grow on common agar and blood agar, but not on MacConkey agar. Its optimal growth temperature is between 35-40 ℃ , and pH 7.0-7.5.

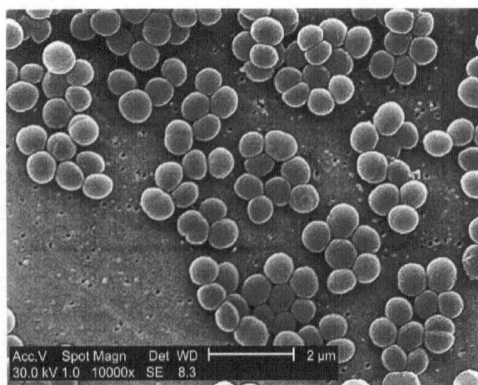

图7-1 电子显微镜下的金黄色葡萄球菌
Figure 7-1 ***Staphylococcus aureus*** **under electron microscope**

金黄色葡萄球菌是葡萄球菌属的一种常见致病菌，简称金葡菌。由于其在适宜的固体培养基、温度，有 O_2 和 CO_2 环境中可以产生金黄色的类胡萝卜素，因此，菌落可呈金黄色。除色素外，金葡菌能够产生 α、β、γ 和 δ 溶血素，在血液琼脂平板上，有的为不完全溶血（α溶血）；有的为明显的完全溶血（β溶血），产溶血素的菌株多为病原菌。金葡菌能够发酵多种糖类，高度耐盐，在普通培养基上生长良好，形成湿润、光滑、隆起的圆形菌落。

金黄色葡萄球菌的细胞壁有两种抗原，一种是荚膜多糖抗原，其成分是 N-乙酰氨基糖醛酸和 N-乙酰岩藻糖胺；另外一种是蛋白质抗原SPA——葡萄球菌蛋白A。SPA是一种单链多肽，与葡萄球菌细胞壁上的肽聚糖共价结合，能与多种哺乳动物IgG的 Fc 片段非特异性结合，具有趋化性、抗吞噬、损伤血小板、促进细胞分裂、引发过敏反应等生物学作用。

金黄色葡萄球菌的毒力因子包括毒素和酶。其中毒素包括α毒素、肠毒素等。α毒素是一种穿孔毒素，可以破坏哺乳动物红细胞膜，插入细胞膜的疏水区，形成微孔，导致红细胞溶解（**图7-2**）。肠毒素按抗原性不同，分为A、B、C1、C2、C3、D和E等20多个型，可破坏肠绒毛，引起以急性胃肠炎为主要症状的食物中毒。金

Staphylococcus aureus, commonly known as "golden staph", is a pathogenic bacterium that belongs to the *Staphylococcus* genus. It is known for its ability to produce a golden-yellow pigment resembling carotenoids on suitable solid media, at appropriate temperatures, in the presence of O_2 and CO_2. In addition to its distinctive pigmentation, *S. aureus* can produce α, β, γ, and δ hemolysins, some of which cause incomplete hemolysis (α-hemolysis), while others cause complete hemolysis (β-hemolysis), with hemolytic strains often being pathogenic. This bacterium can ferment various sugars and is highly salt-tolerant. *S. aureus* grows well on common agar and forms moist, smooth, and convex circular colonies.

The cell wall of *S. aureus* contains two types of antigens. One is the capsule polysaccharide antigen, composed of *N*-acetylglucosaminuronic acid and *N*-acetylgalactosamine. The other is the protein antigen SPA—staphylococcal protein A. SPA is a single-chain polypeptide that covalently binds to the peptidoglycan on the surface of *S. aureus* cell wall. It can non-specifically bind to the Fc fragment of IgG from various mammalian species, and has biological effects such as chemotaxis, anti-phagocytic activity, platelet damage, cell proliferation promotion, and induction of allergic reactions.

The virulence factors of *S. aureus* include toxins and enzymes. Among them, toxins such as α-toxin and enterotoxin are included. α-Toxin is a pore-forming toxin that can destroy mammalian red blood cell membranes by inserting into the hydrophobic region of the cell membrane, forming pores, and causing red blood cells to lyse (**Figure 7-2**). Enterotoxins are classified into more than 20 types, including A, B, C1, C2, C3, D, and E, according to their antigenicity. They can destroy intestinal villi and cause food poisoning with acute gastroenteritis as the main symptom.

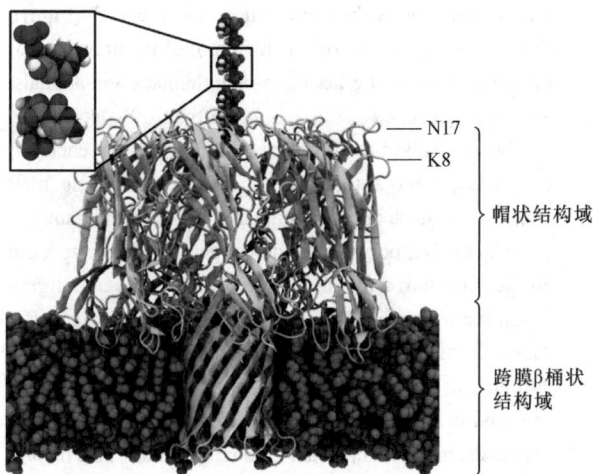

彩图

图7-2 脂质双层中的α毒素（αHL）蛋白，蛋白质通道入口上方为单核苷酸
图像来源见英文。

Figure 7-2 The α-toxin (αHL) protein in lipid bilayer, with the mononucleotide positioned above the vestibule entrance
Source: Manara RM, Tomasio S, Khalid S. 2015. The nucleotide capture region of alpha hemolysin: insights into nanopore design for DNA sequencing from molecular dynamics simulations. Nanomaterials (Basel), 5 (1): 144-153.

黄色葡萄球菌毒力相关的酶包括凝固酶、耐热核酸酶等。凝固酶有游离凝固酶和结合凝固酶两种，其中游离凝固酶分泌于细菌外，类似凝血酶原的作用；结合凝固酶能使含抗凝剂的人或兔血浆凝固。耐热核酸酶能迅速分解感染部位组织细胞和白细胞崩解时释放的核酸，在0.1%甲苯胺蓝平板上呈现紫色。

实验室诊断：金黄色葡萄球菌的分离鉴定程序包括：样本的采集，形态鉴定，分离培养鉴定等。化脓性病灶取脓汁、渗出液，败血症取血液，脑膜炎采取脑脊液，食物中毒则采集食物、呕吐物和粪便等。上述病料涂片、染色、镜检，若在显微镜下可见大量典型的"葡萄串"一样的细菌，则可初步诊断。分离培养，可以采用普通平板、血琼脂平板、高盐培养基，通过菌落形态和溶血特性等进行鉴定。致病性金黄色葡萄球菌能产生凝固酶和耐热核酸酶，菌落金黄色，有溶血性，分解甘露醇等。

The enzymes associated with the virulence of *S. aureus* include coagulases and heat-resistant nucleases. Coagulase has two forms: free coagulase and bound coagulase. Free coagulase is secreted outside the bacteria and acts similarly to prothrombin; bound coagulase can cause anticoagulant-containing human or rabbit plasma to clot. Heat-resistant nucleases can rapidly degrade the nucleic acids released when tissue cells and white blood cells disintegrate at the site of infection, and appear purple on a 0.1% toluidine blue plate.

Laboratory diagnosis: the isolation and identification procedures of *S. aureus* include sample collection, morphological identification, and isolation culture identification. Purulent fluid and exudate are collected from suppurative lesions, blood is collected from sepsis, cerebrospinal fluid is collected from meningitis, and food, vomitus, and feces are collected from food poisoning. The above-mentioned clinical specimens are smeared, stained, and examined under a microscope. If a large number of typical "grape-like clusters" of bacteria are visible, a preliminary diagnosis can be made. Isolation culture can be performed using nutrient agar plates, blood agar plates, and high-salt culture media, and identification can be carried out based on colony morphology and hemolytic characteristics. Pathogenic *S. aureus* can produce coagulase and heat-resistant nucleases, and its colonies are golden in color, show hemolysis, and can ferment mannitol.

第2节 链球菌属
Section 2 *Streptococcus*

链球菌属，顾名思义，因其在显微镜下呈现链条状排列，所以被称为链球菌（**图7-3**），可导致人畜化脓性炎症、肺炎、乳腺炎、败血症等。链球菌属有60多个种，可根据菌体的三种抗原进行分类。这三种抗原分别是：属特异性抗原（为P抗原，即核蛋白抗原）、群特异性抗原（为C抗原，即多糖抗原）和型特异性抗原（为蛋白质抗原，又称表面抗原，多与其毒力与免疫原性有关）。在链球菌细胞壁中也含有一种类似葡萄球菌蛋白A的蛋白质成分，能与人及多种哺乳动物的IgG的Fc结合，但不与禽类的免疫球蛋白结合，称为SPG——链球菌蛋白G。

The *Streptococcus*, as its name suggests, is characterized by its chain-like arrangement under the microscope, and is therefore called "chain cocci" (**Figure 7-3**). It can cause purulent inflammation, pneumonia, mastitis, sepsis, and other diseases in humans and animals. The *Streptococcus* genus has more than 60 species that can be classified based on three antigens on the bacterial cell. These three antigens are: genus-specific antigen (P antigen, *i.e.*, nucleoprotein antigen), group-specific antigen (C antigen, *i.e.*, polysaccharide antigen), and type-specific antigen (protein antigen, also known as surface antigen, which is mostly related to its virulence and immunogenicity). The cell wall of *Streptococcus* also contains a protein component similar to staphylococcal protein A, which can bind to the Fc portion of IgG from humans and many mammals, but not with avian immunoglobulins. This protein is called SPG—streptococcal protein G.

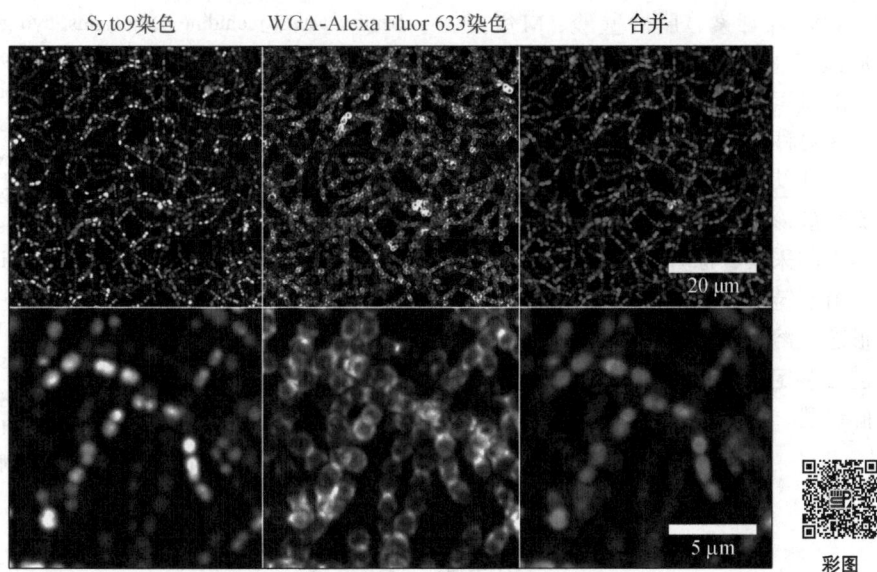

图 7-3 化脓链球菌 GAS JS95 菌株的生物被膜

用染料 Syto9（双链 DNA）和 WGA-Alexa Fluor 633（碳水化合物/胞外聚合物）染色的 GAS JS95 菌株生物被膜 Z-轴视野的投影。图像来源见英文。

Figure 7-3 *Streptococcus pyogenes* GAS JS95 biofilm morphology

Maximum projection of JS95 biofilm Z-stacks stained with Syto9 (dsDNA) and WGA-Alexa Fluor 633 (carbohydrate/extracellular polymeric substance).Source: Matysik A, Kline KA. 2019. *Streptococcus pyogenes* capsule promotes microcolony-independent biofilm formation. J Bacteriol, 201 (18): e00052-19.

链球菌的群特异性抗原又称 C 抗原，即荚膜多糖抗原，是兰氏分群的基础。兰氏分群是美国洛克菲勒大学的微生物学家丽贝卡·兰斯菲尔德建立的一种链球菌分群方法。该方法采用的是抗原和抗体反应原理，用温热的稀盐酸提取 C 多糖抗原，与特异性血清做沉淀反应，将链球菌分为 20 个血清群（A～V，缺 I 和 J）。同群链球菌间，因表面多糖抗原不同被分为不同的型。目前，链球菌的兰氏分群试剂已有商品化的试剂盒可以使用。

链球菌和葡萄球菌一样，在血平板上可以引起溶血。根据溶血能力，可以将链球菌分为 α、β 和 γ 三种溶血类型，其溶血现象分别为不透明草绿色溶血（由血红素转变成胆绿素引起，多为机会致病菌）、完全透明的无色溶血（多为致病菌）和不溶血（一般无致病性）。

链球菌的毒力因子有很多，包括：溶

The group-specific antigen of *Streptococcus*, also known as C antigen or capsule polysaccharide antigen, forms the basis of Lancefield grouping, a method of classifying streptococci established by the microbiologist Rebecca Lancefield at Rockefeller University in the United States. This method uses the principle of antigen-antibody reaction to extract the C polysaccharide antigen with warm dilute hydrochloric acid and precipitate it with specific serum, dividing streptococci into 20 serogroups (A to V, excluding I and J). Within the same serogroup, streptococci are further classified into different types based on their surface polysaccharide antigens. Currently, commercialized kits for Lancefield grouping of streptococci are available.

Similar to *Staphylococcus*, *Streptococcus* can also cause hemolysis on blood agar plates. Based on their hemolytic ability, *Streptococcus* can be grouped into three types: α, β and γ hemolysis. The hemolysis phenomena are represented as opaque greenish hemolysis (caused by conversion of hemoglobin into biliverdin, often seen in opportunistic pathogens), complete transparent hemolysis (often seen in pathogenic bacteria), and no hemolysis (generally non-pathogenic).

There are many virulence factors associated with

血素、致热外毒素、脂磷壁酸、M蛋白、透明质酸酶、链激酶、DNA酶等。其中，链球菌溶血素分为有氧敏感型SLO和氧稳定型SLS两种，SLO是一种胆固醇结合穿孔毒素，可以溶解人和哺乳动物红细胞。

实验室诊断：与金葡菌类似，主要包括：样本的采集，涂片染色，分离培养鉴定等。由于链球菌血清型较多，除了通过菌落形态、溶血特性、细菌形态、生化特性等进行鉴定之外，还可以利用兰氏分群法分群定型。

Streptococcus, including hemolysins, pyrogenic exotoxins, lipoteichoic acids, M protein, hyaluronidase, streptokinase, and DNAase. Among them, *Streptococcus* hemolysins can be divided into two types: oxygen-labile SLO and oxygen-stable SLS. SLO is a cholesterol-binding pore-forming toxin that can lyse human and mammalian red blood cells.

Laboratory diagnosis: similar to those for detecting *Staphylococcus aureus*, laboratory diagnosis involves various procedures, including sample collection, smear staining, isolation and culture identification. Given the wide range of streptococcal serotypes, identification is not limited to evaluating colony morphology, hemolytic properties, bacterial forms, and biochemical characteristics. The Lancefield grouping method can also be utilized for grouping and typing.

思考题　Questions

1. 简述葡萄球菌的基本特征。

2. 有哪些常致牛、羊乳腺炎的链球菌？主要鉴别试验有哪些？

3. 肠球菌属有哪些主要特点？其致病性如何？

1. Please provide a brief overview of the fundamental characteristics of *Staphylococcus*.

2. Which *Streptococcus* spp. normally cause for mastitis in cows and sheep? What are the main diagnostic tests used for identification?

3. What are the major characteristics of *Enterococcus* genus? What is the pathogenicity of *Enterococcus* genus?

第8章 肠杆菌科
Chapter 8 / *Enterobacteriaceae*

内容提要 肠杆菌科细菌广泛存在于自然界，包括正常寄栖菌、机会致病菌和致病菌，种类繁多。其共同点是：革兰氏染色阴性，在显微镜下可见红色细菌，无芽孢，需氧兼性厌氧。肠杆菌科有埃希菌属、沙门菌属、耶尔森菌属等。埃希菌属中最常见的就是大肠杆菌；沙门菌属中常见的是肠道沙门菌，血清型多；耶尔森菌属中常见的有鼠疫、假结核及小肠结肠炎三种病原菌。以上三个属的细菌在显微镜下呈杆状，故称为"肠杆菌"。肠杆菌科中的大肠杆菌和沙门菌可通过食物链或粪口途径感染人和动物，引起发热、腹泻、尿路感染以及脓肿等临床症状。

Introduction The *Enterobacteriaceae* family of bacteria is widely distributed in nature, encompassing commensal, opportunistic, and pathogenic strains with a diverse array of species. Gram-negative staining characterizes these bacteria, and they appear as red rods under the microscope, lacking endospores, and exhibiting facultative anaerobiosis. The most prevalent genera in *Enterobacteriaceae* are *Escherichia*, *Salmonella*, and *Yersinia*. Among these, the most common is *Escherichia coli* in the *Escherichia* genus, while the *Salmonella* genus frequently comprises enteric *Salmonella* with multiple serotypes. *Yersinia*, on the other hand, comprises three pathogenic strains, namely, plague, pseudotuberculosis, and enterocolitis. These three genera exhibit rod-shaped bacteria under the microscope, giving rise to the term "enteric rod bacteria". *Escherichia coli* and *Salmonella* in *Enterobacteriaceae* can cause clinical symptoms, such as fever, diarrhea, urinary tract infections, and abscesses, by infecting humans and animals via the food chain or fecal-oral route.

第1节 埃希菌属
Section 1 *Escherichia*

埃希菌属为革兰氏染色阴性，无芽孢，直杆菌，有菌毛和鞭毛。埃希菌属中的大肠埃希菌，又称大肠杆菌，寄居于温血动物肠道，能为宿主提供微生物源的维生素K，在麦康凯琼脂上培养时，形成红色菌落（**图8-1，图8-2**）。埃希菌属的抗原有三种，分别是O抗原、K抗原和H抗原，其主要成分分别是菌体抗原（含脂多糖）、

The *Escherichia* genus is a group of Gram-negative, non-spore-forming, straight rod-shaped bacteria with pili and flagella. Among them, *Escherichia coli*, also known as *E. coli*, inhabits the intestines of warm-blooded animals and can provide the host with vitamin K of microbial origin. When grown on MacConkey agar, *E. coli* colonies appear as red (**Figure 8-1**, **Figure 8-2**). There are three types of antigens in the *Escherichia* genus: O, K, and H antigens, mainly composed of somatic antigens (including

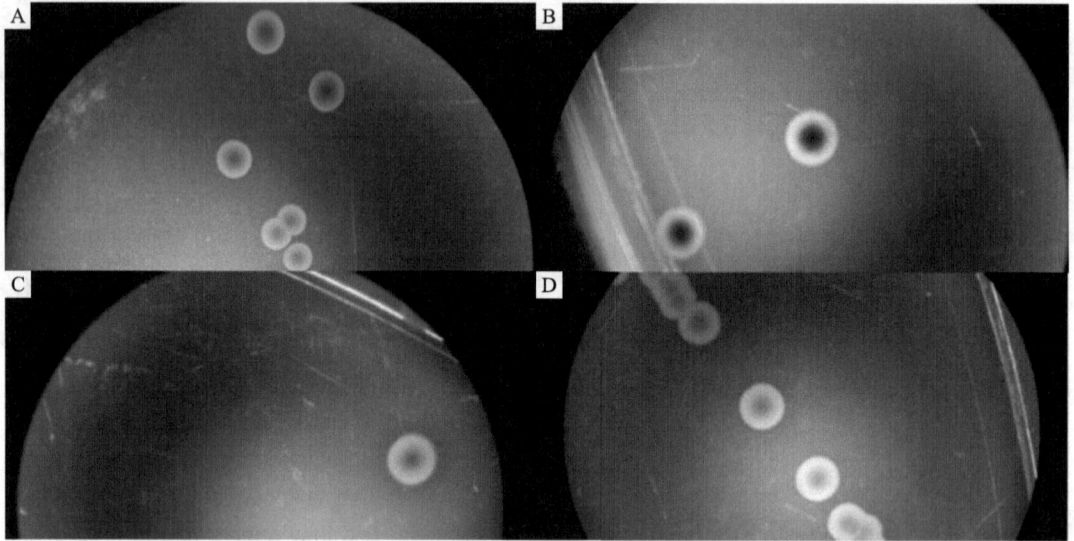

图8-1　产志贺毒素大肠杆菌（STEC）不同菌株在改良麦康凯琼脂平板上的体视显微镜照片
A. STEC O145; B. STEC O26; C. STEC O103; D. STEC O111。图像来源见英文。

Figure 8-1　Stereo-microscopic view of modified MacConkey agar inoculated with Shigatoxin-producing *Escherichia coli* (STEC) strains

A. STEC O145; B. STEC O26; C. STEC O103; D. STEC O111. Source: Verhaegen B, De Reu K, Heyndrickx M, et al. 2015. Comparison of six chromogenic agar media for the isolation of a broad variety of non-O157 Shigatoxin-producing *Escherichia coli* (STEC) serogroups. Int J Environ Res Public Health, 12 (6): 6965-6978.

参考菌株
ATCC 21972　　　D4分离株
血琼脂

麦康凯琼脂

彩图

图8-2　肠杆菌科细菌的分离
大肠杆菌ATCC 21972实验室菌株和D4分离株在血琼脂和麦康凯琼脂平板上的菌落形态、溶血和乳糖发酵特征。图像来源见英文。

Figure 8-2　Isolation of the family *Enterobacteriaceae*

Blood agar and MacConkey agar plates comparing colony morphology, hemolytic properties, and lactose fermentation of *E. coli* lab strain ATCC 21972 and D4 isolated strain. Source: Wolfe AE, Moskowitz JE, Franklin CL, et al. 2020. Interactions of segmented filamentous bacteria (*Candidatus Savagella*) and bacterial drivers in colitis-associated colorectal cancer development. PLoS One, 15 (7): e0236595.

荚膜多糖和鞭毛蛋白。需要特别说明的是在大肠杆菌K抗原中，虽然主要成分是荚膜多糖，但有两个抗原例外，那就是K88和K99，其本质是菌毛蛋白。大肠杆菌的血清型用O: K: H表示，如O8: K23: H19。

根据毒力因子与发病机制的不同，可以将大肠杆菌分为：肠道致病性大肠杆菌（EPEC）和肠道外致病性大肠杆菌（ExPEC）。其中前者包括：肠侵袭型大肠杆菌（EIEC）、肠产毒素型大肠杆菌（ETEC）、肠出血型大肠杆菌（EHEC）、肠致病型大肠杆菌（EPEC）、肠聚集型大肠杆菌（EAEC）、弥散黏附型大肠杆菌（DAEC）共6个类别；肠道外致病性大肠杆菌包括：败血性大肠杆菌（SEPEC）、尿道致病性大肠杆菌（UPEC）、新生儿脑膜炎大肠杆菌（NMEC）、禽致病性大肠杆菌（APEC）共4个类别。

兽医临床上，肠产毒素型大肠杆菌K88是引起仔猪腹泻的主要血清型，K99是引起犊牛和羔羊腹泻的主要血清型，可以引起初生幼畜剧烈水样腹泻所致的脱水死亡；属于肠道外致病的APEC可引起家禽败血症、肠炎、输卵管炎、腹膜炎、心包炎等多种病症。肠出血型大肠杆菌O157: H7是常见的食源性病原，人食用了被大肠杆菌O157: H7污染的食品后，会出现腹部痉挛和血样腹泻，最后发展成溶血性尿毒症，可引起肾衰竭或死亡。

实验室诊断：大肠杆菌的分离鉴定程序包括：取病料接种于培养基进行分离培养和染色镜检，然后再利用纯培养物进一步进行生化试验、血清型鉴定、PCR鉴定等。例如，采集犊牛的腹泻物或直肠拭子，接种选择性培养基麦康凯琼脂或伊红-亚甲蓝琼脂进行初步分离，再将疑似菌落接种三糖铁琼脂斜面。大肠杆菌能分解乳糖和葡萄糖而产酸产气，使三糖铁斜面与底层均呈黄色，且底层有气泡。

lipopolysaccharides), capsular polysaccharides, and flagella proteins, respectively. It should be noted that there are two exceptions to the typical K antigen composition in E. coli, as K88 and K99 are actually pilus proteins. The serotype of *E. coli* is expressed as O:K:H, such as O8:K23:H19.

Based on virulence factors and pathogenesis, *E. coli* can be divided into enteropathogenic *E. coli* (EPEC) and extraintestinal pathogenic *E. coli* (ExPEC). The EPECs can be further classified into 6 types: enteroinvasive *E. coli* (EIEC), enterotoxigenic *E. coli* (ETEC), enterohemorrhagic *E. coli* (EHEC), enteropathogenic *E. coli* (EPEC), enteroaggregative *E. coli* (EAEC), and diffusely adherent *E. coli* (DAEC). The ExPECs include 4 types: septicemic *E. coli* (SEPEC), uropathogenic *E. coli* (UPEC), neonatal meningitis *E. coli* (NMEC), and avian pathogenic *E. coli* (APEC).

In veterinary practice, ETEC K88 is the main cause of diarrhea in piglets, while ETEC K99 is the main cause of diarrhea in calves and lambs, leading to severe dehydration and death in newborn animals. APEC, as an extraintestinal pathogen, can cause various disease such as septicemia, enteritis, salpingitis, peritonitis, and pericarditis in poultry. EHEC O157: H7 is a common foodborne pathogen, and human consumption of food contaminated with EHEC O157: H7 can result in abdominal cramps and hemorrhagic diarrhea, progressing to hemolytic-uremic syndrome, which may lead to renal failure or death.

Laboratory diagnosis: laboratory diagnosis of *E. coli* involves a series of procedures, including isolation and cultivation on culture media, microscopic examination, biochemical tests, serotyping, and PCR identification. For instance, in the case of calf diarrhea, fecal or rectal swab samples are collected and inoculated onto selective media such as MacConkey agar or eosin methylene blue agar for preliminary isolation, followed by inoculation of suspected colonies onto triple sugar iron agar slants. *E. coli* can ferment lactose and glucose with the production of acid and gas, leading to the development of a yellow color with gas bubbles in both the slant and butt of the triple sugar iron agar slant.

第2节 沙门菌属
Section 2 *Salmonella*

沙门菌为革兰氏染色阴性的胞内菌，无芽孢，直杆状，除了鸡沙门菌和鸡白痢沙门菌无鞭毛、不运动外，其余都有菌毛和鞭毛（**图8-3**）。沙门菌属目前有2个种，分别是感染温血动物的肠道沙门菌和感染冷血动物的邦戈沙门菌。肠道沙门菌又分为6个亚种，2500多个血清变种。

沙门菌的抗原有三种，分别是：O抗原、H抗原和Vi抗原，主要成分分别是脂多糖（LPS）、鞭毛蛋白和荚膜多糖。其中，O抗原和H抗原是主要的抗原，H抗原又分为H1和H2。沙门菌的血清型，一般按照O: H1: H2的顺序进行书写。例如，鼠伤寒沙门菌1, 4, [5], 12: i: 1, 2。其中1, 4, [5], 12指的是O抗原，i指的是H1抗原，1, 2指的是H2抗原。在O抗原中的数字1说明这个抗原由溶原噬菌体编码；数字[5]说明该抗原可能不存在。

沙门菌毒力因子主要包括黏附素和毒

Salmonella is a Gram-negative, intracellular bacterium that is non-spore forming and rod-shaped. With the exception of *S. gallinarum* and *S. pullorum*, which lack flagella and motility, all other *Salmonella* have both pili and flagella (**Figure 8-3**). Currently, the *Salmonella* genus comprises two species: *S. enterica*, which infects homeotherms animals, and *S. bongori*, which infects poikilotherms. Within the *S. enterica* species, there are six subspecies and over 2500 serovars.

Salmonella has three types of antigens, namely, O antigen, H antigen, and Vi antigen, which are primarily composed of lipopolysaccharide (LPS), flagellar protein, and capsular polysaccharide, respectively. Among these antigens, O and H antigens are the main antigens, with H antigen further divided into H1 and H2 subtypes. The serotype of *Salmonella* is generally written in the order of O: H1: H2. For instance, *S. typhimurium* 1,4,[5],12:i:1,2 indicates that it has O antigens 1, 4, [5], and 12, H1 antigen i, and H2 antigens 1 and 2. The number 1 in the O antigen indicates that this antigen is encoded by a phage, while the number [5] indicates that this antigen may be absent.

The virulence factors of *Salmonella* mainly include

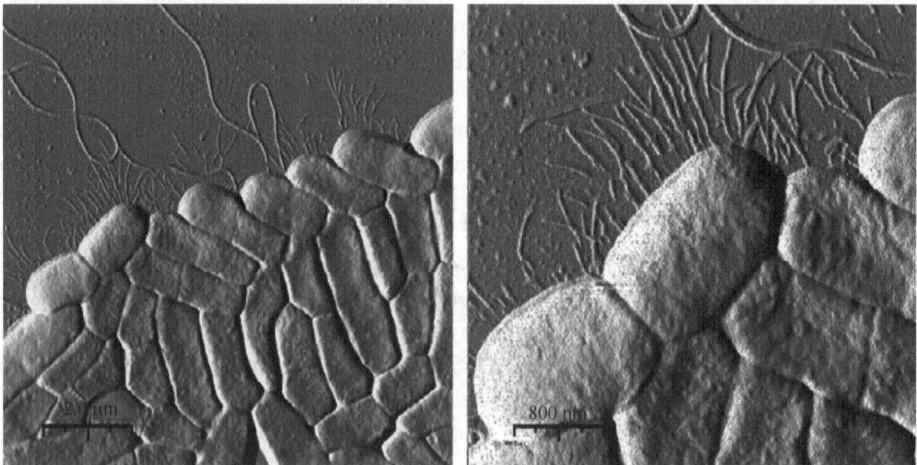

图8-3 鼠伤寒沙门菌MAE14在原子力显微镜两种不同扫描尺寸下后菌毛样结构和鞭毛的高分辨率图像
图像来源见英文。

Figure 8-3 High resolution atomic force microscopy images of pili-like fimbrial structures and flagella in *Salmonella* Typhimurium MAE14 at two different scanning sizes

Source: Jonas K, Tomenius H, Kader A, et al. 2007. Roles of curli, cellulose and BapA in *Salmonella* biofilm morphology studied by atomic force microscopy. BMC Microbiol, 7: 70.

力岛编码的Ⅲ型分泌系统分泌的多种毒力相关蛋白等，其中黏附素包括：鞭毛、菌毛黏附素和非菌毛黏附素等，其作用虽与黏附有关，但有差别。例如，鞭毛可以激活宿主表面Toll样受体TLR5。

实验室诊断： 沙门菌的分离鉴定程序与大肠杆菌类似，可使用普通琼脂、麦康凯平板、伊红-亚甲蓝平板、沙门菌-志贺菌属（SS）平板和三糖铁琼脂进行分离培养。沙门菌可以分解含硫氨基酸产生硫化氢，因此在三糖铁培养基中产生黑色的硫化铁。最后取纯培养物进行生化试验、血清型鉴定、A～F群鉴定和PCR鉴定。

adhesins and various virulence-related proteins secreted by type Ⅲ secretion systems encoded by pathogenicity islands. Adhesins include flagellar adhesins, fimbrial adhesins, and non-fimbrial adhesins, among which their roles in adhesion differ. For example, flagella can activate Toll-like receptor TLR5 on the host surface, although its main function is adhesion.

Laboratory diagnosis: the isolation and identification procedure for *Salmonella* is similar to that of *E. coli*, and can be performed using various media, such as nutrient agar, MacConkey agar, eosin-methylene blue agar, *Salmonella-Shigella* (SS) agar, and triple sugar iron agar. *Salmonella* is capable of decomposing sulfur-containing amino acids to produce hydrogen sulfide, which results in blackening of ferrous sulfate in triple sugar iron agar. Pure cultures are then subjected to biochemical tests, serotyping, group A-F identification, and PCR identification.

第3节 耶尔森菌属
Section 3 *Yersinia*

耶尔森菌属成员包括鼠疫耶尔森菌、假结核耶尔森菌和小肠结肠炎耶尔森菌，为革兰氏阴性胞内菌，无芽孢，除鼠疫耶尔森菌无鞭毛外，其他成员有周身鞭毛。鼠疫耶尔森菌是引起烈性传染病鼠疫的病原菌，同时也是致死性细菌战剂之一，可通过跳蚤在啮齿类动物和人群中传播。欧洲中世纪恐怖的黑死病就是鼠疫，它带走了约2500万人的生命。

鼠疫耶尔森菌可以在野生啮齿类动物—跳蚤—宠物—人群中进行传播。因此，当携带宠物去野外旅行时，应限制宠物与野生啮齿类动物的密切接触，在狩猎和放牧时注意自身防护。小肠结肠炎耶尔森菌是重要的食源性病原，引起人的自限性肠炎，动物多为无症状感染。

实验室诊断： 由于鼠疫可以感染人，处理疑似鼠疫耶尔森菌的病料必须严格执行生物安全相关规定和无菌操作。有关耶尔森菌的分离鉴定与大肠杆菌和沙门菌类似，主要包括：取病料、分离培养、生化和血清学鉴定等步骤。耶尔森菌的培养可

Members of the *Yersinia* genus include *Y. pestis*, *Y. pseudotuberculosis* and *Y. enterocolitica*, which are Gram-negative intracellular bacteria that are non-spore forming and lack endospores. While *Y. pestis* lacks flagella, the other members of the genus have peritrichous flagella. *Y. pestis* is the causative agent of the deadly infectious disease known as plague and is also one of the lethal bacterial warfare agents. It is transmitted through fleas in rodents and humans. The medieval Black Death that ravaged Europe claimed the lives of approximately 25 million people.

Y. pestis can be transmitted through a chain of wild rodents, fleas, pets, and humans. Therefore, when we travel with pets in the wilderness, close contact between pets and wild rodents should be avoided, and personal protective measures should be taken during hunting and grazing. *Y. enterocolitica* is an important foodborne pathogen that causes self-limiting enteritis in humans, while animals are often asymptomatic carriers.

Laboratory diagnosis: due to its potential to infect humans, the handling of suspected *Y. pestis* specimens must strictly adhere to biosafety regulations and aseptic techniques. The isolation and identification of *Yersinia* species are similar to those of *Escherichia coli* and *Salmonella*, and involve the steps of specimen collection, isolation and cultivation, biochemical testing, and

以采用巧克力平板、SS平板、血琼脂平板或麦康凯琼脂平板。其中，耶尔森菌在血琼脂平板上可以形成黏液状的白色菌落（图8-4），不溶血，这是与其他革兰氏阴性和阳性菌的区别。

serological identification. *Yersinia* can be cultured on chocolate agar, SS agar, blood agar, or MacConkey agar. *Yersinia* on blood agar forms white, mucoid colonies (**Figure 8-4**) that do not cause hemolysis, which distinguishes it from other Gram-negative and Gram-positive bacteria.

彩图

图8-4　鼠疫耶尔森菌在绵羊血琼脂上的菌落形态
Figure 8-4　Colonial morphology of *Yersinia pestis* on sheep blood agar

思考题　Questions

1. 简述肠杆菌科的基本特征及其分类概况。

1. Describe the basic characteristics and classification overview of the *Enterobacteriaceae* family.

2. 列举大肠杆菌和沙门菌的实验室诊断程序并指出其鉴定依据。

2. List the laboratory diagnostic procedures for *Escherichia coli* and *Salmonella*, and identify their respective diagnostic criteria.

第9章　弧菌科及气单胞菌科
Chapter 9　Vibrionaceae and Aeromonadaceae

内容提要　为何有人食用被污染的水产品或海鲜后引起食物中毒而导致急性胃肠炎和水样腹泻？这就需要对弧菌科及气单胞菌科相关成员进行了解。

Introduction　Why do some individuals experience food poisoning such as acute gastroenteritis and watery diarrhea after consuming contaminated aquatic products or seafood? This needs an understanding of members belonging to the *Vibrionaceae* and *Aeromonadaceae* families.

第1节　弧　菌　科
Section 1　*Vibrionaceae*

弧菌科成员有霍乱弧菌、副溶血弧菌、创伤弧菌、鳗弧菌等。霍乱弧菌可引起剧烈水样腹泻，曾于1831年发生于英国的伦敦，由污染的水源引起。该公共卫生事件目前已成为现代流行病学的经典案例。副溶血弧菌是海产品中的一种常见污染菌，可引起食用者腹泻；该菌嗜盐，革兰氏染色阴性，在显微镜下呈现弧状、杆状或丝状，一端有鞭毛（**图9-1**）；在绵羊血琼脂平板上不溶血，在TCBS琼脂上菌落呈绿色（**图9-2**），TCBS是硫代硫酸盐-柠檬酸盐-胆盐-蔗糖琼脂培养基的缩写。副溶血弧菌致病菌菌株能产生耐热的热稳定直接溶血毒素（TDH）和不耐热的热不稳定溶血毒素（TLH），裂解人或兔红细胞，可用于区分致病和非致病菌株，即神奈川试验，该试验已被列入我国检验食品中副溶血弧菌的国家标准。

Members of *Vibrionaceae* include *Vibrio cholerae*, *V. parahaemolyticus*, *V. vulnificus* and *V. anguillarum*. *V. cholerae* can cause severe watery diarrhea and was responsible for the cholera outbreak in London in 1831, which has since become a classic example in modern epidemiology. *V. parahaemolyticus* is a common contaminant in seafood and can cause diarrhea in humans. This Gram-negative, salt-tolerant bacterium is curved or rod-shaped, with a single polar flagellum (**Figure 9-1**). It does not hemolyze sheep blood agar but produces green colonies on thiosulfate-citrate-bile-sucrose (TCBS) agar (**Figure 9-2**). TCBS is an abbreviation for thiosulfate-citrate-bile-sucrose agar, a culture medium that contains sulfur compounds, citrate, bile salts, and sucrose. Pathogenic strains of *V. parahaemolyticus* produce a thermostable direct hemolysin (TDH) and a thermolabile hemolysin (TLH), both of which can lyse human or rabbit erythrocytes. The Kanagawa test, which can differentiate pathogenic from non-pathogenic strains, is based on the ability of pathogenic strains to produce TDH and TLH, and has been incorporated into the national standard for detecting *V. parahaemolyticus* in food in China.

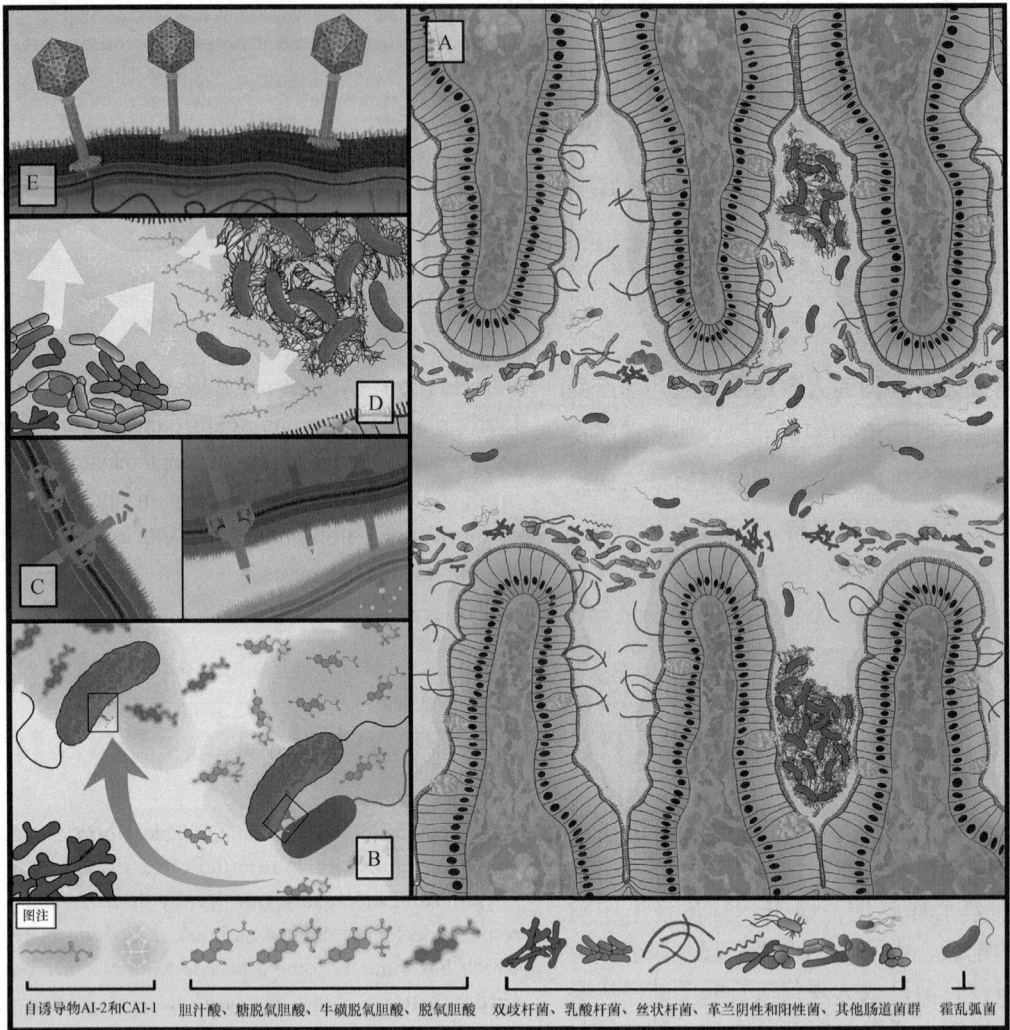

图注

自诱导物AI-2和CAI-1 ｜ 胆汁酸、糖脱氧胆酸、牛磺脱氧胆酸、脱氧胆酸 ｜ 双歧杆菌、乳酸杆菌、丝状杆菌、革兰阴性和阳性菌、其他肠道菌群 ｜ 霍乱弧菌

图 9-1 霍乱弧菌、肠道微生物及其代谢产物与小肠环境之间的互作

A. 霍乱弧菌在酸性胃环境中存活后，进入小肠肠腔，胆汁（绿色）和黏液（黄色）刺激霍乱弧菌表达毒力因子。霍乱弧菌到达肠上皮隐窝后形成生物被膜，黏附于上皮表面；B、C. 霍乱弧菌到达黏液层时，其Ⅵ型分泌系统被激活。该系统可作为收缩性细胞器，从霍乱弧菌刺入邻近的微生物，以转位毒性效应蛋白；D. 自诱导剂AI-2（黄色）由一些共生肠道微生物产生，可在霍乱弧菌中诱导产生群体感应。自诱导物的存在表明霍乱弧菌的密度已经很高，导致能够定植和产生霍乱毒素的毒力基因表达减少，并激活促进离开宿主的基因增加，如鞭毛蛋白；E. 霍乱弧菌特异性噬菌体可以快速感染和裂解霍乱弧菌，从而大幅减少霍乱弧菌的数量。图像来源见英文。 彩图

Figure 9-1 Interactions between the gut microbiota, their metabolites, _V. cholerae_, and the small intestinal environment

A. After _V. cholerae_ survives the acidic gastric environment, the pathogen enters the small intestinal gut lumen, where both bile (green) and mucus (yellow) signal to _V. cholerae_ to express the virulence factors. After reaching the intestinal epithelial crypts, _V. cholerae_ forms biofilms to adhere to the epithelial surface; B and C. When encountering the mucus layer, the _V. cholerae_ type Ⅵ secretion system (T6SS) is activated. This system operates as a contractile organelle that extends from _V. cholerae_ to make contact with neighboring organisms to translocate toxic effectors; D. Autoinducer AI-2 (yellow) is produced by some commensal gut microbes and can induce quorum sensing responses in _V. cholerae_. The presence of autoinducers indicates to _V. cholerae_ a high density of organisms, resulting in reduced expression of virulence genes that enable colonization and cholera toxin production, and activation of genes that promote exit from the host, such as increased flagellar motion; E. Bacteriophage specific to _V. cholerae_ can infect and lyse large numbers of organisms rapidly, drastically reducing _V. cholerae_ populations. Source: Weil AA, Becker RL, Harris JB. 2019. _Vibrio cholerae_ at the intersection of immunity and the microbiome. mSphere, 4 (6): e00597-19.

图9-2 TCBS琼脂上的弧菌菌落

A. 副溶血弧菌（绿松石色）、霍乱弧菌（黄色），以及这二者的混合接种；B. 副溶血弧菌菌落呈青绿色，具有圆形、完整和凸起的形态。图像来源见英文。

Figure 9-2 Colony morphology of *Vibrio* spp. on TCBS agar

A. *V. parahaemolyticus* (turquoise), *V. cholerae* (yellow), and a mixed inoculation of the two species; B. Colonies of *V. parahaemolyticus* appear turquoise with a circular, entire, and convex morphology. Source: Yeung M, Thorsen T. 2016. Development of a more sensitive and specific chromogenic agar medium for the detection of *Vibrio parahaemolyticus* and other *Vibrio* species. J Vis Exp, 117: 54493.

第2节 气单胞菌科
Section 2 *Aeromonadaceae*

嗜水气单胞菌为鱼类常见致病菌，可感染人，引起胃肠炎症状或败血症。该菌革兰氏染色阴性，有鞭毛（**图9-3，箭头**）和菌毛，可以产生气溶素、溶血素和肠毒素，90%人源菌株在绵羊血琼脂上呈现β溶血；能产生细胞色素氧化酶，可以将滤

Aeromonas hydrophila is a common pathogenic bacterium in fish that can also infect humans, causing symptoms of gastroenteritis or sepsis. This Gram-negative bacterium possesses polar flagella (**Figure 9-3, arrows**) and pili, and is capable of producing exotoxins, hemolysins, and enterotoxins. 90% of human isolates display β-hemolysis on sheep blood agar. It is also able

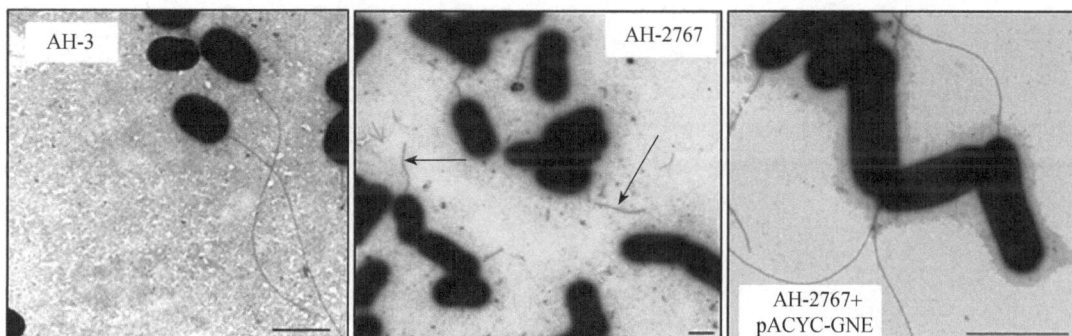

图9-3 三株嗜水气单胞菌细胞的电镜图片

标尺：1 μm。图像来源见英文。

Figure 9-3 Electron microscopy of whole cells from three *Aeromonas hydrophila* strains

The bar represents 1 μm. Source: Merino S, Fulton KM, Twine SM, et al. 2014. *Aeromonas hydrophila* flagella glycosylation: involvement of a lipid carrier. PLoS One, 9 (2): e89630.

纸片上的对苯二胺氧化成红色或紫色的琨类化合物，即为氧化酶试验阳性。气单胞菌与弧菌在低温和不良的环境中，即以活着但不可培养（VBNC）状态存在。

to produce cytochrome oxidase, which can oxidize the substrate p-phenylenediamine on a filter paper to form purple-colored quinone compounds, giving a positive result in the oxidase test. Similar to *Vibrionaceae*, *Aeromonas* and *Vibrio* are capable of existing in a viable but non-culturable (VBNC) state under adverse conditions and low temperatures.

思考题　Questions

1. 某些气单胞菌与弧菌在低温和不良的环境中，以活着但不可培养（VBNC）状态存在，其生物学意义是什么？

2. 副溶血弧菌的主要特征是什么？

3. 嗜水气单胞菌的主要特征是什么？

1. What is the biological significance of certain *Aeromonas* and *Vibrio* species existing in a viable but non-culturable (VBNC) state under adverse conditions and low temperatures?

2. What are the main characteristics of *Vibrio parahaemolyticus*?

3. What are the main characteristics of *Aeromonas hydrophila*?

第10章 巴氏杆菌科和黄杆菌科

Chapter 10 / *Pasteurellaceae* and *Flavobacteriaceae*

内容提要 巴氏杆菌科成员较多，包括巴氏杆菌属、曼氏杆菌属、放线杆菌属、格拉菌属、嗜血杆菌属、禽杆菌属等。黄杆菌科成员有里氏杆菌属等。

Introduction The familiy *Pasteurellaceae* comprises numerous genera, including *Pasteurella*, *Mannheimia*, *Actinobacillus*, *Glaesserella*, *Haemophilus*, *Avibacterium*.The family *Flavobacteriaceae* includes the genus *Riemerella*.

第1节 巴氏杆菌属
Section 1 *Pasteurella*

巴氏杆菌属的典型种为多杀性巴氏杆菌，由法国微生物学家路易斯·巴斯德于1880年从病鸡体内分离获得，并因此命名为多杀性巴氏杆菌。多杀性巴氏杆菌能通过蜱和跳蚤传播，不仅可以感染鸡，还可以感染牛、羊、猪、马、兔等，引起出血性败血症或传染性肺炎（**图10-1**）。多杀性巴氏杆菌为球杆状或短杆状，无鞭毛，无芽孢，革兰氏染色阴性，美蓝或瑞氏染色为两极着色（**图10-2**）；巴氏杆菌在血琼脂平板上生长呈水滴样菌落，不溶血。

多杀性巴氏杆菌的血清型较多，可用大写英文字母和阿拉伯数字表示，其中英文大写字母表示荚膜抗原，阿拉伯数字表示菌体抗原，中间用冒号隔开。例如，多杀性巴氏杆菌血清型A: 3。

实验室诊断： 从患病动物身体上采集新鲜病料（渗出液、血、肝、脾、淋巴结、骨髓等），接种于血琼脂平板和三糖铁琼脂斜面。染色镜检可以采用亚甲蓝染色、瑞氏染色或革兰氏染色。在三糖铁琼

The type species of the *Pasteurella* genus is *Pasteurella multocida*, which was isolated by French microbiologist Louis Pasteur from the body of an infected chicken in 1880 and subsequently named after its ability to cause a variety of diseases in multiple animal species. *P. multocida* is transmitted by ticks and fleas and can infect not only chickens, but also cattle, sheep, pigs, horses, and rabbits, causing septicemia or contagious pneumonia (**Figure 10-1**). This Gram-negative, non-motile, non-spore-forming coccobacillus stains bipolarly with methylene blue or Wright's stain (**Figure 10-2**). On blood agar, *P. multocida* grows as small, glistening, convex, and translucent colonies resembling drops of dew, but does not produce hemolysis.

P. multocida has multiple serotypes that can be identified by uppercase letters and Arabic numerals separated by a colon, with the letter indicating the capsule antigen and the number indicating the somatic antigen. For example, the serotype of *P. multocida* type A with somatic antigen 3 is designated as A:3.

Laboratory diagnosis: fresh pathological specimens such as exudates, blood, liver, spleen, lymph nodes, and bone marrow can be collected from diseased animals and inoculated onto blood agar and triple sugar iron agar slants for isolation and identification. Staining techniques such as methylene blue,

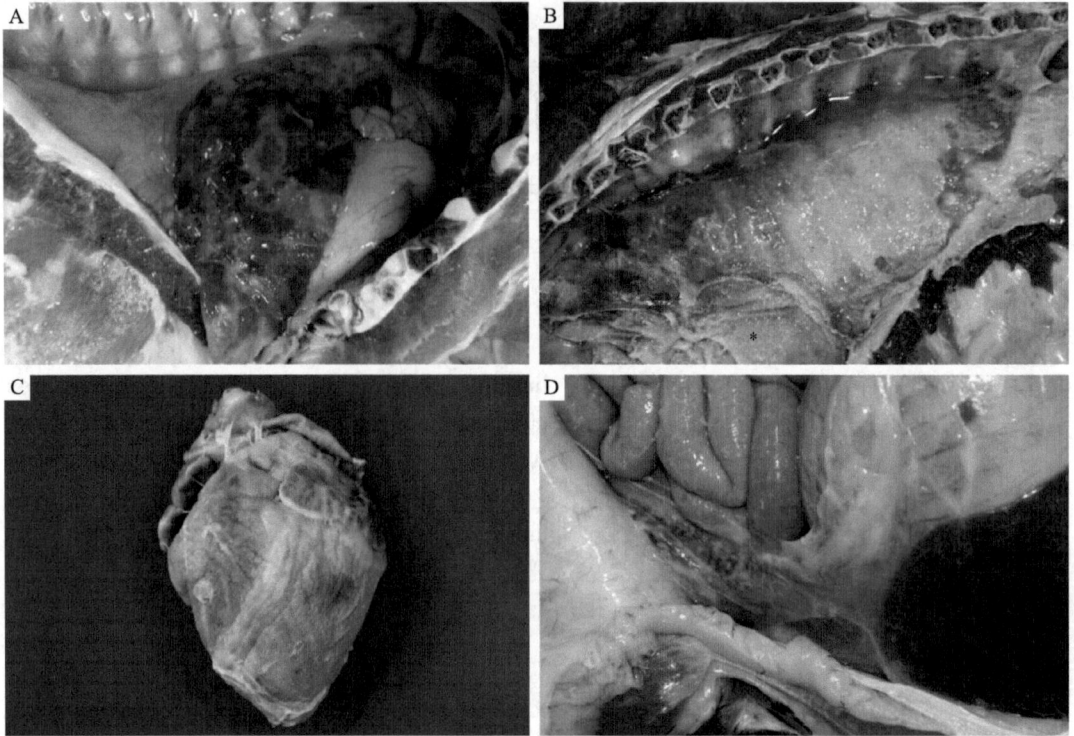

图 10-1　多杀性巴氏杆菌 A 型对实验猪引起的病变

A. 肺：心叶局部广泛出血性胸膜肺炎，胸膜上有纤维蛋白；B. 胸腔：弥漫性纤维蛋白性胸膜炎和心包炎（＊）；C. 心脏：弥漫性纤维蛋白性心包炎；D. 腹腔：纤维蛋白性腹膜炎。

Figure 10-1　Lesions caused by _Pasteurella multocida_ type A in experimentally challenged pigs

A. Lung. Focally extensive hemorrhagic pleuropneumonia in the cardiac lobe with fibrin on the pleura; B. Thoracic cavity. Diffuse fibrinous pleuritis and pericarditis (*); C. Heart. Diffuse fibrinous pericarditis; D. Abdominal cavity. Fibrinous peritonitis.

彩图

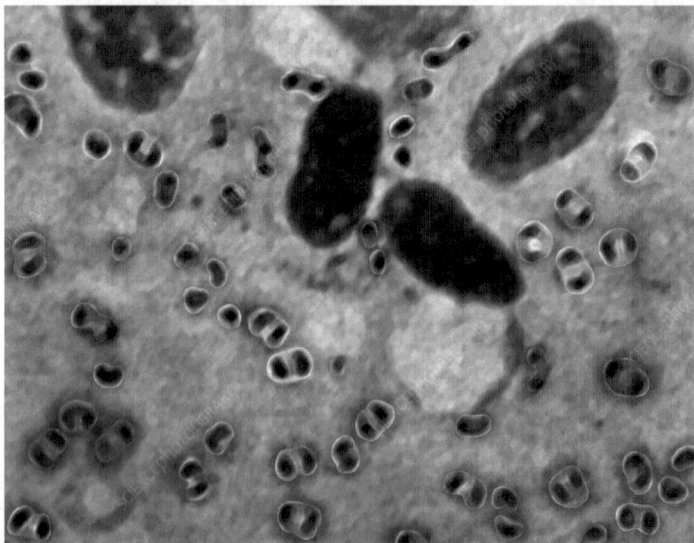

图 10-2　多杀性巴氏杆菌的瑞氏染色图（两极着色）

Figure 10-2　Wright's staining of _Pasteurella multocida_ (bipolar staining)

彩图

脂斜面中，巴氏杆菌能分解乳糖和葡萄糖产酸，使琼脂斜面底层呈黄色。可取单菌落进一步进行生化试验、血清学试验、动物试验、PCR鉴定等。

Wright's stain, or Gram stain can be employed for microscopic examination. On triple sugar iron agar slants, *Pasteurella* species ferment lactose and glucose to produce acid, resulting in a yellow coloration of the lower portion of the agar slant. Further biochemical, serological, animal model, and PCR identification tests can be performed using isolated single colonies.

第2节　曼氏杆菌属
Section 2　*Mannheimia*

溶血性曼氏杆菌为曼氏杆菌属代表成员，可引起牛、羊等反刍动物肺炎，新生羔羊急性败血症。溶血性曼氏杆菌（**图10-3**）

Mannheimia haemolytica, a representative member of the *Mannheimia* genus, is a pathogen that can cause pneumonia in ruminants such as cattle and sheep, as well as acute septicemia in newborn lambs. Under microscope,

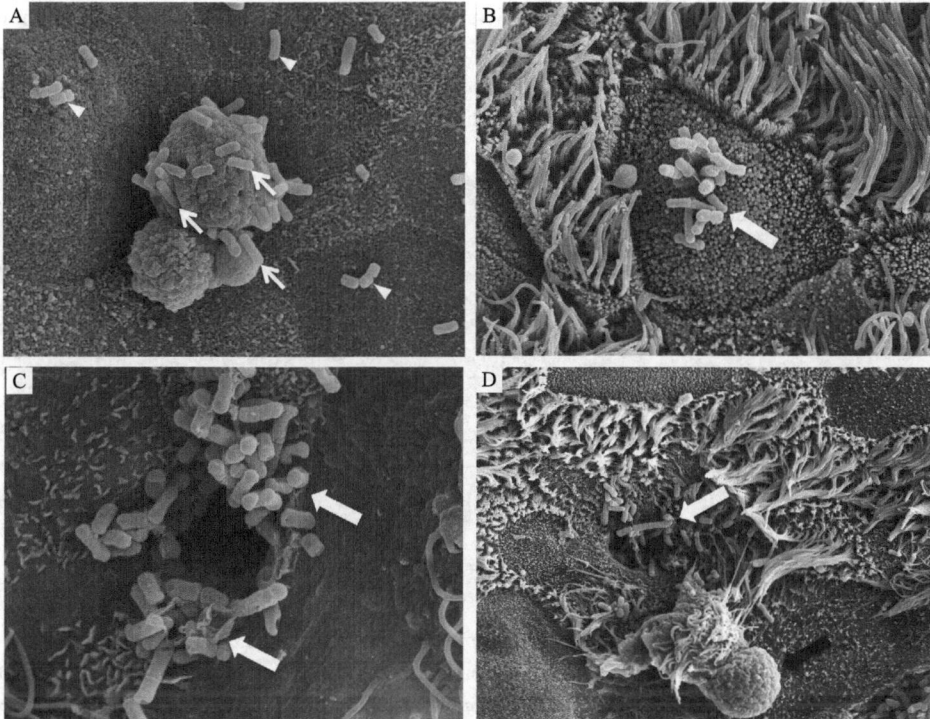

图10-3　感染溶血性曼氏杆菌的牛支气管上皮细胞电镜照片

A. 黏附非纤毛上皮细胞（三角箭头）和黏液（箭头）的细菌；B. 黏附于非纤毛上皮细胞中心的细菌（箭头），但不黏附纤毛；C. 细胞膜内陷处的细菌（箭头）；D. 细菌（白色箭头）与细胞膜内陷相关，这可能由黏液排出引起（黑色箭头）。图像来源见英文。

Figure 10-3　Scanning electron microscopy of differentiated bovine bronchial epithelial cells infected with *Mannheimia haemolytica*

A. Bacteria adhering to nonciliated epithelial cells (arrowheads) and to mucus (arrows); B. Bacteria (arrow) adhering to the center of a nonciliated epithelial cell but not to cilia; C. Large numbers of bacteria (arrows) associated with an invagination of the cell membrane; D. Bacteria (white arrow) associated with an invagination of the cell membrane, which may have been the result of mucus extrusion (black arrow). Source: Cozens D, Sutherland E, Lauder M, et al. 2019. Pathogenic *Mannheimia haemolytica* invades differentiated bovine airway epithelial cells. Infect Immun, 87 (6): e00078-19.

在显微镜下为多形性，有荚膜，无鞭毛，无芽孢，革兰氏染色阴性，瑞氏染色为两极着色；溶血性曼氏杆菌微需氧或兼性厌氧，在牛血琼脂平板上多数为β溶血。

M. haemolytica (**Figure 10-3**) appears pleomorphic, encapsulated, non-motile, non-spore-forming, and Gram-negative, and stains bipolarly with Wright's stain. *M. haemolytica* is a facultative anaerobe or microaerophile that typically displays β-hemolysis on blood agar plates.

第3节　放线杆菌属
Section 3　*Actinobacillus*

胸膜肺炎放线杆菌为放线杆菌属代表成员，可引起猪传染性胸膜肺炎，导致猪的肺部出现淤血坏死，伴有纤维素性胸膜炎（**图10-4**）。该菌革兰氏染色阴性，普通琼脂培养需添加V因子（即辅酶Ⅰ和辅酶Ⅱ）才能生长；兼性厌氧，在10% CO₂中，可长出黏液型菌落；该菌可用巧克力平板培养，在绵羊血琼脂平板上呈β溶血。

Actinobacillus pleuropneumoniae, a representative member of the *Actinobacillus* genus, is a pathogen that can cause contagious pleuropneumonia in pigs, leading to lung congestion, necrosis, and fibrinous pleuritis (**Figure 10-4**). This Gram-negative bacterium requires the addition of V factor (coenzyme Ⅰ and Ⅱ) for growth on routine culture media, and is a facultative anaerobe that can form mucoid colonies in the presence of 10% CO_2. *A. pleuropneumoniae* can be cultured on chocolate agar plates and typically displays β-hemolysis on sheep blood agar plates.

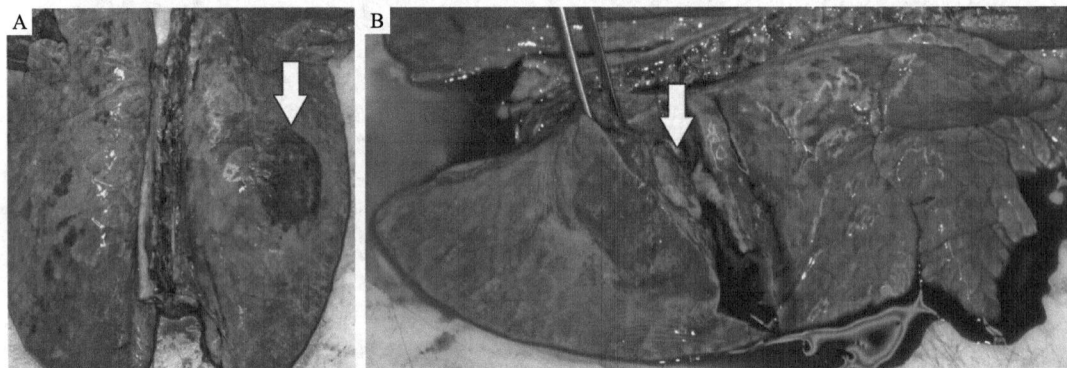

图10-4　攻毒胸膜肺炎放线杆菌第21天后的猪肺部病变
箭头处为包裹在结缔组织中的脓肿样结节（A和B）。图像来源见英文。

Figure 10-4　Lung changes caused by *Actinobacillus pleuropneumoniae* on day 21 post challenge
Arrows show abscess-like nodules encapsulated in connective tissue (A and B). Source: Brauer C, Hennig-Pauka I, Hoeltig D, et al. 2012. Experimental *Actinobacillus pleuropneumoniae* challenge in swine: comparison of computed tomographic and radiographic findings during disease. BMC Vet Res, 8: 47.

V因子需要试验：首先在血平板上用金黄色葡萄球菌划一条线，然后再用胸膜肺炎放线杆菌垂直划若干条线，37℃培养过夜。沿金黄色葡萄球菌划线将出现卫星状菌落，且越靠近金黄色葡萄球菌菌落越多（**图10-5**）。这是因为金黄色葡萄球菌可以裂解红细胞为胸膜肺炎放线杆菌提

The V factor requirement test: the V factor requirement test involves streaking a line of *Staphylococcus aureus* on a blood agar plate, followed by streaking several lines of *A. pleuropneumoniae* perpendicular to the *S. aureus* line. The plate is then incubated overnight at 37 ℃. Along the *S. aureus* line, satellite colonies will appear, with increasing numbers closer to the *S. aureus* line (**Figure 10-5**). This is because *S. aureus* can lyse red blood cells, providing V factor for

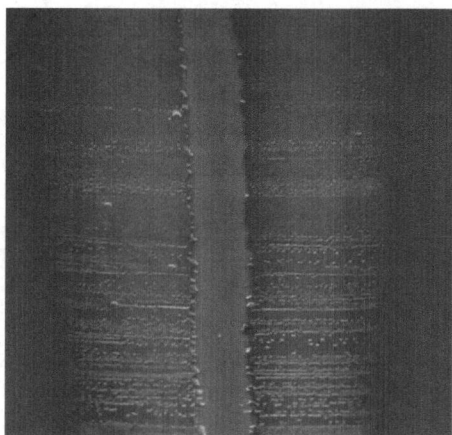

彩图

图 10-5　胸膜肺炎放线杆菌的 V 因子需要试验
Figure 10-5　The V factor requirement test of *A. pleuropneumoniae*

供 V 因子。V 因子需要试验不仅可以用于鉴定胸膜肺炎放线杆菌，还可以鉴定嗜血杆菌。

CAMP 试验： CAMP 是克里斯蒂、阿特金斯和蒙克 - 彼得松三位发明该试验人员的姓氏缩写。其原理是将胸膜肺炎放线杆菌或 B 群链球菌与产生 β 溶血素的金黄色葡萄球菌相互垂直划线但不接触，两条细菌生长线分开的区域出现明显溶血现象，即溶血带变宽。CAMP 试验不仅可以用于鉴定胸膜肺炎放线杆菌，还可以鉴定 B 群链球菌。

the growth of *A. pleuropneumoniae*. The V factor requirement test is not only useful for identifying *A. pleuropneumoniae*, but also for identifying *Glaesserella parasuis*.

CAMP test: the CAMP test is named after the surnames of the three inventors, Christie, Atkins, and Munch-Peterson. The principle of this test is to streak the surface of a blood agar plate with *A. pleuropneumoniae* or group B *Streptococcus* perpendicular to a streak of *Staphylococcus aureus* that produces β-hemolysin, without touching. The area where the two bacterial growth streaks intersect exhibits prominent hemolysis, resulting in an enlarged area of complete lysis called the CAMP reaction zone. The CAMP test is not only used to identify *A. pleuropneumoniae*, but also for the identification of group B *Streptococcus*.

第 4 节　格 拉 菌 属
Section 4　*Glaesserella*

副猪格拉菌，原名为副猪嗜血杆菌，为本属代表种，是引起猪格氏病的病原菌，引起多发性浆膜炎、关节炎、脑膜炎等，病猪出现高热、关节肿胀、呼吸困难及中枢神经症状（**图 10-6，图 10-7**）。副猪格拉菌为革兰氏染色阴性，无芽孢，无鞭毛（**图 10-8**），其生长需要血液中的 V 因子，即辅酶 Ⅰ 和辅酶 Ⅱ。血液红细胞中的 V 因子需要加热 80～90 ℃才能释放出来，因此将血液加热释放出 V 因子后配制的培养基呈褐色，称为巧克力培养基。副猪格拉菌在巧克力琼脂上菌落圆形、光滑。

Glaesserella parasuis, formerly known as *Haemophilus parasuis,* is the representative species of *Glaesserella*, which is a pathogen that causes Glasser's disease in pigs. The infection causes fibrinous polyserositis, arthritis, meningitis, and other clinical symptoms in pigs. Infected pigs exhibit high fever, difficult breathing, joint swelling, and central nervous system symptoms (**Figure 10-6, Figure 10-7**). *G. parasuis* is a Gram-negative bacterium, lacks spores and flagella (**Figure 10-8**). Its growth requires the V factor, which is composed of coenzyme Ⅰ and Ⅱ found in blood. The V factor is released from red blood cells by heating at 80-90 ℃, and the medium containing the V factor is brown and is known as chocolate agar. *G. parasuis* colonies on chocolate agar are circular and smooth.

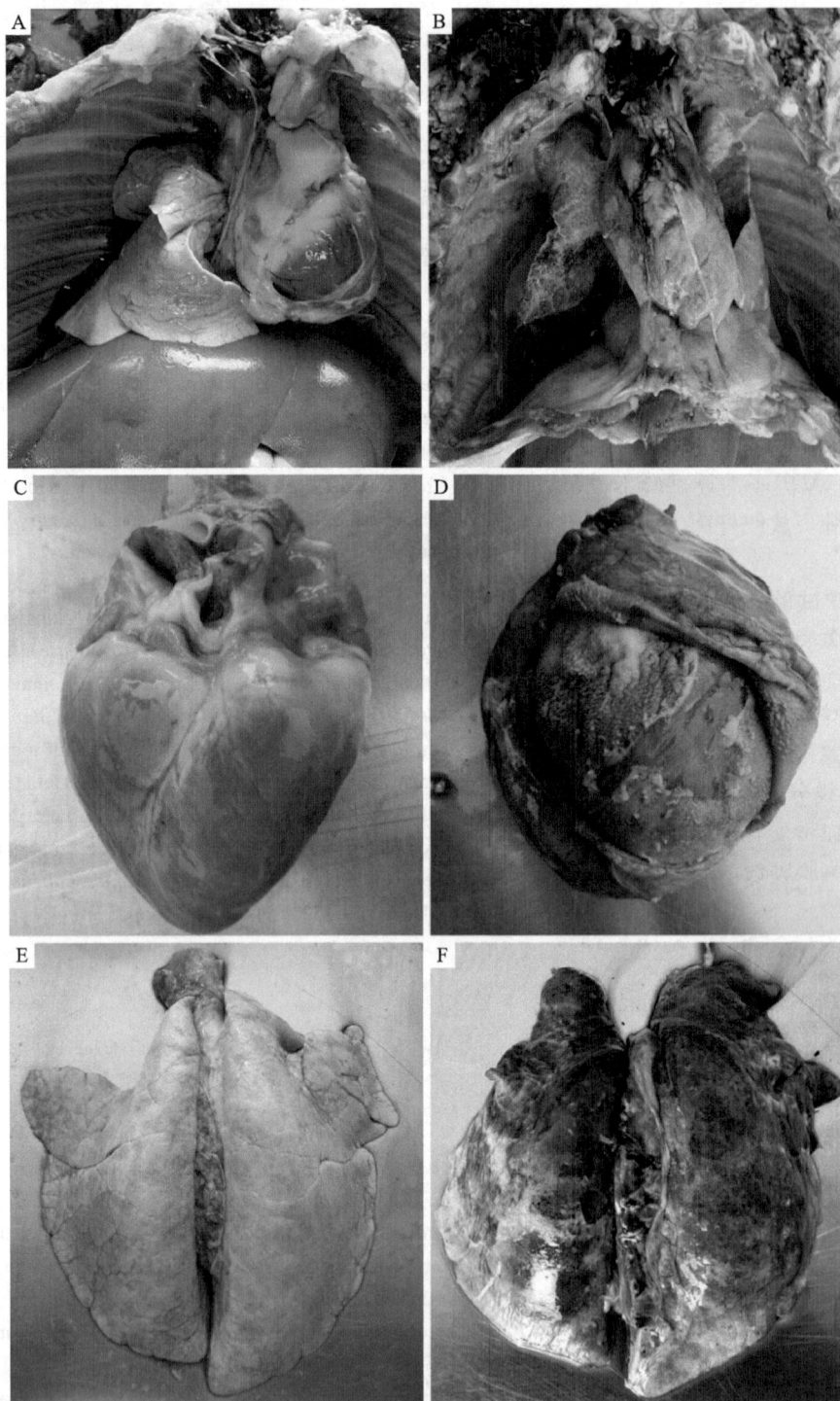

彩图

图 10-6　感染副猪格拉菌仔猪的症状（1）

A、C、E. 对照组；B. 胸腔积液；D. 心包增厚，典型的"绒毛状心脏"；F. 肺充血、出血、坏死。图像来源见英文。

Figure 10-6　Symptoms of piglets infected with *Glaesserella parasuis* (1)

A, C, E. Control group; B. Pleural effusion; D. Pericardial thickening, a typical "fluff heart"; F. Pulmonary congestion, bleeding, necrosis.

Source: Qi B, Li F, Chen K, et al. 2021. Comparison of the *Glaesserella parasuis* virulence in mice and piglets. Front Vet Sci, 8: 659244.

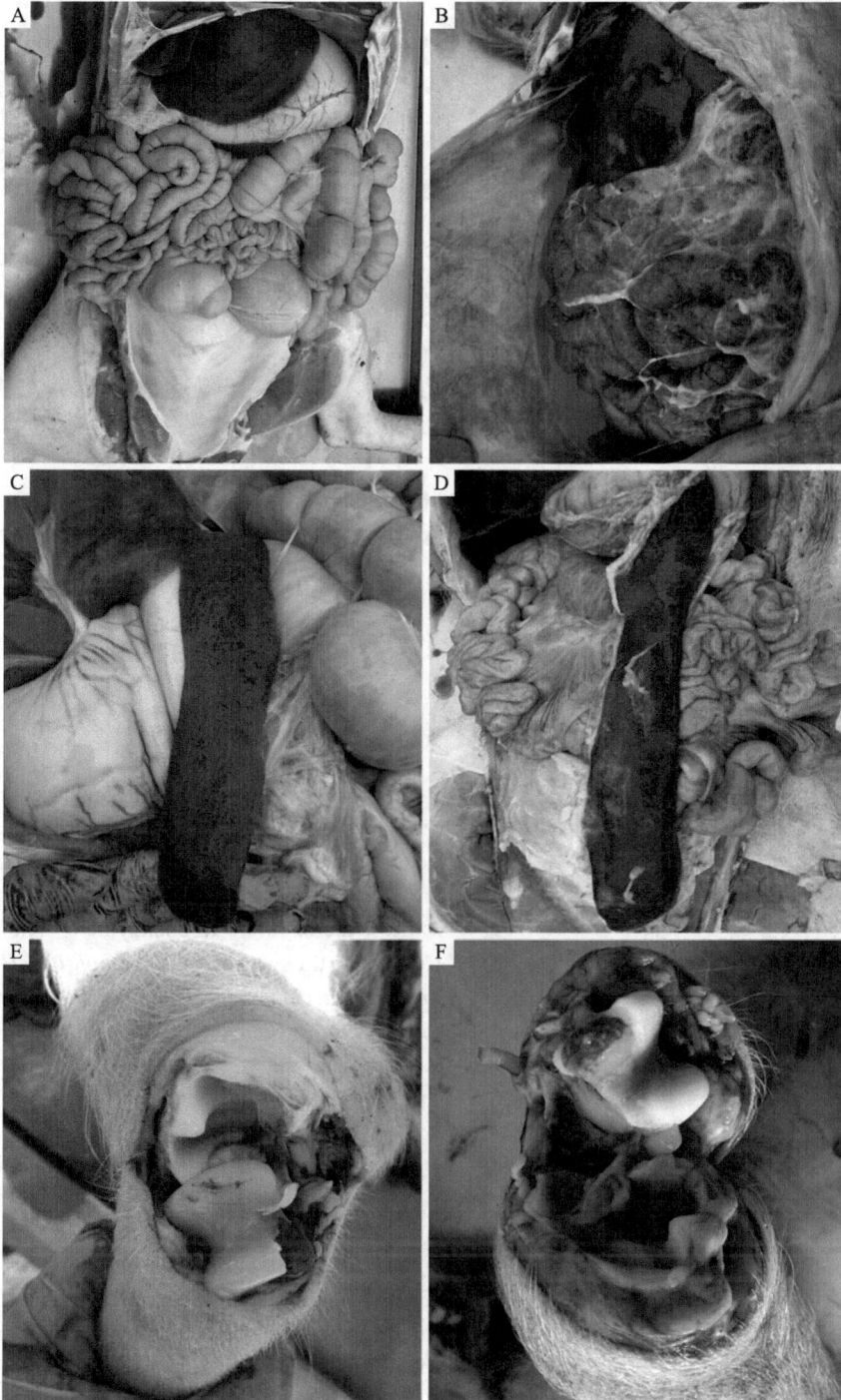

彩图

图 10-7　感染副猪格拉菌仔猪的症状（2）
A、C、E. 对照组；B. 腹膜有纤维渗出、肠壁充血、粘连；D. 脾脏边缘梗死，被纤维状假膜覆盖；F. 关节积液。图像来源见英文。

Figure 10-7　Symptoms of piglets infected with *G. parasuis* (2)
A, C, E. Control group; B. Peritoneal fibrous exudate, intestinal wall congestion, adhesions; D. Spleen edge infarction, covered with fibrous pseudomembrane; F. Joint effusion. Source: Qi B, Li F, Chen K, et al. 2021. Comparison of the *Glaesserella parasuis* virulence in mice and piglets. Front Vet Sci, 8: 659244.

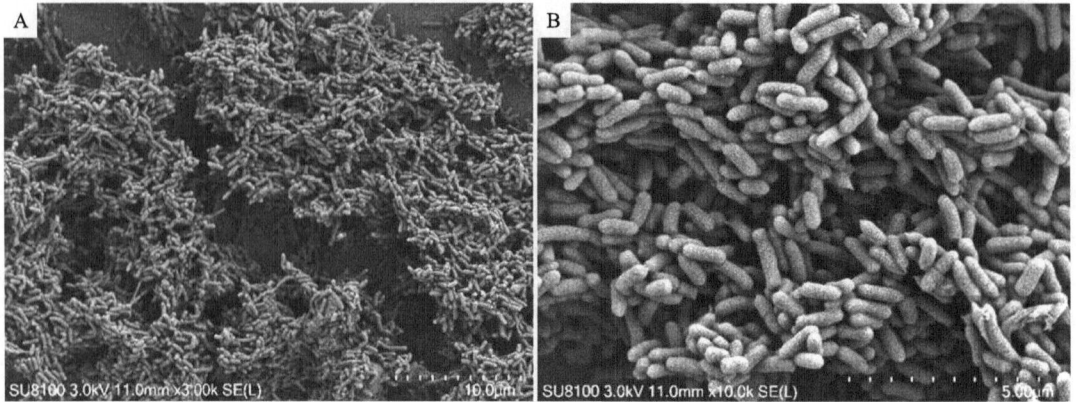

图10-8 副猪格拉菌扫描电镜图像（A和B）
图像来源见英文。

Figure 10-8 *Glaesserella parasuis* biofilms visualized by scanning electron microscope (A and B)

Source: Jiang R, Xiang M, Chen W, et al. 2021. Biofilm characteristics and transcriptomic analysis of *Haemophilus parasuis*. Vet Microbiol, 258: 109073.

第5节 嗜血杆菌属
Section 5 *Haemophilus*

嗜血杆菌是一群酶系统不完全的革兰氏阴性杆菌，培养需要血液中生长因子，尤其是X因子和V因子。X因子即血红素及其衍生物，V因子即辅酶Ⅰ和辅酶Ⅱ。嗜血杆菌主要有三种：猫嗜血杆菌、血红蛋白嗜血杆菌和副兔嗜血杆菌，其中，猫嗜血杆菌和副兔嗜血杆菌培养只需要V因子。

Haemophilus is a group of Gram-negative bacilli with incomplete enzyme systems, which require growth factors, especially X factor and V factor, from blood for cultivation. X factor refers to heme and its derivatives, while V factor refers to coenzyme Ⅰ and Ⅱ. There are mainly three types of *Haemophilus*: *H. felis*, *H. haemoglobinophilus*, and *H. paracuniculus*. Among them, only V factor is required for the cultivation of *H. felis* and *H. paracuniculus*.

第6节 禽杆菌属
Section 6 *Avibacterium*

副鸡禽杆菌属于禽杆菌属，原名为副鸡嗜血杆菌，可引起鸡传染性鼻炎，患鸡眼眶下出现水肿（**图10-9，图10-10**）。副鸡禽杆菌为革兰氏染色阴性杆菌，生长需要辅酶Ⅰ（NAD，因菌株而异），有A、B、C三个血清群，外膜蛋白中的血凝素可凝集鸡红细胞。在巧克力琼脂平板上菌落呈现圆形、光滑、灰白半透明。

Avibacterium paragallinarum, formerly known as *Haemophilus paragallinarum*, belongs to the genus *Avibacterium* and is the causative agent of infectious coryza in chickens. The disease is characterized by swelling beneath the eye pit (**Figure 10-9, Figure 10-10**). *A. paragallinarum* is a Gram-negative rod-shaped bacterium that requires coenzyme Ⅰ (NAD, depending on the variants) as a growth factor. It has three serogroups (A, B, and C) based on the outer membrane protein hemagglutinin, which can agglutinate chicken red blood cells. On chocolate agar, the colonies appear round, smooth, and grayish-white translucent.

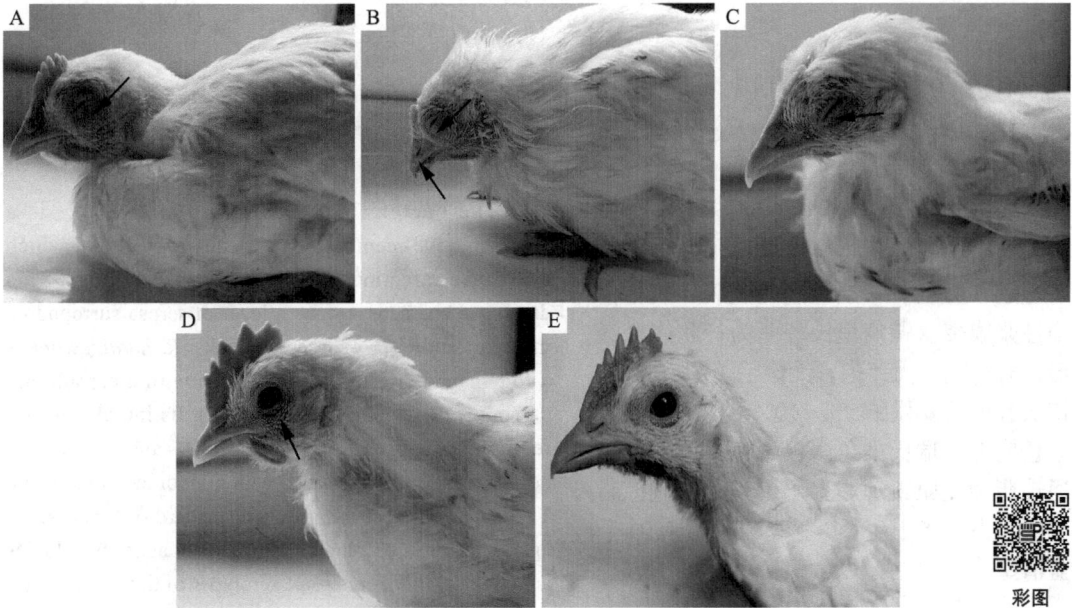

图10-9 副鸡禽杆菌感染鸡的临床症状

攻毒后第1天（A）、第3天（B）、第5天（C）、第7天（D）面部肿胀（箭头）。E为对照组健康鸡。图像来源见英文。

Figure 10-9 Clinical symptoms of chickens infected with *Avibacterium paragallinarum*

Facial swelling (arrow) at 1 dpi (A), 3 dpi (B), 5 dpi (C), and 7 dpi (D). E represents the healthy chicken in the control group. Source: Guo M, Liu D, Chen X, et al. 2022. Pathogenicity and innate response to *Avibacterium paragallinarum* in chickens. Poult Sci, 101 (1): 101523.

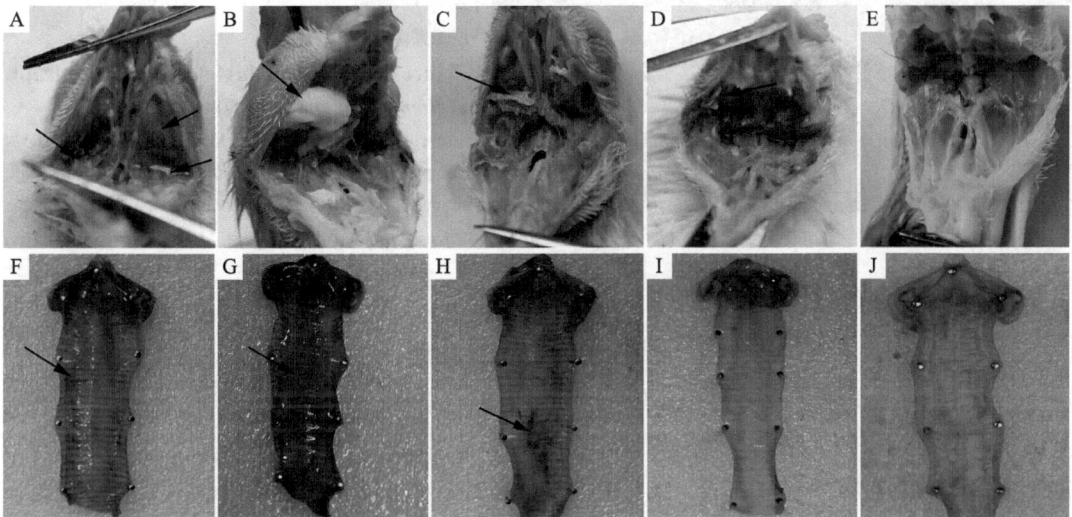

图10-10 副鸡禽杆菌感染鸡的大体病变

攻毒后第1天（A）、第3天（B）、第5天（C）、第7天（D）鼻腔和眶下窦大体病变（箭头）。攻毒后第1天（F）、第3天（G）、第5天（H）、第7天（I）气管大体病变（箭头）。E和J为对照组健康鸡。图像来源见英文。

Figure 10-10 Gross lesions of chickens infected with *A. paragallinarum*

The gross lesions in the nasal cavity and infraorbital sinus (arrow) at 1 dpi (A), 3 dpi (B), 5 dpi (C), and 7 dpi (D). The gross lesions in the trachea (arrow) at 1 dpi (F), 3 dpi (G), 5 dpi (H), and 7 dpi (I). E and J represent the healthy chicken in the control group. Source: Guo M, Liu D, Chen X, et al. 2022. Pathogenicity and innate response to *Avibacterium paragallinarum* in chickens. Poult Sci, 101 (1): 101523.

第7节 里氏杆菌属
Section 7 *Riemerella*

鸭疫里氏杆菌为里氏杆菌属代表成员，可引起雏鸭传染性浆膜炎，即患病鸭的心脏和肝脏有一层包裹的浆膜，故称传染性浆膜炎。鸭疫里氏杆菌为杆状或椭圆形，有荚膜，无芽孢，无鞭毛（**图10-11**），初次分离需要供给5%～10% CO_2，革兰氏染色阴性，瑞氏染色两极着色。该菌在普通琼脂和麦康凯琼脂上不生长，在血琼脂和巧克力琼脂上生长，菌落圆形，有21个血清型。

The bacterium *Riemerella anatipestifer* is a representative member of the genus *Riemerella*, which can cause infectious polyserositis in ducklings. This disease is characterized by a layer of serosa surrounding the heart and liver of infected ducks. *R. anatipestifer* is a rod-shaped or elliptical bacterium with a capsule but no spores or flagella (**Figure 10-11**). Its initial isolation requires a supply of 5% to 10% CO_2, and it is Gram-negative and polarly stained with a modified silver stain. The bacterium does not grow on standard or MacConkey agar, but grows on blood and chocolate agar. Its colonies are circular, and there are 21 serotypes of this bacterium.

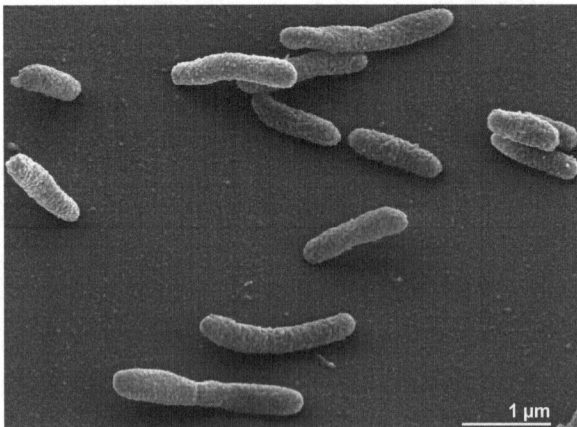

图10-11　鸭疫里氏杆菌ATCC 11845^{T}的扫描电子显微照片
图像来源见英文。

Figure 10-11　Scanning electron micrograph of *Riemerella anatipestifer* ATCC 11845^T
Source: Mavromatis K, Lu M, Misra M, et al. 2011. Complete genome sequence of *Riemerella anatipestifer* type strain (ATCC 11845). Stand Genomic Sci, 4 (2): 144-153.

思考题　Questions

1. 同是革兰氏阴性菌，为何大肠杆菌可以在麦康凯培养基上生长而巴氏杆菌却不能？

2. 巧克力琼脂平板是将巧克力融化后与普通营养琼脂混合配制的吗？

3. 如何分离和鉴定鸭疫里氏杆菌？

1. Despite both being Gram-negative bacteria, why is it that *E. coli* can grow on MacConkey agar while *Pasteurella* cannot?

2. Is chocolate agar a type of nutrient agar that is mixed with melted chocolate?

3. What are the methods for isolating and identifying *Riemerella anatipestifer*?

第11章 革兰氏阴性需氧杆菌
Chapter 11 / Gram-negative aerobic bacilli

内容提要 革兰氏阴性需氧杆菌成员较多，包括布氏杆菌属、假单胞菌属、伯克霍尔德菌属、博德特菌属、弗朗西斯菌属、不动杆菌属等，多数为人兽共患病原菌。

Introduction There are members of Gram-negative aerobic bacilli including the genera *Brucella*, *Pseudomonas*, *Burkholderia*, *Bordetella*, *Francisella* and *Acinetobacter*. Most of them are zoonotic pathogens.

第1节 布氏杆菌属
Section 1 *Brucella*

布氏杆菌又名布鲁菌，1887年由英国医生大卫·布鲁斯从马耳他死亡士兵脾脏中最先分离，故名布氏杆菌。该菌为球形或短杆形，无芽孢、无荚膜、无鞭毛（**图11-1**），为革兰氏染色阴性胞内菌。柯兹罗夫斯基染色（改良的齐-尼抗酸染色）为红色。本菌专性需氧，但初次分离需5%～10% CO_2环境、泛酸钙和赤藓糖醇；且生长缓慢，5～10 d或更久才形成肉眼可见的菌落，实验室长期传代菌株可在2～3 d出现菌落。布氏杆菌在血琼脂平板上不溶血，有光滑型、黏液型和粗糙型三种菌落形态。

赤藓糖醇，一种多元醇，可以促进布氏杆菌的生长。该物质主要聚积于牛、羊、猪等动物的胎盘位置，也存在于乳腺和附睾处，成为布氏杆菌在体内的定居处，这就是该菌引起不孕不育和流产的主要原因。

Brucella, also known as Bruce bacteria, was first isolated in 1887 by British physician David Bruce from the spleen of a deceased soldier in Malta, hence its name. This bacterium is spherical or short rod-shaped, lacks spores, capsules, and flagella (**Figure 11-1**), and is a Gram-negative intracellular pathogen. Kuznetsov staining (modified Ziehl-Neelsen acid-fast staining) results in a red color. The bacterium is obligately aerobic, but requires 5% to 10% CO_2, panthothenate, and erythritol for initial isolation. Its growth is slow, taking five to ten days or longer to form colonies visible to the naked eyes. However, laboratory strains that have been passaged for an extended period can form colonies in two to three days. *Brucella* does not hemolyze on blood agar plates and has three colony morphologies: smooth, mucoid, and rough.

Erythritol, a polyol, has been shown to stimulate the growth of *Brucella*. This substance primarily accumulates at the placentae of animals such as cattle, sheep, and pigs, as well as in the mammary glands and epididymis, serving as a colonization site for Brucella within the host. Such bacterial colonization is considered the primary cause of infertility and abortion associated with *Brucella* infection.

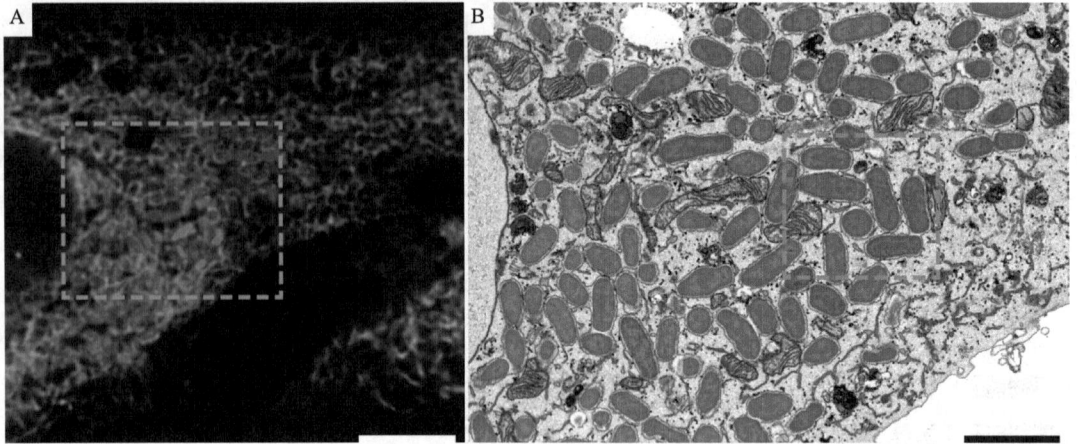

图 11-1 囊泡中复制的布氏杆菌

A. 感染 HeLa 细胞 24 h 后的流产布氏杆菌激光共聚焦图像：HeLa 细胞表达 Emerald-Sec61β（绿色），布氏杆菌分别表达 BFP（蓝色）和 dsRed（粉色）；B. 离子束/扫描电镜断层扫描 A 中标记位置的图像。标尺：6 μm（A）；2 μm（B）。图像来源见英文。

Figure 11-1 Replicative *Brucella*-containing vacuoles

A. Confocal microscope image of HeLa cell expressing Emerald-Sec61β (green), co-infected with *B. abortus* expressing BFP (blue) and *B. abortus* expressing dsRed (pink) at 24 h post infection. B. Single image from a focused ion beam/scanning electron microscopic tomography (FIB/SEM) tomogram of the site marked in A. Scale bars: 6 μm (A); 2 μm (B). Source: Sedzicki J, Tschon T, Low SH, et al. 2018. 3D correlative electron microscopy reveals continuity of Brucella-containing vacuoles with the endoplasmic reticulum. J Cell Sci, 131 (4): jcs210799.

布氏杆菌引起的疾病称为布病，人兽共患。细菌在生殖器官、乳腺和网状内皮系统内定居，引起母畜流产、附睾炎、脓肿等。患附睾炎的感染公畜的睾丸和附睾肿大（**图 11-2**）。另外，在关节部位出现皮下囊肿，虽然囊肿出现在关节附近，但囊肿和关节之间没有连通，关节的正常功能不受影响，这与链球菌感染引起的关节炎不同。囊肿内有积液和纤维状物质。

布氏杆菌的抗原主要有两类，分别是属内抗原和属外抗原。属内抗原是在布氏杆菌属内存在的抗原，属外抗原是布氏杆菌属与其他属的细菌共有的类似抗原，在血清学试验中可发生非特异性交叉反应。属内抗原有 A、M、C、R 四种表面抗原，即锚定在细胞膜表面的脂多糖最外侧的特异性多糖侧链。该多糖侧链又称 O 抗原，在光滑型菌株中 O 抗原表现为 A、M、C 三种抗原类型；在粗糙型菌株中 O 抗原表现为 R 型抗原。布氏杆菌在培养过程中可发

Brucellosis, a disease caused by *Brucella*, is a zoonotic infection. The bacteria colonize the reproductive organs, mammary glands, and reticuloendothelial system, causing abortion, epididymitis, and abscesses in female livestock. In males, the infection leads to swelling of the testicles and epididymis (**Figure 11-2**). Additionally, subcutaneous nodules may appear near joints. While these nodules are located near joints, they are not connected to the joint itself, and normal joint function is not affected. This is distinct from joint inflammation caused by *Streptococcus*. The nodules contain fluid and fibrous material.

Brucella has two main types of antigens: genus-specific antigens and non-genus-specific antigens. Genus-specific antigens are present within the *Brucella* genus, while non-genus-specific antigens are similar to antigens shared between *Brucella* and other bacterial genera, and can cause non-specific cross-reactions in serological tests. There are four types of surface antigens in intracellular antigens, namely A, M, C, and R. These are specific polysaccharide side chains that are anchored to the outermost layer of lipopolysaccharides on the cell membrane. The polysaccharide side chains are also known as O antigens. In smooth strains of *Brucella*, O antigens appear in three types, A, M, and C. In rough strains, O

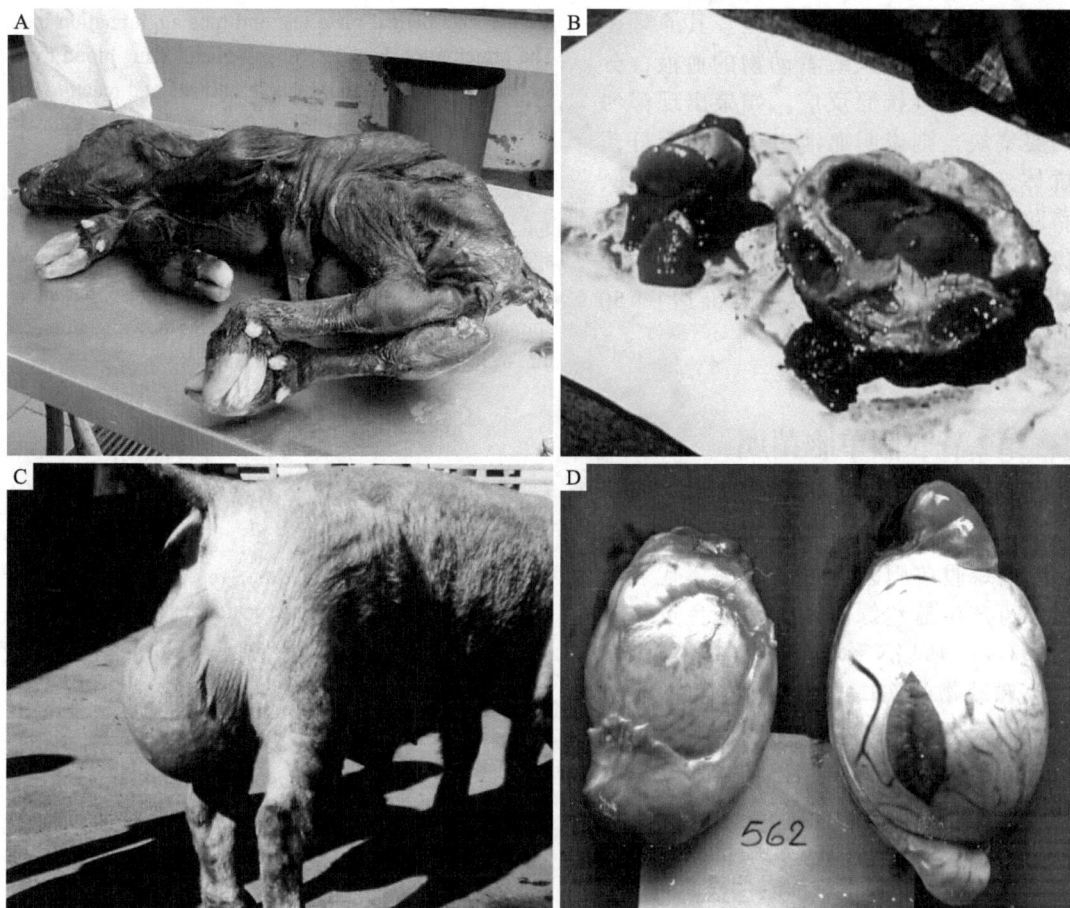

图11-2　布氏杆菌感染动物的临床症状和大体病变
A. 布氏杆菌生物变种1感染而流产的牛胎儿；B. 流产布氏杆菌感染公牛的睾丸坏死（单侧睾丸炎，右）；C. 猪布氏杆菌感染导致单侧睾丸增大（野猪）；D. 羊布氏杆菌感染引起公羊附睾炎。注意右侧附睾的增大。图像来源见英文。

Figure 11-2　Clinical signs and gross lesions of *Brucella* infected animals
A. Aborted fetus due to *B. abortus* biovar 1 infection; B. Necrosis in the testicle (right) of a *B. abortus* infected bull with unilateral orchitis; C. Unilateral testicular enlargement due to *B. suis* infection (boar); D. Epididymitis in rams due to *B. ovis* infection. Note the increase in size of the right epididymis. Source: Megid J, Mathias LA, Robles CA. 2010. Clinical manifestations of brucellosis in domestic animals and humans. The Open Veterinary Science Journal, 4: 119-126.

生光滑型→粗糙型（S→R）变异，即S型的A、M、C缺失后暴露了R抗原。属外抗原中的S型菌株C/Y表位与小肠结肠炎耶尔森菌O: 9、土拉热弗朗西斯菌、霍乱弧菌和大肠杆菌O157存在共同抗原成分；R型菌株与驹放线杆菌、绿脓杆菌和多杀性巴氏杆菌存在共同抗原成分。

实验室诊断： 由于布病为人兽共患病，因此其检测具有非常重要的公共卫生学意义。目前常用的金标准检测方法

antigens appear as R antigens. Brucella can undergo smooth-to-rough (S→R) variation during culture, where A, M, and C antigens are lost, exposing the R antigen. The C/Y epitopes of extracellular antigens in S-type strains share antigenic components with O: 9 of *Yersinia enterocolitica*, *Francisella tularensis*, *Vibrio cholerae*, and *Escherichia coli* O157. R-type strains share antigenic components with *Actinobacillus equuli*, *Pseudomonas aeruginosa*, and *Pasteurella multocida*.

Laboratory diagnosis: as brucellosis is a zoonotic disease, its detection has significant public health importance. The currently preferred gold standard detection methods

是虎红平板或试管凝集试验。其原理是采集疑似感染的人或者动物的血液，分离血清与虎红抗原反应。如果出现浑浊的凝集块，则说明血清中存在布氏杆菌抗体，为凝集试验阳性。人和牛、马、骆驼等大动物的试管凝集试验和平板凝集试验的血清凝集价达1∶100时判为阳性；羊、猪、犬血清的凝集价达1∶50时判为阳性。

are the rose bengal plate test and tube agglutination test. The principle of these tests involves collecting blood from suspected infected individuals or animals and reacting the serum with rose bengal antigen. The appearance of a turbid agglutination indicates the presence of *Brucella* antibodies in the serum, leading to a positive agglutination test. For humans and large animals such as cattle, horses, and camels, a serum agglutination titre of 1∶100 or above in tube agglutination and plate agglutination tests is considered positive. For small animals such as sheep, pigs, and dogs, a serum agglutination titre of 1∶50 or above is considered positive.

第2节 假单胞菌属
Section 2 *Pseudomonas*

铜绿假单胞菌为假单胞菌属代表成员，可产生蓝色绿脓素和蓝绿色荧光素（**图11-3**），所以又称绿脓杆菌。感染该菌的人和动物的伤口或感染部位可呈现绿色。铜绿假单胞菌为革兰氏染色阴性，无芽孢，有端生单或丛鞭毛。在马血琼脂平板上，菌落的四周有溶血环。该菌有O抗原、H抗原、黏液抗原（荚膜抗原）、R抗原和菌毛抗原。其中O抗原包括共同多糖抗原和O特异性抗原有两种，前者由鼠李糖均聚物组成；后者由3～5种不同糖类构成的异聚物重复单元组成，是用于该菌血清分型的基础。

Pseudomonas aeruginosa, a representative member of the *Pseudomonas* genus, produces pyocyanin and fluorescein (**Figure 11-3**), giving it the nickname "green pus bacillus". Infections caused by this bacterium in humans and animals can result in green discoloration of wounds or infected areas. *P. aeruginosa* is Gram-negative, non-spore forming, and has polar, single or multiple flagella. Hemolysis is observed around colonies on horse blood agar. The bacterium has O, H, slime (capsular), R, and fimbrial antigens. The O antigen includes two types of antigens, namely common polysaccharide antigens and O-specific antigens. The former consists of rhamnolipid aggregates, while the latter consists of repeating units of three to five different sugars and is the basis for the serotyping of this bacterium.

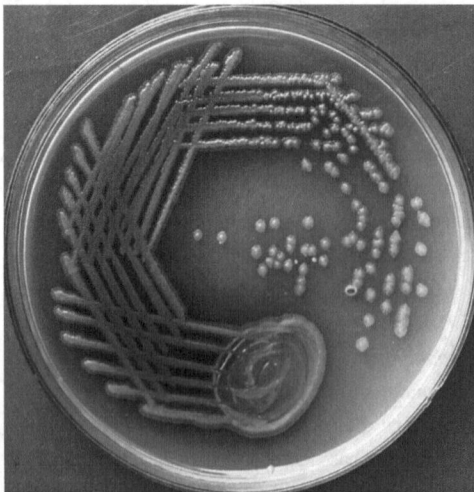

彩图

图11-3 营养琼脂上的铜绿假单胞菌（产生蓝色绿脓素和蓝绿色荧光素）
图像来源见英文。

Figure 11-3 *Pseudomonas aeruginosa* **on nutrient agar (producing blue pyocyanin and a fluorescent pigment called as pyoverdine)**
Source: Rath S, Padhy RN. 2015. Surveillance of acute community acquired urinary tract bacterial infections. Journal of Acute Disease, 4 (3): 186-195.

第3节 伯克霍尔德菌属
Section 3 *Burkholderia*

鼻疽伯克霍尔德菌又称鼻疽杆菌，人兽共患，引起单蹄兽鼻疽（**图11-4**），在历史上曾被用作生物武器。该菌为革兰氏阴性胞内菌，无芽孢，无鞭毛，有荚膜多糖，专性需氧，在添加5%绵羊血或1%甘油的营养琼脂平板上生长良好，有溶血，麦康凯琼脂上可生长。存在S→R变异。诊断和检疫可用鼻疽菌素点眼（变态反应），观察3~24 h，眼结膜红肿并排出脓性分泌物者为阳性。

Burkholderia mallei, also known as glanders bacillus, is a zoonotic bacterium that causes glanders in solipeds (**Figure 11-4**). It has been historically used as a biological weapon. The bacterium is Gram-negative, intracellular, non-spore forming, non-flagellated, and has a polysaccharide capsule. It is strictly aerobic and grows well on nutrient agar plates containing 5% sheep blood or 1% glycerol, and shows hemolysis. It can also grow on MacConkey agar. S→R variation is observed. Diagnosis and quarantine can be conducted via dropping mallei in the eyes (a hypersensitivity reaction). A positive reaction is indicated by the development of conjunctiva redness with purulent discharge within 3 to 24 hours.

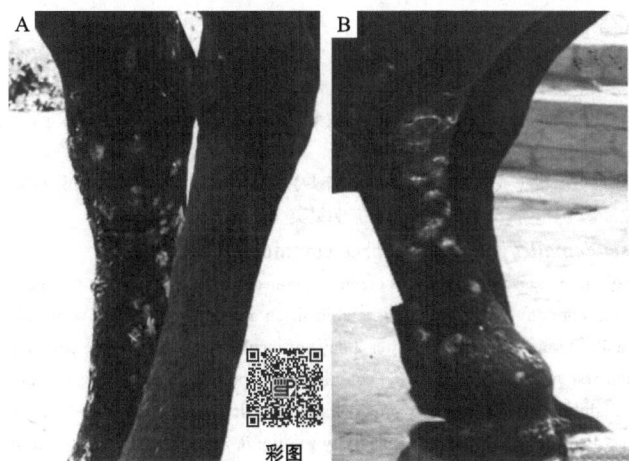

彩图

图11-4 马鼻疽的静脉注射实验治疗
A. 开始治疗前右后腿出现溃疡（Ⅳ疗程）；B. 治疗第二周后溃疡愈合结痂。图像来源见英文。

Figure 11-4 Response of a glandered horse to intravenous course of experimental treatment
A. Note developing ulcers on right hind leg before the start of treatment (Ⅳ course of therapy); B. Healing ulcers and scar after the termination of second week of Ⅳ course of therapy. Source: Khan I, Wieler LH, Melzer F, et al. 2013. Glanders in animals: a review on epidemiology, clinical presentation, diagnosis and countermeasures. Transbound Emerg Dis, 60 (3): 204-221.

伪鼻疽伯克霍尔德菌为伯克霍尔德菌属的另一种细菌，因其所致的临床症状与鼻疽非常相似而得名，引起类鼻疽，人兽共患。与鼻疽杆菌不同的是，伪鼻疽伯克霍尔德菌有鞭毛。该菌为革兰氏染色阴性胞内菌，呈现两极着色；在普通琼脂上生长迅速，部分菌株可在麦康凯琼脂上生长；在巧克力琼脂上菌落粗糙，继续培养可变成橙色的黏液状（**图11-5**），这一菌落形态与耻垢分枝杆菌类似。

Burkholderia pseudomallei, another bacterium in the *Burkholderia* genus, is named for its clinical symptoms, which are very similar to glanders, and causes melioidosis, a zoonotic disease. Unlike *B. mallei*, *B. pseudomallei* has flagella. The bacterium is Gram-negative, intracellular, and exhibits bipolar staining. It grows rapidly on common agar, and some strains can grow on MacConkey agar. On chocolate agar, the colonies are rough, and with continued growth, they can become orange, mucoid (**Figure 11-5**), and resemble those of *Mycobacterium smegmatis*.

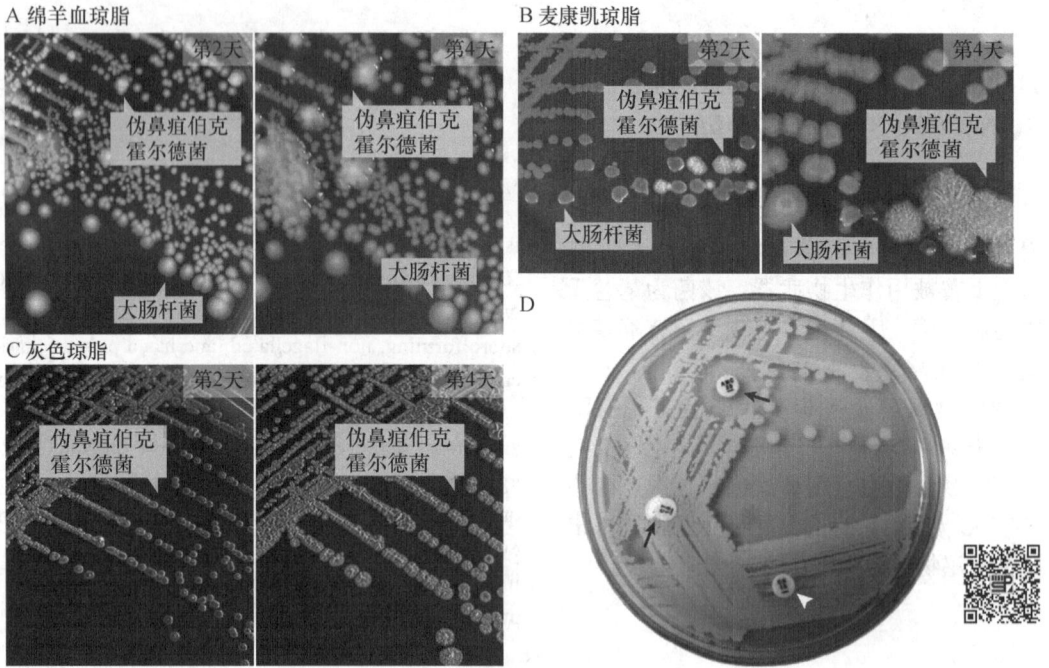

A 绵羊血琼脂

B 麦康凯琼脂

C 灰色琼脂

D

图 11-5　伪鼻疽伯克霍尔德菌在三种常见培养基上的菌落形态

伪鼻疽伯克霍尔德菌和大肠杆菌的典型外观。A. 伪鼻疽伯克霍尔德菌培养第 2 天形成类似大肠菌群的奶油状非溶血菌落；到第 4 天，菌落被轻微的金属光泽覆盖，干燥起皱；B. 培养第 2 天，伪鼻疽伯克霍尔德菌菌落类似于无色、不发酵乳糖的大肠菌群；到第 4 天，菌落干燥起皱；C. 培养第 2 天，出现针尖大小的伪鼻疽伯克霍尔德菌菌落，颜色从透明到淡粉色；到第 4 天，菌落呈深粉色到紫色，扁平，略微干燥，有一定的金属光泽。大肠杆菌不能生长，因琼脂含庆大霉素抑制大肠杆菌；D. 抗生素敏感性纸片扩散试验：伪鼻疽伯克霍尔德菌对黏菌素（或多黏菌素）（黑色箭头）和庆大霉素（三角箭头）具有耐药性（尽管某些地区存在敏感菌株），对复方阿莫西拉霉素（红色箭头）敏感。图像来源见英文。

Figure 11-5　Identification of _Burkholderia pseudomallei_ colonies on three common types of agar

Typical appearance of *B. pseudomallei* and *Escherichia coli*. A. *B. pseudomallei* forms creamy, non-haemolytic colonies that resemble a coliform after 2 days of incubation; by day 4, the colonies are covered by a slight metallic sheen and become dry and wrinkled. B. *B. pseudomallei* colonies resemble a colorless, non-lactose fermenting coliform after 2 days of incubation; by day 4, the colonies appear dry and wrinkled. C. After 2 days of incubation, the first visible *B. pseudomallei* colonies are pinpoint with a clear to pale pink color; by day 4, they become darker pink to purple, flat, slightly dry and wrinkled with a definite metallic sheen. *E. coli* fails to grow because it is inhibited by gentamicin in the agar. D. Three-disc diffusion antibiotic sensitivity testing: *B. pseudomallei* is resistant to colistin (or polymyxin) (black arrow) and gentamicin (arrowhead) (although sensitive isolates exist in some areas) and sensitive to co-amoxiclav (red arrow). Source: Wiersinga WJ, Virk HS, Torres AG, et al. 2018. Melioidosis. Nat Rev Dis Primers, 4: 17107.

第 4 节　博德特菌属
Section 4　*Bordetella*

支气管败血博德特菌为博德特菌属代表成员，可引起支气管炎、鼻炎等，是猪萎缩性鼻、犬传染性支管炎的病原体（**图 11-6**）。该菌为革兰氏染色阴性，无芽

Bordetella bronchiseptica, a representative member of the *Bordetella* genus, can cause bronchitis, rhinitis, and other respiratory diseases, and is the causative agent of atrophic rhinitis in pigs and infectious canine tracheobronchitis (**Figure 11-6**). The bacterium is

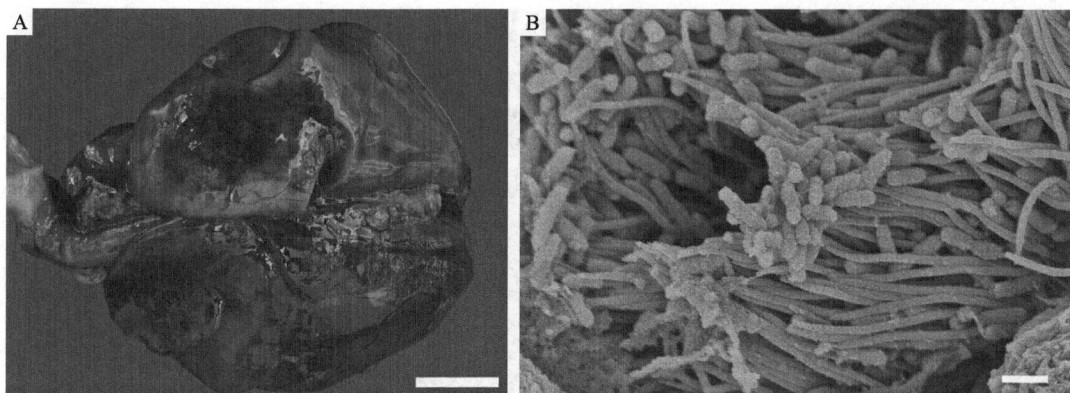

图11-6 支气管败血博德特菌引起的幼犬支气管肺炎的病理学和电镜观察
A. 肺部弥漫性肿胀，有暗红色区域。标尺：1 cm；B. 支气管黏膜扫描电镜，可见众多球杆菌附着于支气管黏膜纤毛上端。标尺：10 μm。图像来源见英文。

Figure 11-6 Pathological findings and electron microscopy of _B. bronchiseptica_ bronchopneumonia in puppies
A. The lung is diffusely swollen with dark red areas. Scale bar: 1 cm; B. SEM of the bronchial mucosa. Numerous coccobacilli are attached to the cilia of bronchial mucosa. Scale bar: 10 μm. Source: Wiersinga WJ, Virk HS, Torres AG, et al. 2018. Melioidosis. Nat Rev Dis Primers, 4: 17107.

彩图

孢；在血琼脂上呈现不明显的β溶血，存在不溶血变异菌落；在麦康凯琼脂上菌落为灰蓝色；在史密斯-巴斯克维尔选择性培养基上为灰色透明菌落，有狭窄红色环。支气管败血博德特菌有O、K和H抗原。

Gram-negative, non-spore forming, and exhibits slight β-hemolysis on blood agar, with non-hemolytic variant colonies present. On MacConkey agar, the colonies appear gray-blue, and on Smith-Baskerville selective medium, the colonies are transparent gray with a narrow red ring. _B. bronchiseptica_ possesses O, K, and H antigens.

第5节 弗朗西斯菌属
Section 5 _Francisella_

土拉热弗朗西斯菌为弗朗西斯菌属成员，引起土拉热，又称野兔热。弗朗西斯菌为人兽共患病原菌，曾被用作生物武器研究。该菌宿主范围广，可以感染哺乳类、鸟类、两栖类和爬行类，而且能在昆虫体内繁殖。土拉热弗朗西斯菌为革兰氏阴性胞内菌，两极染色，菌体玫瑰色，无芽孢，无鞭毛（**图11-7**）；在葡萄糖半胱氨酸血琼脂上菌落灰色，光滑隆起，边缘有绿色环。

Francisella tularensis, a member of the _Francisella_ genus, causes tularemia, also known as rabbit fever. It is a zoonotic pathogen that has been studied as a potential biological weapon. The bacterium has a wide host range and can infect mammals, birds, amphibians, reptiles, and can reproduce in insects. _F. tularensis_ is Gram-negative, intracellular, and exhibits bipolar staining, with the bacterial body appearing rose-colored. It is non-spore forming and non-flagellated (**Figure 11-7**). On cysteine-glucose-blood agar, the colonies appear gray, smooth, and raised, with a greenish edge.

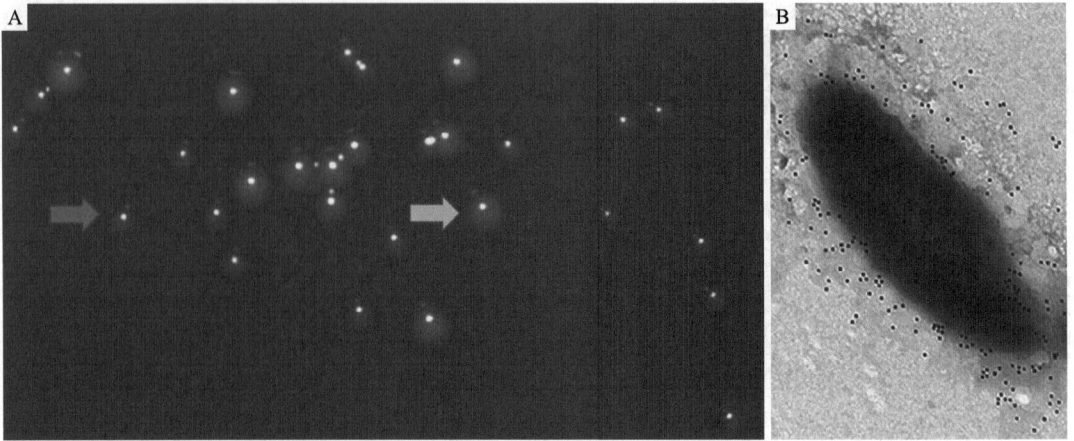

图11-7　土拉热弗朗西斯菌的蓝/灰色菌落和透射电镜照片

A. 在血琼脂上，土拉热弗朗西斯菌菌落较小，带有蓝色光泽（蓝色箭头）。O抗原自发丢失后形成更较大的灰色菌落（灰色箭头）。丢失O抗原的土拉热弗朗西斯菌对血清敏感且失去毒性，对小鼠无致病性；B. 土拉热弗朗西斯菌O抗原荚膜和荚膜状复合物的透射电镜照片。在土拉热弗朗西斯菌细胞周围可见O抗原荚膜和荚膜状复合物。图像来源见英文。

Figure 11-7　Blue/gray colonies and transmission electron microscopy of *Francisella tularensis*

A. On blood agar, virulent *F. tularensis* colonies are smaller, with a blueish sheen (blue arrow). Spontaneous loss of the O antigen results in formation of larger, gray colonies (gray arrow) that are serum sensitive and avirulent in the mouse model; B. Transmission electron microscopy of the O antigen capsule and the capsule-like complex (CLC) of *F. tularensis*. Both the O antigen capsule and the CLC can be seen directly or indirectly surrounding *F. tularensis* cells. Source: Freudenberger Catanzaro KC, Inzana TJ. 2020. The *Francisella tularensis* polysaccharides: what is the real capsule? Microbiol Mol Biol Rev, 84 (1): e00065-19.

第6节　不动杆菌属
Section 6　*Acinetobacter*

鲍氏不动杆菌为不动杆菌属代表成员，为人兽共患病原机会致病菌，引起呼吸道炎症、手术感染、新生驹败血症、牛乳腺炎等，高度耐药。革兰氏阴性，有荚膜，无鞭毛（**图11-8**）。在显微镜下为球形、链条状、成双或成丝。不同的形状代表该菌不同的生长状态，处于生长期的不动杆菌为杆状，生长停滞时为球状或丝状。专性需氧，营养要求不高，菌落S型或M型，奶油色或灰白色，在血琼脂平板上不溶血。

Acinetobacter baumannii, a representative member of the *Acinetobacter* genus, is a zoonotic pathogen that causes respiratory inflammation, surgical infections, septicemia in newborn foals, bovine mastitis, and is highly resistant to antibiotics. The bacterium is Gram-negative, encapsulated, and non-flagellated (**Figure 11-8**). Under the microscope, it appears as spheres, chains, pairs, or filaments, depending on its growth state. When actively growing, *A. baumannii* appears rod-shaped, and in a dormant state, it appears spherical or filamentous. It is strictly aerobic, with low nutritional requirements, and the colonies are S- or M-shaped, cream-colored, or gray-white, and do not exhibit hemolysis on blood agar plates.

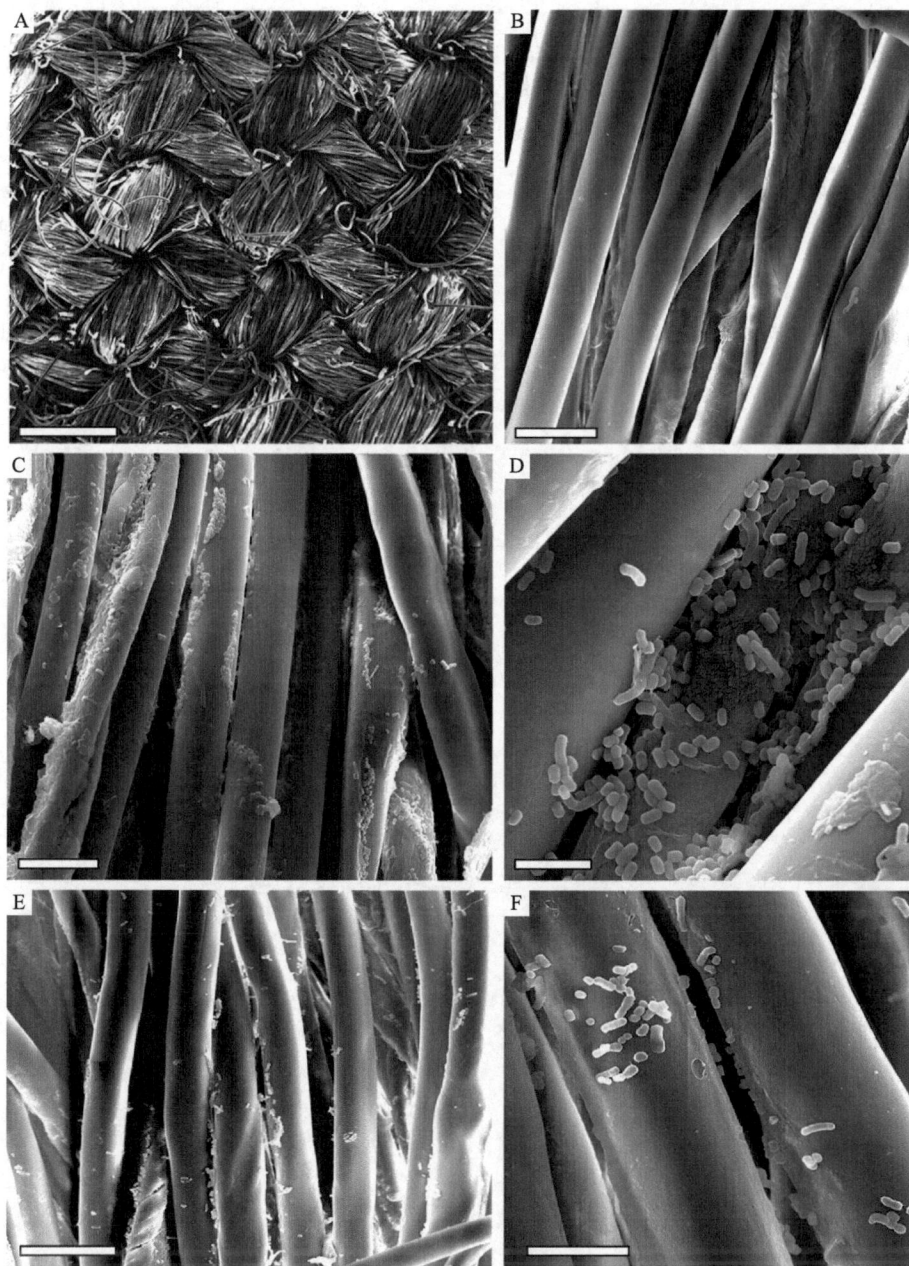

图 11-8　用扫描电镜分析实验室白色外套碎片上的鲍氏不动杆菌

各组为不同放大倍率下观察到的随机显微视野。A、B 为对照样品；C、D 为实验开始时（第 0 天）感染鲍氏不动杆菌的样品；E、F 为 22℃下 60 d 后的感染样本。原始放大倍数：A×100；B、C、E×2 500；D×10 000；F×5 000。标尺：A 为 0.5 mm；B 为 20 μm；C、E 为 25 μm；D 为 5 μm；F 为 10 μm。图像来源见英文。

Figure 11-8　Scanning electron microscopy analysis of the morphology of *Acinetobacter baumannii* cells maintained on white lab coat fragments

Panel shows random microscopy fields observed at different magnifications. A, B. Control samples; and C, D. Infected samples at the beginning of the experiments (day 0); E, F. Infected samples after 60 days at 22℃. Original magnification: A ×100; B, C, E ×2 500; D×10 000; F ×5 000. Scale bars: A, 0.5 mm; B, 20 μm; C, E, 25 μm; D, 5 μm; F, 10 μm. Source: Chapartegui-González I, Lázaro-Díez M, Bravo Z, et al. 2018. *Acinetobacter baumannii* maintains its virulence after long-time starvation. PLoS One, 13 (8): e0201961.

思考题　Questions

1. 布氏杆菌之所以存在属内抗原，是因为其细胞壁脂多糖的哪个结构发生了变化？

2. 在虎红平板凝集中被凝集的是IgG还是IgM，为什么？

1. The reason behind the existence of antigenic variation among the *Brucella* species lies in the alteration of which structure of their cell wall lipopolysaccharides?

2. Which antibody, IgG or IgM, is responsible for the agglutination observed in tiger red plate test, and why?

第12章 革兰氏阴性微需氧菌和厌氧菌

Chapter 12 / Gram-negative microaerobes and anaerobes

内容提要 革兰氏阴性微需氧和厌氧菌中的人兽共患病原菌代表是弯曲菌和螺杆菌。微需氧菌是介于厌氧和需氧之间的细菌,其生存环境中的氧气含量低于空气中的氧含量。例如,弯曲菌属中的空肠弯曲菌栖居于氧气含量较低的人和动物肠道。

Introduction Representative zoonotic pathogens among Gram-negative microaerophilic and anaerobic bacteria are *Campylobacter* and *Helicobacter* species. Microaerophilic bacteria are those that require low levels of oxygen to grow, occupying a niche between anaerobic and aerobic organisms. For instance, *Campylobacter jejuni*, a member of the *Campylobacter* genus, thrives in the lower-oxygen environment of the intestines of humans and animals.

第1节 弯 曲 菌 属
Section 1 *Campylobacter*

空肠弯曲菌为弯曲菌属的代表成员。该菌为人兽共患病原菌,革兰氏染色阴性,呈多形性,无芽孢,一端或两端无鞘单鞭毛(**图12-1**)。可在CCD培养基(炭头孢哌酮脱氧胆酸琼脂)培养,有两种菌落,一种为大菌落,边缘不整齐;一种为小菌落,圆形隆起。该菌微需氧,血平板上不溶血,麦康凯培养基上生长微弱或不生长。具有耐热O抗原和不耐热H、K抗原;可引起食物中毒以及人和动物腹泻、发热、呕吐和血便等症状。目前无有效的疫苗可以预防,健康的犬和猫可携带本菌并感染人。

Campylobacter jejuni is a representative member of the *Campylobacter* genus. This Gram-negative, pleomorphic bacterium lacks spores and possesses one or two unsheathed flagella at one or both poles (**Figure 12-1**). It is a zoonotic pathogen that can cause diarrhea, fever, vomiting, and bloody stools in humans and animals. *C. jejuni* is microaerophilic and can grow on CCD agar (charcoal cefoperazone deoxycholate agar), forming two types of colonies: large colonies with an irregular edge and small, circular, raised colonies. It does not hemolyze on blood agar and may grow weakly or not at all on MacConkey agar. This bacterium has a thermostable O antigen and thermodisable H and K antigens. Currently, there is no effective vaccine available for prevention, and healthy dogs and cats may carry and transmit the organism to humans.

图12-1　具有完整鞭毛的空肠弯曲菌81-176野生型菌株负染透射电镜图像
图像来源见英文。

Figure 12-1　Negatively stained transmission electron microscopy images of the helical *C. jejuni* 81-176 wild-type strain with intact flagella

Source: Frirdich E, Biboy J, Pryjma M, et al. 2019. The *Campylobacter jejuni* helical to coccoid transition involves changes to peptidoglycan and the ability to elicit an immune response. Mol Microbiol, 112 (1): 280-301.

第2节　螺 杆 菌 属
Section 2　*Helicobacter*

　　幽门螺杆菌为螺杆菌属代表成员。该菌为人兽共患，革兰氏阴性，呈螺旋形、弯曲或直杆形，无芽孢，鞭毛有鞘（**图12-2**）；微需氧，最佳培养气体环境为5% O_2、5% CO_2和90% N_2；营养要求高，在脑心汤血琼脂上菌落无色半透明；该菌能产生尿素酶，将尿素分解成CO_2，幽门螺杆菌呼气试验就是基于这一反应原理。该菌引起人胃溃疡和胃癌；同属的猪螺杆菌引起猪胃溃疡、胃炎等，人也可经口感染，如食用被污染且未煮熟的猪肉。

　　Helicobacter pylori is a representative member of the *Helicobacter* genus. This Gram-negative, spiral, curved, or rod-shaped bacterium possesses sheathed flagella and lacks spores (**Figure 12-2**). It is a zoonotic pathogen that can cause gastric ulcers and gastric cancer in humans. *H. pylori* is microaerophilic and thrives in an environment with 5% O_2, 5% CO_2, and 90% N_2. It has high nutritional requirements and forms colorless, translucent colonies on brain heart infusion blood agar. This bacterium produces urease, which breaks down urea into CO_2, and the *H. pylori* breath test is based on this principle. Related species, such as *H. suis*, can cause gastric ulcers and gastritis in pigs, and humans can also become infected by ingesting contaminated, undercooked pork.

思考题　Questions

1. 空肠弯曲菌的主要特征是什么？

2. 幽门螺杆菌的主要特征是什么？

1. What are the main characteristics of *Campylobacter jejuni*?

2. What are the main characteristics of *Helicobacter pylori*?

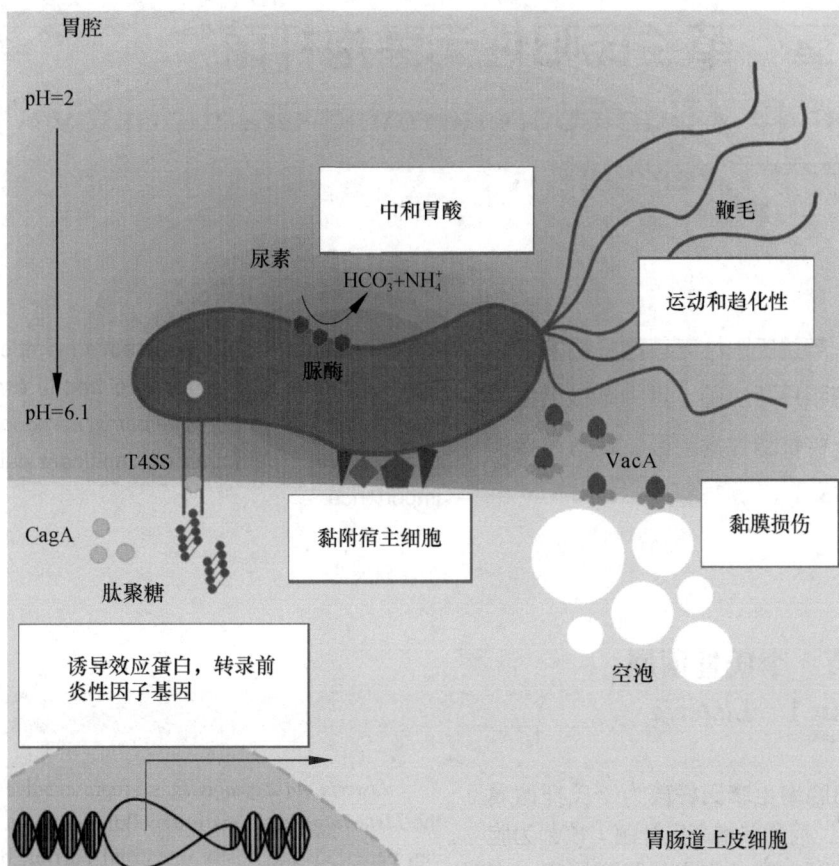

图12-2 幽门螺杆菌毒力因子

幽门螺杆菌具有高度遗传变异性，有些菌株与胃病相关，最突出的是表达毒力因子VacA和Cag-PAI毒力岛蛋白的菌株。VacA诱导空泡的形成，从而对胃上皮产生损伤。Cag-PAI毒力岛包含大约30个编码基因，这些基因编码蛋白共同构成Ⅳ型分泌系统。分泌系统将毒力因子CagA和肽聚糖等分子泌入胃黏膜上皮细胞质内。一旦进入细胞质，CagA被磷酸化，从而触发下游一系列细胞内事件。尿素酶将尿素转化为氨和二氧化碳对幽门螺杆菌在胃中的存活至关重要。鞭毛对幽门螺杆菌在胃黏膜的定植中起重要作用。鞭毛可根据胃腔pH和尿素等化合物的浓度产生运动，从而使幽门螺杆菌能够穿透黏膜层到达上皮内部。图像来源见英文。

Figure 12-2 _Helicobacter pylori_ virulence factors

H. pylori is genetically highly variable, and some strains are more strongly associated with gastric pathologies. The most prominent are those that express the virulence factors VacA and Cag-PAI. VacA induces the formation of vacuoles, thus generating damage in the gastric epithelium. The Cag-PAI contains approximately 30 genes that code for proteins, that together make up a T4SS. The secretion system introduces a number of molecules, including the virulence factor CagA and peptidoglycans, into the cytoplasm of epithelial cells of the gastric mucosa. Once in the cytoplasm, CagA is phosphorylated, and this triggers downstream intracellular events. The conversion of urea into ammonium and carbon dioxide by urease is essential to the survival of _H. pylori_ in the stomach. Flagella play an important role in the colonization of the gastric mucosa, as they produce differential motility depending on the pH of the stomach lumen and the concentration of compounds such as urea, thus enabling _H. pylori_ bacteria to swim across the mucous layer towards the epithelial lining. Source: Molina-Castro S, Ramírez-Mayorga V, Alpízar-Alpízar W. 2018. Priming the seed: _Helicobacter pylori_ alters epithelial cell invasiveness in early gastric carcinogenesis. World J Gastrointest Oncol, 10 (9): 231-243.

第13章 革兰氏阳性无芽孢杆菌
Chapter 13 / Non-spore-forming Gram-positive bacilli

内容提要 革兰氏阳性无芽孢杆菌的代表是李氏杆菌和丹毒丝菌，其中单核增生李氏杆菌是一种食源性病原菌，具有重要的公共卫生意义。

Introduction The representative members of non-spore-forming Gram-positive bacilli are *Listeria monocytogenes* and *Erysipelothrix*. *L. monocytogenes* is a foodborne pathogen of significant public health importance.

第1节 李氏杆菌属
Section 1 *Listeria*

单核细胞增生李氏杆菌为李氏杆菌属的代表成员，简称单增李氏杆菌。该菌为胞内寄生菌，可以突破宿主的肠道屏障、血脑屏障和胎盘屏障；通过食用污染的食物感染，因此被称为食源性病原菌（**图13-1**）。其所致疾病称为李氏杆菌病，病症多样，包括肠炎、脑膜炎、流产、败血症等，死亡率可高达30%；反刍动物李氏杆菌病也可出现类似病症。该菌为短杆状（**图13-2**），革兰氏染色阳性，无荚膜，无芽孢；在20～25 ℃培养可产生多根周鞭毛，37 ℃及以上温度不产生鞭毛；生长温度范围1～45 ℃，普通琼脂上可生长，在血液培养基上可见其周围有狭窄的β溶血环，CAMP试验阳性。

单增李氏杆菌在不同鉴别培养基上的菌落形态和颜色不一样，在BCM或ALOA鉴别培养基上，菌落蓝绿色光泽（**图13-3**）；在李氏杆菌快速专用培养基上为绿色菌落；在七叶苷琼脂上为黑色菌落；在半固

Listeria monocytogenes is a representative member of the *Listeria* genus. This intracellular parasitic bacterium can penetrate the host's intestinal barrier, blood-brain barrier, and placental barrier. It is known as a foodborne pathogen, transmitted through the consumption of contaminated food (**Figure 13-1**). The resulting disease is called listeriosis, which manifests in various forms such as enteritis, meningitis, miscarriage, sepsis, and other symptoms. The mortality rate can reach up to 30%, and similar symptoms can also appear in ruminant animals. This bacterium is a short rod (**Figure 13-2**) with Gram-positive staining, no capsule, and no endospore. It can produce multiple peritrichous flagella when cultured at 20-25 ℃, but no flagella are produced at temperatures above 37 ℃. It can grow in a temperature range of 1-45 ℃ and can grow on nutrient agar. In blood culture, a narrow β-hemolysis zone can be observed around it, and the CAMP test is positive.

The colony morphology and color of *L. monocytogenes* can vary on different identification culture media. On BCM or ALOA identification media, the colonies exhibit a blue-green luster (**Figure 13-3**). On *Listeria* rapid specialized media, the colonies appear green, while on esculin hydrate agar, they form black colonies. When

单增李氏杆菌污染的食物

肝脏

脑

血流

淋巴结

肠道

脾脏

胎盘

胎儿

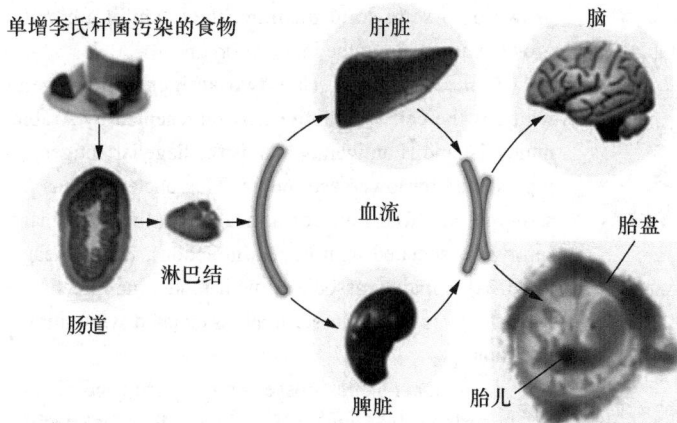

图13-1 单增李氏杆菌通过污染的食物感染

图像来源见英文。

Figure 13-1 The infection by *Listeria monocytogenes* by ingestion of contaminated food

Source: Cossart P. 2011. Illuminating the landscape of host-pathogen interactions with the bacterium *Listeria monocytogenes*. Proc Natl Acad Sci U S A，108 (49): 19484-19491.

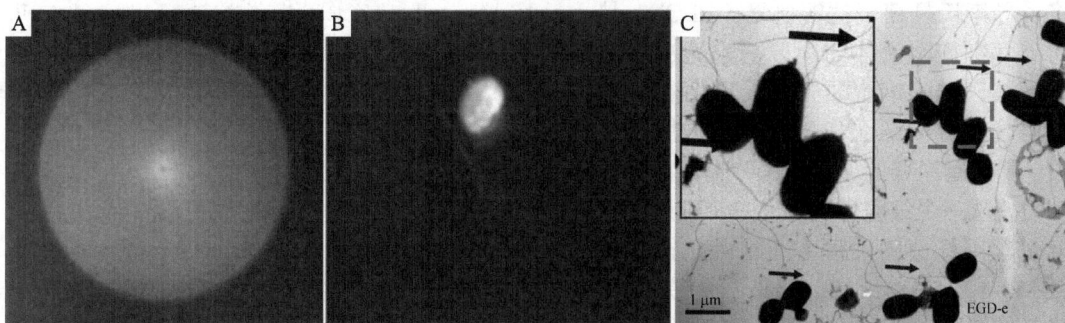

图13-2 单增李氏杆菌的运动和鞭毛形成

单增李氏杆菌运动试验（A、B）和透射电镜图像（C）；单增李氏杆菌EGD-e分别在软琼脂（0.25%）于30℃（A、C）或37℃（B）培养16 h。鞭毛为箭头所示。标尺：1 μm。图像来源见英文。

Figure 13-2 *L. monocytogenes* motility and flagellar formation

A motility assay (A, B) and TEM (C) were performed by growing the *L. monocytogenes* strain EGD-e on soft agar (0.25%) at 30℃ (A, C) or 37 ℃ (B) for 16 h. Flagella are shown by arrows. Scale bar: 1 μm. Source: Cheng C, Wang H, Ma T, et al. 2018. Flagellar basal body structural proteins FlhB, FliM, and FliY are required for flagellar-associated protein expression in *Listeria monocytogenes*. Front Microbiol, 9: 208.

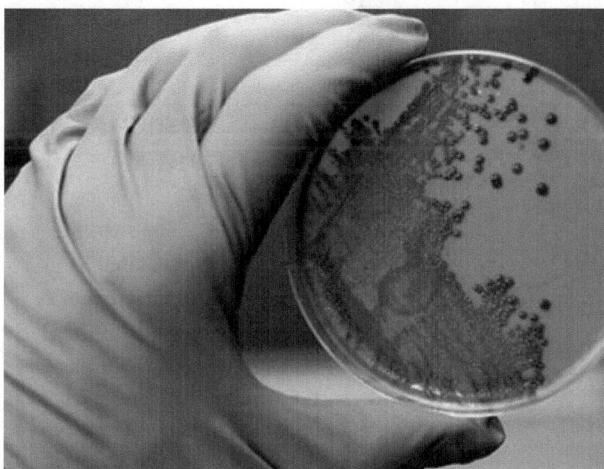

彩图

图13-3 单增李氏杆菌在BCM鉴别培养基上可见蓝绿色光泽菌落

图像来源见英文。

Figure 13-3 Blue-green colonies of *L. monocytogenes* on BCM differential culture media

Source: Maxwell A. 2016. *Listeria*: the science of adhesion and food safety. Thermo Fisher Scientific.

体培养基中沿穿刺线云雾状生长。

　　单增李氏杆菌有两种抗原，分别是：O抗原，即菌体抗原，用阿拉伯数字表示；H抗原，即鞭毛抗原，用小写英文字母表示。该菌有13个血清型4个谱系，人源主要为谱系Ⅰ（4b），食源主要为谱系Ⅱ（1/2a），动物源主要为谱系Ⅲ。

　　单增李氏杆菌表面有两个特殊的毒力因子内化素A（InlA）和内化素B（InlB），可特异性黏附宿主上皮细胞，从而通过内化方式进入细胞内部，被吞噬进入溶酶体中。溶酶体内部的酸性环境将诱导李氏杆菌表达溶血素（LLO），从而破坏细胞溶酶体的膜，并逃逸出吞噬溶酶体，进入细胞质。李氏杆菌在细胞质内，利用ActA蛋白劫持宿主的肌动蛋白，形成彗星样的尾巴为菌体提供前进的动力，从一个细胞拱进另外一个细胞，这就是李氏杆菌感染宿主细胞的拉链机制（**图13-4**）。在李氏杆菌感染宿主或37 ℃培养的细胞时，H抗原即鞭毛抗原受温度调控，不再表达。

grown on a semi-solid medium, they exhibit a cloudy growth pattern along the inoculation line.

　　L. monocytogenes has two antigens: O antigen, which is the cell wall antigen, is represented by Arabic numerals; and H antigen, which is the flagellar antigen, is represented by lowercase letters. This bacterium has 13 serotypes and four lineages. Lineage Ⅰ (4b) is the main serotype associated with human infections, while lineage Ⅱ (1/2a) is mainly associated with foodborne infections. Lineage Ⅲ is the main serotype associated with animal infections.

　　L. monocytogenes possesses two virulence factors on its surface, InlA and InlB, which allow for specific adherence to host epithelial cells and subsequent entry into these cells via internalization. Once inside the cell, the bacterium is engulfed into a lysosome, where the acidic environment induces the expression of listeriolysin O (LLO). LLO destroys the lysosomal membrane, allowing the bacterium to escape into the cytoplasm. Once inside the cytoplasm, *L. monocytogenes* uses the ActA protein to hijack the host's actin filaments, forming a comet tail that propels the bacterium forward and aids in cell-to-cell spread, known as the "zipper" mechanism of *L. monocytogenes* infection (**Figure 13-4**). During infection of the host or when cultured at 37 ℃, the H antigen or flagellar antigen is temperature-regulated and no longer expressed.

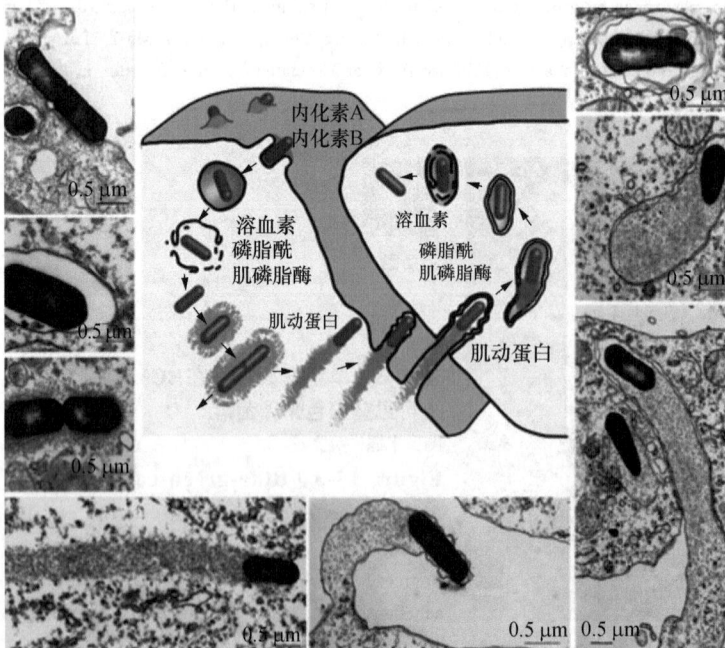

图13-4　单增李氏杆菌主要毒力因子介导的感染
图像来源见英文。

Figure 13-4　The steps of the infection mediated by major virulence factors of *L. monocytogenes*
Source: Cossart P. 2011. Illuminating the landscape of host-pathogen interactions with the bacterium *Listeria monocytogenes*. Proc Natl Acad Sci U S A, 108 (49): 19484-19491.

实验室诊断：单增李氏杆菌在公共卫生和食品安全领域具有重要意义，其微生物学诊断程序包括：增菌培养、纯培养（取纯培养物进行革兰氏染色，溶血性、运动性、生化特性、血清学、PCR鉴定等）、动物试验（通过小鼠等动物模型检测致病性）等。

Laboratory diagnosis: *L. monocytogenes* is of significant importance in the fields of public health and food safety. The microbiological diagnostic procedures for this bacterium include culture, purification (Gram staining, hemolytic activity, motility, biochemical characteristics, serology, PCR identification, *etc.*), and animal testing (testing virulence using animal models such as mice).

第2节 丹毒丝菌属
Section 2 *Erysipelothrix*

猪丹毒丝菌为丹毒丝菌属代表成员，引起动物全身或局部感染，如急性败血症、亚急性钻石状皮肤红斑、慢性关节炎和心内膜。因皮肤表面出现红疹斑块而得名"丹毒"（**图13-5**）。猪丹毒丝菌可感染哺乳动物、禽和鱼类，其中猪分离率最

Erysipelothrix rhusiopathiae, the representative member of the *Erysipelothrix* genus, can cause systemic or localized infections in animals, such as acute septicemia, subacute diamond-shaped skin erythema, chronic arthritis, and endocarditis. It is named "erysipelas" due to the appearance of red rash on the skin surface (**Figure 13-5**). *E. rhusiopathiae* can infect mammals, birds, and fish,

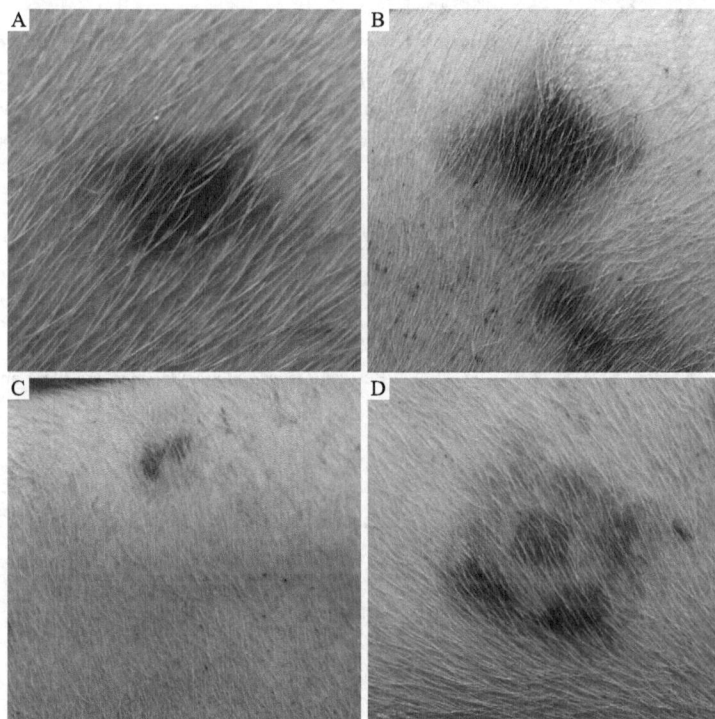

彩图

图13-5 从鱼和海豹中分离的猪丹毒丝菌在猪皮肤引起的局部荨麻疹病变

A. 海豹分离株；B. 鲱鱼分离株；C. 大鲱鱼分离株；D. 毛鳞鱼分离株。图像来源见英文。

Figure 13-5 Local urticarial lesions in pigs induced by selected *Erysipelothrix* strains isolated from fish and harbour seal

A. Isolate from a harbour seal; B. Isolate from a herring; C. Isolate from a large herring; D. Isolate from a capelin. Source: Opriessnig T, Shen HG, Bender JS, et al. 2013. *Erysipelothrix rhusiopathiae* isolates recovered from fish, a harbour seal (*Phoca vitulina*) and the marine environment are capable of inducing characteristic cutaneous lesions in pigs. J Comp Pathol, 148 (4): 365-372.

高；人可经外伤感染，发生皮肤病变，称"类丹毒"，要注意与化脓链球菌引起的丹毒区分。猪丹毒丝菌为革兰氏染色阳性，细杆菌，两端钝圆易成丝状（**图13-6**），微需氧菌，有荚膜，无鞭毛，无芽孢；血平板培养呈现灰白色露滴状透明小菌落，有狭窄的α溶血环，菌落光滑（致病菌）或粗糙（非致病菌）；明胶穿刺呈试管刷状生长，但不液化。

with pigs having the highest isolation rate. In humans, it can cause skin lesions through traumatic infection, which is called "erysipeloid". It is important to differentiate it from erysipelas caused by pyogenic streptococcus. *E. rhusiopathiae* is a Gram-positive, rod-shaped bacterium, with blunt ends that easily form filaments (**Figure 13-6**). It is a facultative anaerobe, encapsulated, non-motile, and non-spore-forming. On blood agar plates, it forms small, transparent droplet-shaped colonies with narrow alpha-hemolysis zones. The colonies can be smooth (pathogenic strains) or rough (non-pathogenic strains). On gelatin stab cultures, it grows in a tube-brush-like pattern but does not cause liquefaction.

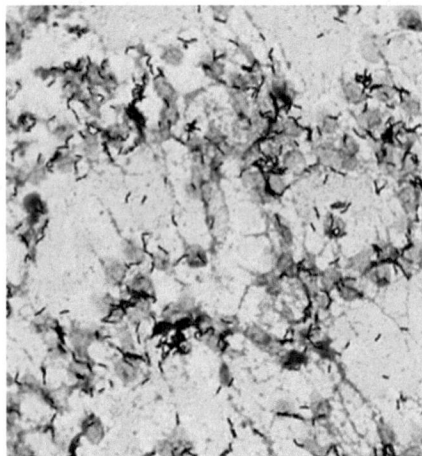

彩图

图13-6　猪丹毒丝菌革兰氏染色（丝状）
图像来源见英文。

Figure 13-6　Gram staining of *Erysipelothrix rhusiopathiae* (filamentous)
Source: Kobayashi KI, Kawano T, Mizuno S, et al. 2019. *Erysipelothrix rhusiopathiae* bacteremia following a cat bite. IDCases, 18: e00631.

思考题　Questions

1. 除了单增李氏杆菌CAMP试验阳性之外，还有其他细菌吗？

2. 如果将单增李氏杆菌接种于半固体培养基置于37℃培养，能看到沿穿刺线云雾状生长的细菌吗？为什么？

3. 如何对猪丹毒丝菌进行微生物学诊断？

1. Besides *Listeria monocytogenes*, are there any other bacteria that yield a positive CAMP test?

2. If *Listeria monocytogenes* is inoculated onto a semi-solid medium and incubated at 37℃, can you observe a cloudy pattern of growth along the inoculation line? Why or why not?

3. What is the microbiological diagnosis of *Erysipelothrix rhusiopathiae*?

第14章 革兰氏阳性产芽孢杆菌
Chapter 14 / Spore-forming Gram-positive bacilli

内容提要 产芽孢的细菌是一群差异很大的细菌，大多数是革兰氏阳性并能运动的杆菌，包括芽孢杆菌属、芽孢乳杆菌属、梭菌属、脱硫肠状菌属、芽孢八叠球菌属和颤螺菌属等。革兰氏阳性产芽孢杆菌中的芽孢杆菌属和梭菌属对兽医学及公共卫生具有重要意义。炭疽芽孢杆菌是芽孢杆菌属最重要的代表，其芽孢具有极强的抵抗力。梭菌属大多数成员专性厌氧，产气荚膜梭菌、肉毒梭菌和破伤风梭菌是梭菌属的代表性致病梭菌，三者均因产生毒素而致病。

Introduction The spore-forming bacteria are a diverse group of bacteria that mostly include Gram-positive and motile rods, such as *Bacillus, Sporolactobacillus, Clostridium, Desulfotomaculum, Sporosarcina* and *Oscillospira*. *Bacillus* and *Clostridium* are of significant importance in veterinary medicine and public health. *Bacillus anthracis*, the causative agent of anthrax, is the most important representative of the *Bacillus* genus and has extremely resistant spores. Most members of the *Clostridium* genus are obligate anaerobes, and *Clostridium perfringens, C. botulinum* and *C. tetani* are representative members of pathogenic *Clostridium* species that cause diseases by producing toxins.

第1节 芽孢杆菌属
Section 1 *Bacillus*

芽孢杆菌属细菌革兰氏阳性，需氧或兼性厌氧；菌体呈杆状，菌端钝圆或平直；每个细胞只形成一个芽孢，能极强地抵抗不良环境条件；大多以周鞭毛运动；某些种可在一定条件下产生荚膜。不同种的菌落形态和大小多变，在某些培养基上可产生色素；大多数种产生触酶，氧化酶阳性或阴性。

本属种类繁多，包括34个正式种和至少200个位置不定的种，其中炭疽芽孢杆菌是重要的人和动物的病原菌，绝大多数为自然界广泛分布的非病原菌。

Bacillus genus are Gram-positive bacteria that are aerobic or facultatively anaerobic. They are rod-shaped, with blunt or straight ends, and produce a single spore per cell that can withstand harsh environmental conditions. Most *Bacillus* species are motile with peritrichous flagella, and some can produce capsules under certain conditions. The colony morphology and size of *Bacillus* species vary greatly, and some can produce pigments on certain culture media. Most species produce catalase and may be positive or negative for oxidase.

This genus encompasses a wide range of species, including 34 formally recognized species and at least 200 putative species. Among them, *Bacillus anthracis* is an important pathogen that cause diseases in humans and animals, while the vast majority are non-pathogens and widely distributed in nature.

炭疽芽孢杆菌

炭疽芽孢杆菌是引起人类、各种家畜和野生动物炭疽的病原，在兽医学和医学上均占有相当重要的地位。炭疽是世界动物卫生组织（WOAH）规定的通报疫病，自第一次世界大战以来就被列为首选的生物武器，在美国曾发生投送炭疽粉末的生物恐怖事件。

该菌革兰氏阳性大杆菌，大小为（1.0～1.2）μm×（3～5）μm，无鞭毛，不运动；芽孢椭圆形，位于菌体中央；可形成荚膜和S层（**图14-1**）；DNA（G＋C）含量为32.2%～33.9%。

Bacillus anthracis

Bacillus anthracis is a pathogen that causes anthrax in humans, various domesticated animals, and wildlife, and holds significant importance in both veterinary and human medicine. Anthrax is a notifiable disease designated by the World Organisation for Animal Health (WOAH) and has been listed as the highest priority bioterrorism agent since World War Ⅰ. A bioterrorism incident involving the dissemination of anthrax spores through powder occurred in the United States.

This bacterium is a Gram-positive, large rod-shaped bacterium, with a size of (1.0-1.2) μm×(3-5) μm, and it is non-motile and lacks flagella. The spores are oval-shaped and located in the center of the cell. It can form a capsule and an S-layer (**Figure 14-1**). The DNA (G＋C) content is 32.2%-33.9%.

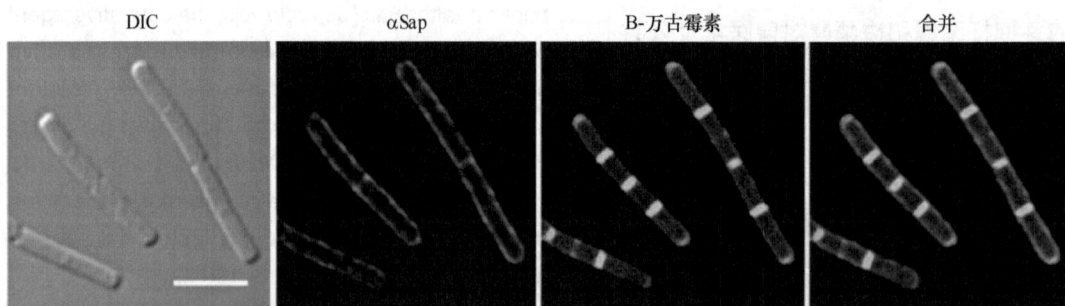

图14-1　炭疽芽孢杆菌中S层和S层相关蛋白的分布
炭疽芽孢杆菌样本用BODIPY-万古霉素（B-万古霉素）或兔抗Sap蛋白血清（αSap）进行染色，并通过相差显微镜（DIC）或荧光显微镜观察图像并合并。标尺：5 μm。图像来源见英文。

Figure 14-1　Distribution of S-layer and S-layer-associated proteins in *B. anthracis*
Samples were stained with BODIPY-vancomycin (B-vancomycin) or rabbit antiserum raised against Sap (αSap) and viewed via differential interference contrast (DIC) or fluorescence microscopy, and images were acquired and merged. Scale bar: 5 μm. Source: Oh SY, Lunderberg JM, Chateau A, et al. 2016. Genes required for *Bacillus anthracis* secondary cell wall polysaccharide synthesis. J Bacteriol, 199 (1): e00613-16.

在动物组织和血液中，菌体单在或呈2～5个相连的短链，相连的菌端平截而呈竹节状，荚膜丰厚。炭疽杆菌只有在离开机体，接触空气中的氧气之后，方能形成芽孢。

该菌为需氧菌，但在厌氧条件下也可生长。可生长温度范围为15～40 ℃，最适生长温度30～37 ℃。最适pH为7.2～7.6。营养要求不高，普通培养基中即能良好生长，培养24 h后，强毒菌株形成灰白色、

In animal tissues and blood, the bacterial cells are typically found singly or in short chains of 2-5 cells, with the connected ends appearing flat and bamboo-shaped, and a thick capsule surrounding them. Anthrax bacilli are only able to form spores after leaving the host and coming into contact with oxygen in the air.

The bacterium is an aerobic microbe, but it can also grow under anaerobic conditions. It can grow in a temperature range of 15-40 ℃, with an optimal temperature of 30-37 ℃. The optimal pH is 7.2-7.6. Its nutritional requirements are not high, and it can grow well in common culture media. After 24 hours of incubation,

表面干燥、边缘呈卷发状的粗糙（R）型菌落，无毒或弱毒菌株形成稍小而隆起、表面较为光滑湿润、边缘比较整齐的光滑（S）型菌落（**图14-2**）。

highly virulent strains form rough (R) colonies that are gray-white, dry on the surface, and have curled edges, while non-virulent or weakly virulent strains form smooth (S) colonies that are slightly smaller, raised, moist on the surface, and have neat edges (**Figure 14-2**).

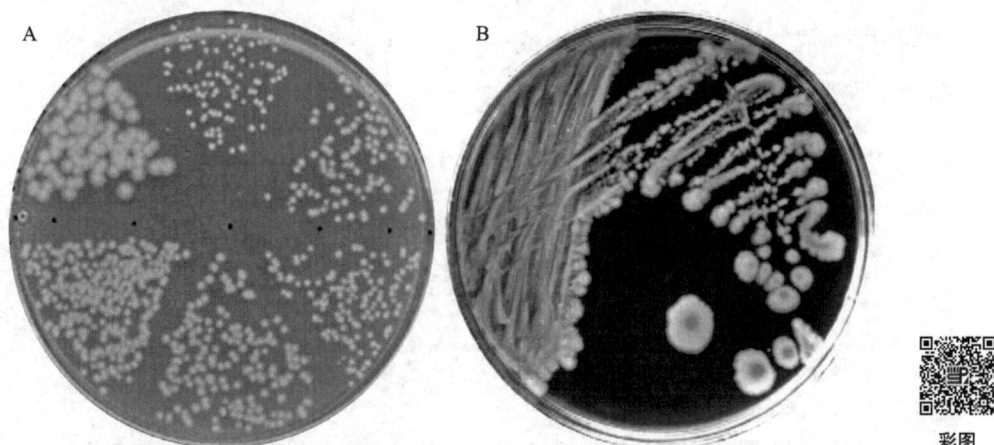

图14-2 不同琼脂平板上的炭疽芽孢杆菌

A. BHI琼脂平板上的炭疽芽孢杆菌菌落，30 ℃生长24 h；B. ACA显色琼脂平板上的炭疽芽孢杆菌菌落，37 ℃生长24 h，菌落呈奶油色，中心蓝绿色不明显。图像来源见英文。

Figure 14-2 Growth of *B. anthracis* on agar plates

A. *B. anthracis* were inoculated on BHI agar and grown at 30 ℃ for 24 h; B. After 24 h, the *B. anthracis* colonies have developed cream-colored colonies without discernable teal centers on anthracis chromogenic agar (ACA). Source: Figure 14-2A: Oh SY, Lunderberg JM, Chateau A, et al. 2016. Genes required for *Bacillus anthracis* secondary cell wall polysaccharide synthesis. J Bacteriol, 199 (1): e00613-16. Figure 14-2B: Juergensmeyer MA, Gingras BA, Restaino L, et al. 2006. A selective chromogenic agar that distinguishes *Bacillus anthracis* from *Bacillus cereus* and *Bacillus thuringiensis*. J Food Prot, 69 (8): 2002-2006.

在含青霉素0.5 IU/mL的培养基中，由于细胞壁的肽聚糖合成受到抑制，原生质体相互连接成串，此特点可与其他需氧芽孢杆菌鉴别，称为"串珠试验"。

该菌与菌体相关的抗原有荚膜抗原、菌体抗原及芽孢抗原。

该菌可引致各种家畜、野兽和人类的炭疽（**图14-3**），牛、绵羊、鹿等易感性最强，马、骆驼、猪、山羊等次之，犬、猫、食肉兽等则有相当大的抵抗力，禽类一般不感染。

死于炭疽的病畜尸体严禁剖检，只能自耳根部采取血液，必要时可切开肋间采取脾脏。皮肤炭疽可采取病灶水肿液或渗出物，肠炭疽可采取粪便。

In a culture medium containing 0.5 IU/mL of penicillin, the synthesis of peptidoglycan in the cell wall is inhibited, and the protoplasts are connected in chains. This characteristic, which can be used to differentiate it from other aerobic spore-forming bacilli, is called the "string-of-pearls test".

The bacterium has capsule antigen, cell wall antigen, and spore antigen related to the bacterial cell.

The bacterium can cause anthrax in various domestic and wild animals as well as humans (**Figure 14-3**). Cattle, sheep, and deer are the most susceptible, followed by horses, camels, pigs, and goats, while dogs, cats, and carnivores have relatively strong resistance, and poultry generally do not become infected.

The carcasses of animals that died of anthrax should not be dissected, and blood samples can only be taken from the ear root. If necessary, the spleen sample can be collected by cutting between the ribs. For cutaneous anthrax, the fluid from the lesion can be collected, and for intestinal anthrax, feces can be collected.

图14-3　从纳米比亚埃托沙国家公园沙斑马的角度看炭疽的传播

A. 斑马在放牧时感染炭疽，大约一周内死亡，并立即吸引食腐动物；B. 食腐动物迅速打开尸体，炭疽芽孢散落于地面。在尸体位置处，食草动物在早期可以相当容易地识别和避免部分腐烂的尸体；C. 经过多年植被恢复，尸体慢慢融入环境。D. 食草动物在草原再次受到感染。图像来源见英文。

Figure 14-3　The life cycle of anthrax in Etosha National Park, Namibia, viewed from the perspective of zebra

A. Zebra become infected while grazing, dying within approximately a week and immediately attracting scavengers; B. The scavengers quickly open a carcass, depositing spores into the ground. During the early stages of a carcass site, herbivores can fairly easily identify and avoid partially decomposed carcasses; C. Carcasses slowly blend into the environment over a period of years and vegetation returns; D. Herbivores return and once again become infected. Source: Carlson CJ, Getz WM, Kausrud KL, et al. 2018. Spores and soil from six sides: interdisciplinarity and the environmental biology of anthrax (*Bacillus anthracis*). Biol Rev Camb Philos Soc, 93 (4): 1813-1831.

炭疽痊愈动物可获得坚强的免疫，再次感染者很少。抗炭疽血清也可用于治疗和紧急预防。我国目前应用的菌苗有两种：Ⅱ号炭疽芽孢苗和无毒炭疽芽孢苗。

Animals that recover from anthrax can acquire strong immunity and are rarely re-infected. Anthrax antiserum can also be used for treatment and emergency prevention. Currently in China, there are two types of vaccines available: the Type Ⅱ anthrax spore vaccine and the avirulent anthrax spore vaccine.

第2节　梭　菌　属
Section 2　*Clostridium*

该属细菌菌体呈杆状、厌氧呼吸、不还原硫酸盐产生硫化物，这3项特性可与其他革兰氏阳性产芽孢杆菌相区别。该菌革兰染色阳性，常单在、成双、或呈链状排列，运动或不运动，运动者具周鞭

Bacteria in *Clostridium* genus are rod-shaped, anaerobic, and do not produce sulfides from sulfate reduction, distinguishing them from other Gram-positive spore-forming bacteria. They are Gram-positive and often occur singly, in pairs, or in chains, and can be motile or non-motile, with those that are motile possessing peritrichous flagella. They

毛。可形成卵圆形或圆形的芽孢，常使菌体膨大。当芽孢位于菌体中央时形如梭状，故名梭菌。但并非梭菌属所有成员均为梭形，许多梭菌的芽孢位于菌体近端或末端。

绝大多数种专性厌氧，对氧的耐受差异很大。梭菌在自然界分布广泛，如土壤、污水、海洋沉积物、腐烂的植物、动植物产品、人和其他脊椎动物及昆虫的肠道、人与动物的伤口及软组织感染灶。本属有74个正式种，100多个不定种。多为非病原菌，致病菌中较常见的约10种，多为人兽共患病病原。可分解碳水化合物或蛋白胨产生有机酸和醇类，有些种是解糖菌，有些是解胨菌，有的两者皆可分解或都不分解。

肉毒梭菌

该菌是一种腐生性细菌，广泛分布于土壤、海洋和湖泊的沉积物，哺乳动物、鸟类和鱼类的肠道，以及饲料和食品中。当有适宜营养和厌氧环境时，即可生长繁殖并产生肉毒毒素。人食入含此毒素的食品，家畜摄入含此毒素的饲料，即可发生肉毒中毒症。

菌体多呈直杆状，仅解胨菌株有的可略弯曲；单在或成双，革兰氏阳性。着生周鞭毛，能运动。易在液体或固体培养基上形成芽孢（**图14-4**）。芽孢卵圆，位于菌体近端，大于菌体直径，使细胞膨大。

该菌专性厌氧；对温度的要求因菌株不同而异，一般最适生长温度为30～37 ℃，多数菌株在25 ℃和45 ℃可生长，产毒素的最适温度为25～30 ℃；6.5% NaCl、20%胆汁和pH 8.5抑制其生长。在血琼脂平板上，可形成直径1～6 mm不规则菌落，β溶血；在巧克力琼脂平板上，解胨者能使菌落周围培养基变为半透明。在庖肉培养基中生长良好。芽孢的抵抗力极

are able to form oval or round spores, which often cause the bacterial cell to become swollen. When the spore is located in the center of the bacterial cell, it appears spindle-shaped, hence the name "clostridium". However, not all members of the *Clostridium* genus are spindle-shaped; many have spores that are located near the proximal or distal ends of the bacterial cell.

The majority of species in this genus are obligate anaerobes with varying degrees of oxygen tolerance. They are widely distributed in nature, found in soil, sewage, marine sediments, decaying plants, animal and insect intestines, and infectious sites in human and animal wounds and soft tissues. The genus comprises 74 formally recognized species and more than 100 indeterminate species. Most of them are non-pathogenic, but about 10 species are known to cause diseases in both humans and animals. They can break down carbohydrates or peptones to produce organic acids and alcohols. Some species are saccharolytic, some proteolytic, and some can utilize both or neither.

Clostridium botulinum

Clostridium botulinum is a saprophytic microorganism that is widely distributed in the sediment of soil, marine and lake, intestinal tract of mammals, birds and fish, as well as feed and food. It can grow and reproduce and produce botulinum toxin when suitable nutrition and anaerobic environment are present. Human ingestion of food containing the toxin or livestock ingestion of feed containing the toxin can lead to botulism.

The bacterial cells are generally straight rods, only slightly curved in some protein decomposing strains; they occur singly or in pairs, and are Gram-positive. They possess peritrichous flagella and are motile. They are capable of forming endospores in liquid or solid culture media (**Figure 14-4**). The endospores are oval and located near the proximal end of the cell; they are larger than the diameter of the bacterial cell, causing the cell to swell.

The bacterium is an obligate anaerobe, and its temperature requirements vary depending on the strain. Generally, the optimal temperature for growth is 30-37 ℃, and most strains can grow at 25-45 ℃. The optimal temperature for toxin production is 25-30 ℃. The growth of the bacterium is inhibited by 6.5% NaCl, 20% bile, and a pH of 8.5. On blood agar plates, irregular colonies with diameters ranging from 1-6 mm can be formed, and the bacterium exhibits β-hemolysis. On chocolate agar plates, the colonies of some strains make the surrounding medium appear translucent. The bacterium grows

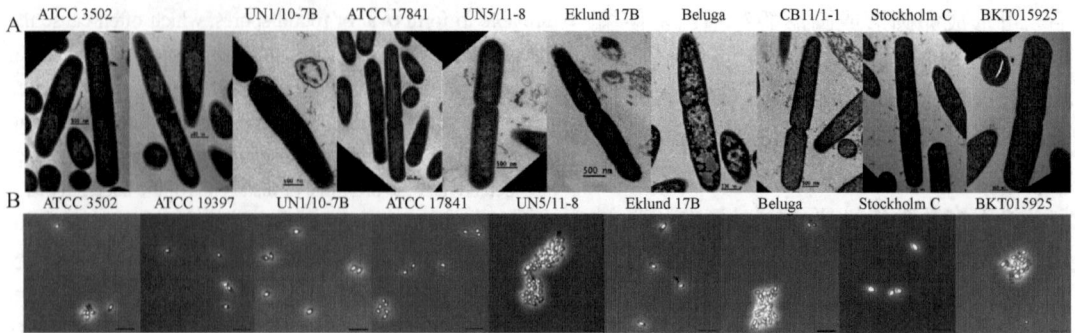

图14-4 肉毒梭菌芽孢萌发和自凝集图像

A. 电镜下的肉毒梭菌超薄切片芽孢超微结构；B. 每个肉毒梭菌芽孢的悬浮液沉淀物。图像来源见英文。

Figure 14-4　Images of sporulation and auto-aggregation of *Clostridium botulinum*

A. Electron Images of thin sections of sporulating *C. botulinum* ultrastructure; B. Sedimentation in spore suspension of each strain. Source: Portinha IM, Douillard FP, Korkeala H, et al. 2022. Sporulation strategies and potential role of the exosporium in survival and persistence of *Clostridium botulinum*. Int J Mol Sci, 23 (2): 754.

强，121 ℃处理15 min可被灭活。肉毒毒素在pH 3～6范围内其毒性不减弱，但对碱敏感，在pH 8.5以上即被破坏。此外，0.1%高锰酸钾或100 ℃加热20 min均能破坏毒素。

　　根据毒素抗原性的差异，可将该菌分为A、B、C（Cα和Cβ）、D、E、F、G7个型，A、B型菌存在于土壤，其余各型则常见于湿地环境中。用各型毒素或类毒素免疫动物，只能获得中和相应型毒素的特异性抗毒素。

破伤风梭菌

　　破伤风梭菌又名破伤风杆菌，是引起人畜破伤风的病原。破伤风以骨骼肌发生强直性痉挛的症状为特征，故又称为强直症，因此该菌也被称为强直梭菌。

　　该菌为两端钝圆、细长、直或略弯曲的杆菌（**图14-5**），大小为（0.5～1.7）μm×（2.1～18.1）μm，长度变化较大。严格厌氧，接触氧后很快死亡；最适生长温度为37 ℃，在25 ℃和45 ℃生长微弱或不生长；最适pH 7.0～7.5。营养要求不高，在普通培养基中即能生长；在血琼脂平板上生长可形成直径4～6 mm的菌落，常常伴

well on meat infusion agar. The spores of the bacterium are highly resistant and can be inactivated by treatment at 121 ℃ for 15 minutes. The toxicity of the botulinum toxin is not reduced in the pH range of 3-6, but it is sensitive to alkaline conditions and is destroyed at a pH above 8.5. In addition, 0.1% potassium permanganate or heating at 100 ℃ for 20 minutes can also destroy the toxin.

Based on differences in toxin antigenicity, the bacterium can be divided into seven types: A, B, C (Cα and Cβ), D, E, F, and G. Types A and B are found in soil, while the other types are commonly found in wetland environments. Immunization of animals with each type of toxin or toxoid only results in the production of specific antitoxins that neutralize the corresponding type of toxin.

Clostridium tetani

Clostridium tetani, also known as tetanus bacillus, is a pathogen that causes tetanus in humans and animals. Tetanus is characterized by the symptom of skeletal muscle stiffness and spasms, hence it is also known as "lockjaw", and the bacterium is also referred to as spasms-causing *Clostridium*.

The bacterium is a straight or slightly curved rod-shaped bacterium with blunt ends, elongated shape, and a size of (0.5-1.7) μm×(2.1-18.1) μm (**Figure 14-5**). It is strictly anaerobic and dies quickly upon exposure to oxygen. The optimum growth temperature is 37 ℃, and it grows weakly or not at all at 25 ℃ and 45 ℃. The optimum pH range for growth is 7.0-7.5. Its nutritional requirements are not high, and it can grow in common culture media. On

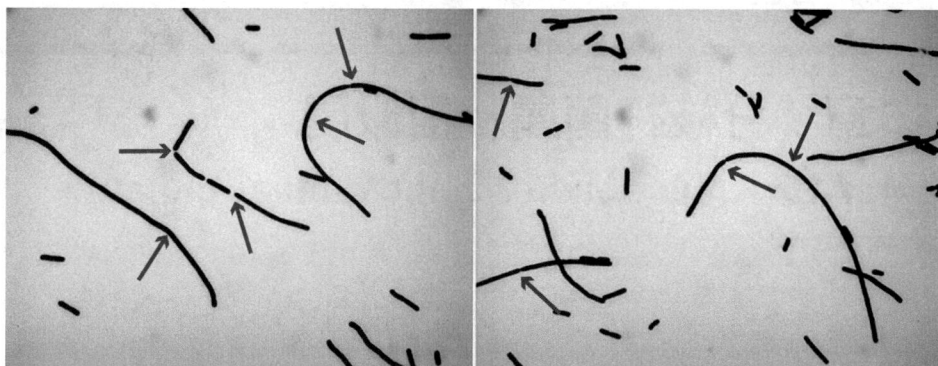

图14-5　破伤风梭菌的形态学变化

在生长中期直至自溶，部分破伤风梭菌变成长链细丝状。箭头表示在一些形成长链细丝的梭菌中可见间隔凹陷。图像来源见英文。

Figure 14-5　*Clostridium tetani* morphological changes

During mid-growth stage until autolysis, a subpopulation of cells formed long-chained filaments. Arrows indicate the septal invaginations that can be seen in some of the cells that formed long-chained filaments. Source: Orellana CA, Zaragoza NE, Licona-Cassani C, et al. 2020. Time-course transcriptomics reveals that amino acids catabolism plays a key role in toxinogenesis and morphology in Clostridium tetani. J Ind Microbiol Biotechnol, 47 (12): 1059-1073.

有狭窄的β溶血环。

通常在外伤、断脐、去势、断角等时候，芽孢通过伤口侵入机体，即可在其中萌发和繁殖，产生强烈神经毒素，引发破伤风。造成此病发生的创伤必须具有厌氧环境。

该菌产生破伤风痉挛毒素和破伤风溶血素。破伤风类毒素能诱导产生坚强的免疫力，可以有效地预防本病的发生。主动免疫预防可用明矾沉淀破伤风类毒素，注射后1个月产生免疫力，免疫期1年。若第2年再注射一次，则免疫力可持续4年。破伤风治疗或紧急预防可用破伤风抗毒素血清，迅速产生被动免疫力，能维持14～21 d。

blood agar plates, it forms colonies with a diameter of 4-6 mm and often accompanied by a narrow β-hemolysis zone.

The spores of the bacterium can enter the body through wounds, umbilical stumps, castration, dehorning, and other types of trauma. Once inside the body, they can germinate and grow under anaerobic conditions, producing a potent neurotoxin that triggers tetanus. The type of injury that causes the disease must have an anaerobic environment.

This bacterium produces tetanus neurotoxin and tetanolysin. Tetanus toxoid can induce strong immunity and effectively prevent the occurrence of tetanus. Active immunization can be achieved by injecting precipitated tetanus toxoid. Immunity develops after one month of injection and lasts for one year. If a second injection is given in the second year, the immunity can last for 4 years. Passive immunization for the treatment or emergency prevention of tetanus can be achieved with tetanus antitoxin serum, which quickly produces passive immunity that lasts for 14-21 days.

思考题　Questions

1. 炭疽芽孢杆菌的生物学特性是什么，为何被用作生物武器？

1. What are the biological characteristics of *Bacillus anthracis*, and why is it used as a biological weapon?

2. 破伤风感染的特点，如何防治？

2. What are the characteristics of tetanus infection and how to prevent and treat it?

第15章 分枝杆菌属及相似属
Chapter 15 / *Mycobacteria* and similar genera

内容提要 分枝杆菌属是革兰氏阳性菌中抗酸菌的代表,特殊的细胞壁影响了其染色特性。本章主要介绍分枝杆菌属、放线菌属和嗜皮菌属。

Introduction *Mycobacterium* is a representative of acid-resistant bacteria among Gram-positive bacteria, and its unigue cell wall affects its staining characteristics. This chapter mainly introduces *Mycobacterium*, *Actinomyce* and *Dermatophilus*.

第1节 分枝杆菌属
Section 1 *Mycobacterium*

分枝杆菌属归属于放线菌目、分枝杆菌科,因其革兰氏染色着色不良,常用抗酸染色法进行染色,成红色(**图15-1**),故又称抗酸杆菌。

形态及细胞壁组成:分枝杆菌属多为细长而稍弯曲杆菌,大小(1.5~4.0)μm×(0.2~0.5)μm,由于细菌与细菌之间常相互聚集,形成树枝样形态,故称作分枝杆菌。分枝杆菌属无鞭毛,无芽孢,无荚膜。细胞壁不仅含有肽聚糖,还含有特殊的糖脂。因糖脂的存在,致使分枝杆菌革兰氏染色不易着色,而抗酸染色为红色。分枝杆菌表面还含有由糖脂类覆盖的分枝菌酸构成的索状因子,使分枝杆菌相互粘连,在液体培养基中呈索状排列。

抵抗力:由于分枝杆菌富含类脂和蜡脂,因此对外界环境抵抗力强,对阴湿、干燥、低温、酸(3% HCl)或碱(4% NaOH)具有抵抗力;但对乙醇、湿热和紫外线敏感,62~63℃加热15 min、煮沸或日光直射2 h可杀死该菌。

The genus *Mycobacterium* belongs to the order of *Actinomycetales* and the family of *Mycobacteriaceae*. Due to its poor staining with Gram staining, acid-fast staining is commonly used, which results in a red coloration (**Figure 15-1**), hence the name acid-fast bacilli.

Morphology and cell wall composition: *Mycobacterium* genus is typically comprised of slender, slightly curved rods ranging in size from 1.5-4.0 μm in length and 0.2-0.5 μm in width. Due to the tendency of bacteria to aggregate, branches are formed, thus earning the name "branching bacteria". Members of the genus *Mycobacterium* are devoid of flagella, spores, and capsules. Additionally, their cell walls not only comprise peptidoglycan but also contain distinct glycolipids. The presence of glycolipids in *Mycobacterium* species renders them Gram-stain resistant and hence, difficult to stain. Acid-fast staining, on the other hand, imparts a reddish hue to these bacteria. The surface of *Mycobacterium* species also harbors pili-like appendages composed of arabinogalactan-bound mycolic acids, which enable these bacteria to adhere to one another and form characteristic palisade arrangements in liquid cultures.

Resistance: owing to the abundance of lipids, including wax esters and mycolic acids, *Mycobacterium* species exhibit robust resistance to various environmental stresses, such as dampness, dryness, low temperature, and extremes of acidity (3% HCl) or alkalinity (4% NaOH). However, these bacteria are susceptible to ethanol, high humidity, and ultraviolet

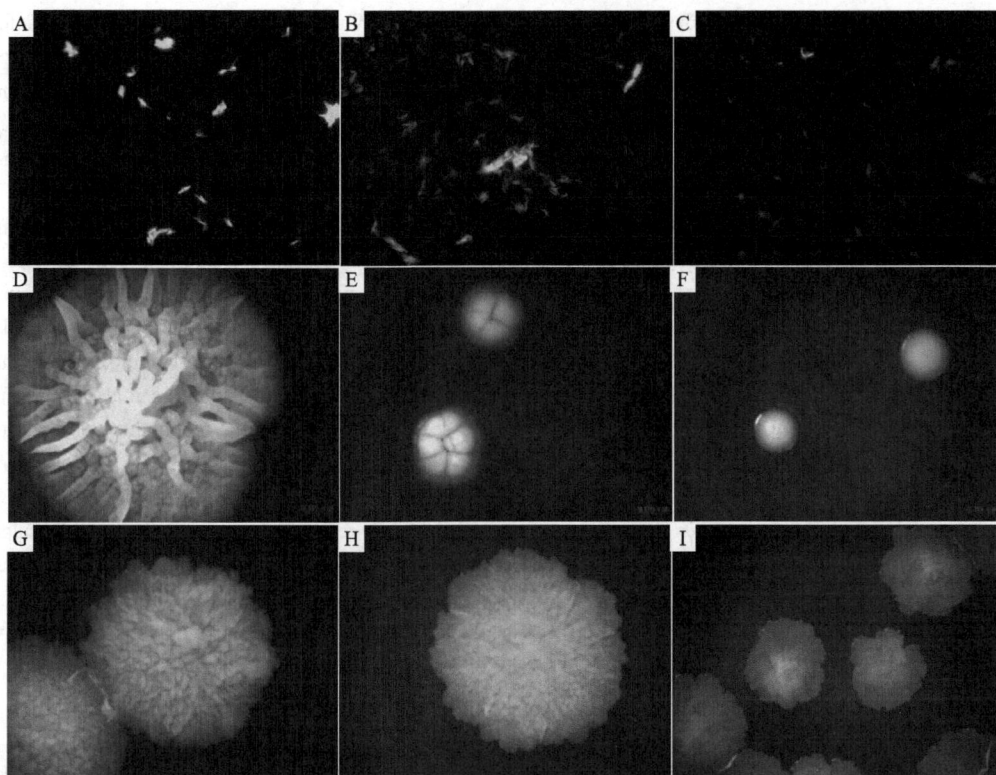

图 15-1 显微镜下的分枝杆菌

玻片上的耻垢分枝杆菌（A～C），7H10培养基琼脂平板上的耻垢分枝杆菌单菌落（D～F），7H10-OADC培养基琼脂平板上的结核分枝杆菌单菌落（G～I）。注：菌落大小因培养时间不同而异。图像来源：图片由浙江农林大学宋厚辉提供。

Figure 15-1 Microscopy of *Mycobacteria*

彩图

Microscopy of *M. smegmatis* on slides (A-C) and single colonies of *M. smegmatis* (D-F) on 7H10 agar plates, and single colonies of *M. tuberculosis* (G-I) on 7H10 agar plated supplemented with OADC. Note that the sizes of colonies are determined by the incubation times on agar plates. Source: courtesy of Houhui Song，Zhejiang A&F University.

radiation. *Mycobacterium* can be effectively eliminated by subjecting them to temperatures of 62-63 ℃ for 15 minutes, boiling, or direct exposure to sunlight for 2 hours.

结核分枝杆菌及牛分枝杆菌

两者许多特性相近，本节一并介绍。

培养特性：专性需氧，最适温度为37 ℃，需要特殊营养，在培养基上生长缓慢，每分裂1代需时18～24 h。固体培养基上，3～4周可见菌落生长，菌落呈颗粒、结节或花菜状，乳白色或米黄色，不透明（**图 15-1**）。液体培养基中生长时，因菌体含有索状因子而使得细菌呈长链状缠绕，

Mycobacterium tuberculosis and *Mycobacterium bovis*

Many characteristics of these two organisms are similar, and thus will be introduced in this section.

Cultivation characteristics: they are obligate aerobes with an optimum growth temperature of 37℃. They require specialized nutrients, and their growth on culture media is slow, with a generation time of 18-24 hours. On solid media, colony growth can be observed after 3-4 weeks, appearing as granular, nodular, or cauliflower-like structures with a milky-white or beige color and an opaque appearance

沿容器壁索状生长，形成有褶皱的菌膜浮于液面，需在培养基中添加Tween-80、Tyloxapol等表面活性剂后才能形成均一培养物。

致病性和免疫性：牛分枝杆菌是人兽共患结核病的主要病原菌。胞内寄生，疾病进程较缓慢，且以形成局部病灶为特点（**图15-2**），肺结核是主要类型，但也可发生肺外结核，如乳腺结核、肠结核、淋巴结核、生殖器结核等。牛分枝杆菌卡介苗是目前人结核病的唯一官方认可疫苗，为牛分枝杆菌的传代致弱株。结核病的免疫呈现细胞免疫和体液免疫分离的现象，细胞免疫随病情的加重而减弱，体液免疫随病情的加强而增强。当细菌在机体消失或者死亡后，其免疫也随之终止。

(**Figure 15-1**). When grown in liquid media, the bacteria form long chains due to the presence of pili-like appendages, which entwine and grow along the walls of the container, forming wrinkled pellicles that float on the liquid surface. To obtain a homogeneous culture, surfactants such as Tween-80 and Tyloxapol must be added to the culture medium.

Pathogenicity and immunogenicity: *Mycobacterium bovis* is the primary causative agent of zoonotic tuberculosis in humans. It is a cytozoic parasite, and the disease progresses slowly, with the formation of localized lesions as a characteristic feature (**Figure 15-2**). Pulmonary tuberculosis is the main type, but extrapulmonary tuberculosis can also occur, such as breast tuberculosis, intestinal tuberculosis, lymph node tuberculosis, and genital tuberculosis. Bacille Calmette-Guérin (BCG) vaccine, derived from attenuated strains of *M. bovis*, is the only officially recognized vaccine for human tuberculosis. Immunity to tuberculosis exhibits a separation between cellular and humoral immunity, with cellular immunity weakening as the disease progresses and humoral immunity strengthening as the disease worsens. The immune response terminates upon the disappearance or death of the bacteria in the body.

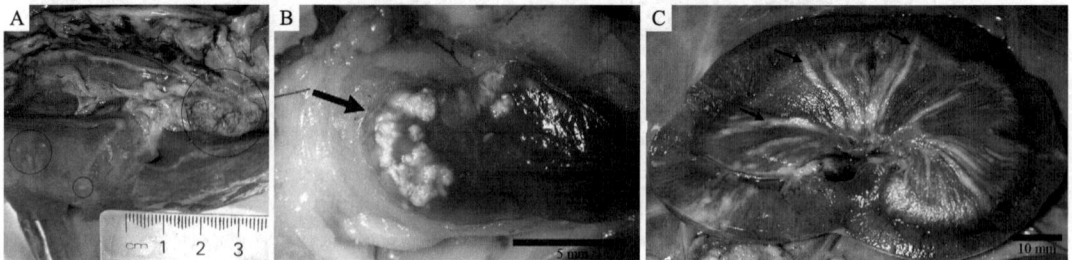

图15-2 自然感染肺结核獾的可见病变
A. 肺结核和右气管支气管淋巴结的大体病变（圆圈）；B. 腹股沟淋巴结内可见的结核结节（箭头）；C. 肾结核大体病变，注意肉芽肿呈放射状分布（箭头）。图像来源见英文。

Figure 15-2　Visible lesions of tuberculosis of a naturally-infected badger
A. Gross lesions of tuberculosis in the lungs and right tracheobronchial lymph node (circle); B. Small visible lesions of tuberculosis (arrow) in an inguinal lymph node; C. Gross lesions of tuberculosis in the kidney. Note the radial orientation of granulomas (arrows). Source: Corner LA, Murphy D, Gormley E. 2010. *Mycobacterium bovis* infection in the Eurasian badger (*Meles meles*): the disease, pathogenesis, epidemiology and control. J Comp Pathol, 144 (1): 1-24.

彩图

诊断：可采用抗酸染色后进行显微镜检查，若发现红色杆菌，可做出初步诊断。进一步进行细菌的分离培养，病料经6%硫酸或4%氢氧化钠处理14 min后接种于罗氏培养基斜面进行培养，培养阳性后根据培养特性和生化特征进一步鉴定，也可用PCR或实时荧光定量PCR方法对样本进行核酸检测。

Diagnosis: acid-fast staining can be used for initial detection by observing red acid-fast bacilli under a microscope. Further isolation and cultivation of the bacteria can be achieved by treating the specimen with 6% sulfuric acid or 4% sodium hydroxide for 14 minutes, then inoculating it onto a slant of Löwenstein-Jensen medium for culture. Positive growth can be identified based on culture and biochemical characteristics, or through nucleic acid detection using PCR or real-time fluorescent quantitative PCR methods.

第2节 放 线 菌 属
Section 2 *Actinomyces*

放线菌属是一类能形成分枝菌丝的细菌，菌丝直径小于1 μm，革兰氏染色呈阳性，不含分枝菌酸，不产生芽孢和分生孢子。主要寄居在人和动物的口腔、上呼吸道、胃肠道和泌尿生殖道，多数无致病性。在少数有致病性的放线菌中，较为常见的包括对人致病性较强的衣氏放线菌，以及引起牛、猪等放线菌病的牛放线菌。

"大颌病"和"木舌病"是两种症状相似的疾病，症状为口咽肿胀，但大小不一。"大颌病"的实质是一种由牛放线菌引起的严重骨髓炎。该菌可穿越口咽黏膜，随后穿透下颌骨骨膜，最终导致下颌骨骨髓炎。"木舌病"的实质是由林氏放线杆菌引起的舌部感染。该菌可致舌头变得异常肿大和疼痛，并可能因创面感染扩大而形成溃疡（**图15-3**）。

Actinomyces is a genus of bacteria capable of forming branched filaments with a diameter smaller than 1 μm. These bacteria are Gram-positive and lack mycolic acids, endospores, and reproductive spores. They commonly reside in the oral cavity, upper respiratory tract, gastrointestinal tract, and urogenital tract of humans and animals, with the majority being non-pathogenic. Among the pathogenic *Actinomyces*, the most commonly encountered species include *Actinomyces israelii*, which is pathogenic to humans, and *Actinomyces bovis*, which causes actinomycosis in cattle, pigs, and other animals.

"Lumpy jaw" and "wooden tongue" are two diseases with similar symptoms, characterized by swelling of the oral cavity and pharynx, albeit with varying degrees of severity. "Lumpy jaw" is a severe form of osteomyelitis caused by the bacterium *Actinomyces bovis*. This bacterium can penetrate the mucous membranes of the oral cavity, subsequently breaching the periosteum of the mandible, and ultimately leading to osteomyelitis of the mandible. "Wooden tongue" is an infection of the tongue caused by *Actinobacillus lignieresii*. This bacterium can cause abnormal swelling and pain in the tongue, and may lead to ulceration due to the expansion of the wound infection (**Figure 15-3**).

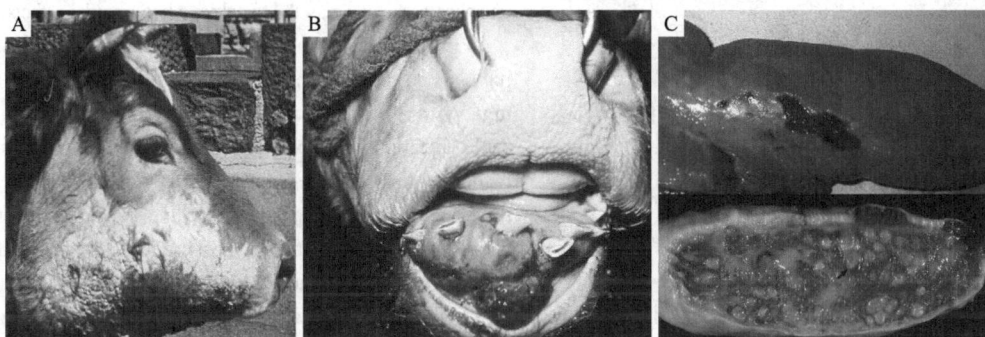

图15-3 "大颌病"和"木舌病"
A. 牛头部侧视图：下颌明显肿胀，伴有浅表皮肤溃疡和渗出；B. 牛口腔视图：下颌切牙明显变形，伴有严重下颌骨骨髓炎；C. 舌头明显肿大，边缘有多处慢性溃疡，舌横切面可见大量脓肿（箭头）。图像来源见英文。

Figure 15-3 "Lumpy jaw" and "wooden tongue"

彩图

A. Lateral view of marked swelling with superficial cutaneous ulceration and exudation; B. Rostral view of severe mandibular osteomyelitis with marked distortion of the mandibular incisors; C. The tongue is markedly enlarged with several areas of chronic ulceration along its lateral edge. Cross section through a case of wooden tongue with numerous pyogranulomas (arrow) scattered throughout the tongue. Source: Njaa BL, Panciera RJ, Clark EG, et al. 2012. Gross lesions of alimentary disease in adult cattle. Vet Clin North Am Food Anim Pract, 28 (3): 483-513.

牛放线菌

形态和培养特性：形态随所处环境而异，在培养基上呈短杆状或棒状，老龄培养物常呈分枝丝状。在病灶中形成硫磺样颗粒状黄白色小菌块，显微镜下呈菊花状，菌丝末端膨大，向周围呈放射状排列，颗粒中央革兰氏染色阳性，外围为革兰氏染色阴性，无抗酸特性。厌氧或微需氧，初代培养时需厌氧；最适pH为7.2～7.4，最适生长温度为37 ℃；在含有甘油、葡萄糖和血清的培养基中生长良好。在血琼脂上，37 ℃厌氧培养2 d可见半透明、乳白色、不溶血的粗糙菌落，紧贴培养基，呈小米粒状，无气生菌丝。无运动性，无荚膜和芽孢。

抵抗力：对石炭酸抵抗力较强，对高温抵抗力较弱，80 ℃经5 min或0.1%升汞经5 min可将其杀死。对青霉素、链霉素、四环素、头孢菌素、林可霉素、锥黄素和磺胺类药物敏感，但因药物难渗透到脓灶中，故达不到杀菌目的。

致病性：牛、羊、马、猪易感，主要侵害牛和猪，奶牛发病率较高。牛感染牛放线菌后主要侵害颌骨、唇、舌、咽、齿龈、头颈皮肤及肺，尤以颌骨缓慢肿大、脓性肉芽肿为多见。猪感染后病变多局限于乳房。

微生物学诊断：取病料（如脓汁）少许，用蒸馏水稀释，找到其中的硫磺样颗粒，在水中洗净，置于载玻片上加一滴5%氢氧化钾溶液，覆以盖玻片用力按压，置显微镜下观察，或将压片加热固定后革兰氏染色镜检，结合临床特征即可做出诊断。必要时可做病原的分离，分离培养时用血液琼脂，初次培养生长比较缓慢，可观察1周以上。

Actinomyces bovis

Morphological and cultural characteristics: the morphology of this organism varies depending on its environment, appearing as short or rod-shaped bacilli on culture media, with older cultures often exhibiting branched filaments. In the lesion, it forms sulfur granule-like yellow-white small bacterial masses with a characteristic "daisy" appearance under the microscope, with the filamentous tips swelling and radiating outward in a radial pattern. The central portion of the granules is Gram-positive, while the periphery is Gram-negative and lacks acid-fast properties. It is an obligate or facultative anaerobe, requiring anaerobic conditions during primary culture. The optimal pH range for growth is 7.2-7.4, and the optimal growth temperature is 37 ℃. It grows well in media containing glycerol, glucose, and serum. On blood agar, after two days of anaerobic incubation at 37 ℃, it forms opaque, rough, non-hemolytic colonies that are tightly adherent to the culture medium, resembling millet seeds and lacking aerial mycelium. It is non-motile and lacks capsules and spores.

Resistance: this organism exhibits relatively strong resistance to carbolic acid, but is weakly resistant to high temperatures; exposure to 80 ℃ for 5 minutes or 0.1% mercuric chloride for 5 minutes is sufficient to kill it. It is sensitive to antibiotics such as penicillin, streptomycin, tetracycline, cephalosporins, lincomycin, kanamycin, and sulfonamides. However, due to poor penetration of these drugs into abscesses, they may fail to achieve bactericidal effects.

Pathogenicity: this organism primarily affects cattle, pigs, sheep, horses, and is highly prevalent in cattle and pigs, with a higher incidence observed in dairy cows. In cattle, infection with *Actinomyces bovis* primarily affects the mandible, lips, tongue, pharynx, gums, skin of the head and neck, and lungs, with mandibular swelling and the formation of purulent granulomas being the most common presentation. In pigs, the lesions are typically localized to the udder.

Microbiological diagnosis: to diagnose *Actinomyces* infection, a small amount of infected material (such as pus) is diluted with distilled water, and the sulfur granules within are isolated and washed with water. A drop of 5% potassium hydroxide solution is added to the cleaned granules on a glass slide, covered with a cover slip, and pressed down firmly. The sample is then examined under a microscope or Gram-stained after heat fixation. The diagnosis can be made based on the clinical features combined with microscopic observation. If necessary, the pathogen can be isolated by culturing on blood agar, although initial growth may be slow and may require observation for over a week.

第3节 嗜皮菌属
Section 3 *Dermatophilus*

嗜皮菌属是一类需氧、革兰氏阳性球杆菌。能形成分枝状细丝，产生具有运动力的游动孢子，具有球杆菌 - 菌丝 - 游动孢子的形态发育过程。该属的成员通常不产生气生菌丝体，通常在含有血液、血清、葡萄糖的培养基中生长最好。不抗酸，触酶阳性。在本属成员中，刚果嗜皮菌可导致多种动物发生疾病，称为嗜皮菌病。

刚果嗜皮菌

1915年，刚果嗜皮菌首次在刚果发病的牛体中分离到。我国于1969年在甘肃的牦牛中首次发现，并相继在其他省份的水牛、山羊中被发现。该菌主要感染牛、羊和马，导致皮肤表层发生渗出性皮炎，并形成痂块。刚果嗜皮菌病传染性强，可造成患病动物的死亡；由于该病为人兽共患病，人亦可被感染，因此存在公共卫生隐患。

培养特性：需氧、兼性厌氧的革兰氏阳性细丝状球杆菌，能产生孢子，多为4个或4个以上一排（**图15-4**）。在固体培养基中会形成圆形有波浪边缘的菌落，呈白色、金黄色和金色。在血琼脂平板出现β溶血，可能出现多种形态不一的菌落（**图15-4**）。在半固体培养基中，会形成类似放线菌的菌落。在普通肉汤、0.1%葡萄糖肉汤等液体培养基中生长较好。接种24 h后培养基呈现轻度浑浊，48 h后出现白色絮片状的生长物悬浮在培养基中。

抵抗力：刚果嗜皮菌抵抗力较强，菌株可存活的时间较长（2～5年），其活力不受贮藏、温度、培养基等环境因素的影响。其孢子对干燥环境有较强的抵抗力，在干燥皮肤结痂中存活时间长达42个月。

Dermatophilus is an aerobic, Gram-positive genus of cocci that can form branching filaments and produce motile spores, undergoing a morphological development process from cocci to filaments to motile spores. Members of this genus typically do not produce aerial mycelium and grow best in media containing blood, serum, and glucose. They are non-acid-fast and positive for catalase. Among the members of this genus, *Dermatophilus congolensis* can cause a variety of diseases in animals, known as dermatophilosis.

Dermatophilus congolensis

In 1915, *Dermatophilus congolensis* was first isolated from a diseased cow in the Congo. In China, it was first discovered in yaks in Gansu province in 1969, and subsequently found in water buffalo and goats in other provinces. This bacterium primarily infects cattle, sheep, and horses, causing exudative dermatitis and the formation of scabs on the skin surface. *Dermatophilus congolensis* infections are highly contagious and can result in the death of infected animals. Moreover, as this disease can be transmitted from animals to humans, it poses a potential public health threat.

Cultivation characteristics: this Gram-positive filamentous coccus is an obligate or facultative anaerobe that can produce spores, typically arranged in rows of four or more (**Figure 15-4**). On solid media, it forms circular colonies with wavy edges, appearing white, golden, or yellow. On blood agar plates, it exhibits β-hemolysis and may produce colonies of various morphologies (**Figure 15-4**). On semi-solid media, colonies resembling those of *Actinomyces* can form. It grows well in liquid media such as ordinary and 0.1% glucose broth. After 24 hours of inoculation, the medium appears slightly turbid, and after 48 hours, white flocculent growth is suspended in the medium.

Resistance: *D. congolensis* exhibits strong resistance, with strains remaining viable for a prolonged period of 2-5 years, and its vitality is not affected by storage, temperature, or culture media. Its spores possess strong resistance to dry environments and can survive for up to 42 months in dried scabs on the skin. This bacterium can survive exposure to 75% ethanol or 2% lysol for 30 minutes, and 2%

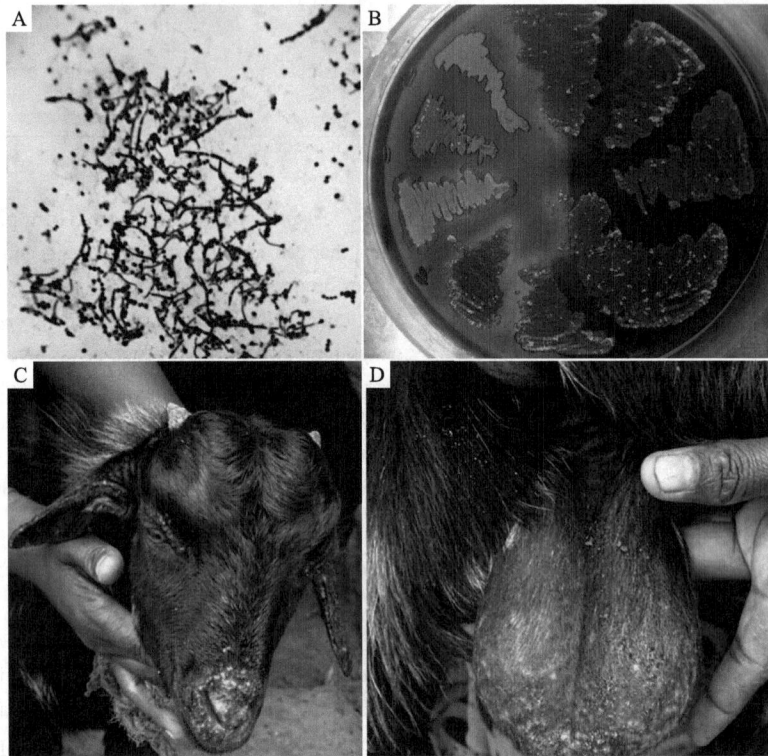

图 15-4　刚果嗜皮菌的特征

A. 革兰氏染色可见分裂成球状的刚果嗜皮菌分枝细丝；B. 刚果嗜皮菌在血琼脂平板上有溶血，菌落形态各异；C. 患有嗜皮菌皮肤病的山羊面部、耳部有干燥结痂皮炎；D. 患有嗜皮菌病的山羊阴囊处可见干燥结痂皮炎。图像来源见英文。

Figure 15-4　Characteristics of *D. congolensis*

A.Gram's staining of a colony shows fragmentation of branching filaments into cocci; B. Colonies of *D. congolensis* isolates on blood agar show hemolysis and varied colony morphology; C. Dermatitis with dried scabs on the face, ears of a goat with dermatophilosis; D. Dermatitis with dried scabs on the scrotum of a goat with dermatophilosis. Source: Chitra MA, Jayalakshmi K, Ponnusamy P, et al. 2017. *Dermatophilus congolensis* infection in sheep and goats in Delta region of Tamil Nadu. Vet World, 10 (11): 1314-1318.

75% 乙醇或 2% 来苏尔作用 30 min，2% 甲醛或 0.1% 新洁尔灭作用 10 min，该菌依然存活。0.2% 新洁尔灭作用 10 min，60 ℃ 加热 10 min，80 ℃ 加热 5 min 或煮沸 1 min 均能有效杀灭该菌。

流行病学：嗜皮菌病的传染源是带菌的动物，该类动物的病变皮肤中存在大量的菌丝或孢子，随着雨水或者渗出物扩散在环境中。在气候炎热地区的多雨季节，宿主长期淋雨、被毛潮湿的情况下，大量孢子可从感染皮肤中释放出来，发病率有升高趋势，因此该病有一定的季节性和地区性。此外，该菌在全球都有分布，除了动物和人直接接触感染外，还可通过昆虫

formaldehyde or 0.1% dodecyl dimethyl benzyl ammonium bromide for 10 minutes. However, it can be effectively killed by 0.2% dodecyl dimethyl benzyl ammonium bromide for 10 minutes, heating at 60 ℃ for 10 minutes, heating at 80 ℃ for 5 minutes, or boiling for 1 minute.

Epidemiology: the source of infection for dermatophilosis is carrier animals, in which large amounts of hyphae or spores exist in the affected skin, spreading into the environment with rainwater or exudate. In hot and humid regions during the rainy season, the host is exposed to prolonged rain and wet fur, and a large number of spores can be released from the infected skin, leading to an increasing incidence of the disease. Therefore, this disease has a certain seasonality and regional distribution. Moreover, the bacteria are distributed worldwide, and in addition to direct contact between animals and humans, they can also be indirectly

间接水平传播。

致病性：动物感染后体温一般不会明显升高，潜伏期为2~14 d。病程初期，在病灶皮肤表面有小块的充血症状，而后形成丘疹，出现浆液性渗出物，与被毛、皮肤杂屑粘连在一起，形成圆形结节，呈灰色或黄褐色，突出于表皮，大小不等（**图15-4**）。发病的位置起初在背部，随后可沿肋骨向两侧延伸，甚至可波及头颈部、乳房后部，病畜可自愈，但严重者可衰竭而死亡。死亡的动物极度消瘦，有大范围的皮炎，也有继发性肺炎或其他并发症。

诊断及防治：当患畜皮肤出现渗出性皮炎和结痂，体温无明显变化时可初步判断为刚果嗜皮菌感染。确诊需要进行实验室诊断。在痂皮、刮屑涂片或培养物中，镜检结果显示有革兰氏阳性分枝的菌丝及球菌状孢子时，结合临床症状，可确诊。

该病尚无商业化疫苗，多使用抗生素、皮质类固醇、杀虫剂联合治疗。防止该病的主要措施为严格隔离病畜，尽可能防止家畜淋雨或被蜱和吸血蝇类叮咬，加强对集市贸易检疫和家畜运输检疫。

transmitted through insects.

Pathogenicity: after infection, the body temperature of animals generally does not increase significantly, and the incubation period is 2-14 days. In the early stage of the disease, there is small area of congestion on the surface of the affected skin, followed by the formation of papules, which exude serous exudate and adhere to the hair and skin debris, forming round nodules that protrude from the epidermis, ranging in size and color from gray to yellow-brown (**Figure 15-4**). The initial site of the disease is on the back, which can then extend along the ribs to both sides, and even involve the head and neck, posterior udder, and other parts. Affected animals can recover spontaneously, but severe cases can lead to emaciation and death. Dead animals are extremely emaciated, with a large area of skin inflammation, as well as secondary pneumonia or other complications.

Diagnosis and control: when livestock exhibit exudative dermatitis and crusts on the skin without an obvious change in body temperature, preliminary diagnosis of *Dermatophilus congolensis* infection can be made. Laboratory diagnosis is required for confirmation. Diagnosis can be confirmed by examining crusts, scrapings, or cultures under a microscope, which show Gram-positive branched filaments and coccoid spores, combined with clinical symptoms.

Currently no commercial vaccine is available for this disease, and treatment mainly relies on a combination of antibiotics, corticosteroids, and insecticides. The main measures to prevent this disease include strict isolation of infected animals, prevention of exposure to rain and bites from ticks and blood-sucking flies, as well as strengthening quarantine inspections for livestock trade and transportation.

思考题　Questions

1. 分枝杆菌属多采用何种染色法？

2. 如何科学预防牛放线菌病？

3. 嗜皮菌病的主要特征是什么？

4. 如何科学预防嗜皮菌病？

1. What staining method is commonly used for *Mycobacterium*?

2. How to scientifically prevent bovine actinomycosis?

3. What are the main features of dermatophilosis?

4. How to scientifically prevent dermatophilosis?

第16章 螺 旋 体
Chapter 16 / *Spirochaete*

内容提要 螺旋体的基本结构与细菌类似，具有轴鞭毛而能活泼运动，多数为厌氧培养，有致病性的属包括疏螺旋体属、短螺旋体属、密螺旋体属和钩端螺旋体属。引致莱姆病的伯氏疏螺旋体可通过硬蜱传播，外膜脂蛋白在致病、检测和免疫方面均具有重要意义；猪痢短螺旋体对幼龄猪具有肠致病性；钩端螺旋体对人和动物可致发热、黄疸等，诊断主要是检测血清抗体水平升高的情况。

Introduction The basic structure of spirochaetes is similar to that of bacteria. They have axial filaments which enable them to move vigorously. Most of them are anaerobic. The pathogenic genera include *Borrelia*, *Brachyspira*, *Treponema* and *Leptospira*. *Borrelia burgdorferi*, the causative agent of Lyme disease, can be transmitted by hard ticks. The lipoproteins on their outer membranes are of great significance in pathogenicity, detection and immunity. *Brachyspira hyodysenteriae* is an intestinal pathogen of piglets. *Leptospira* can cause fever and jaundice in humans and animals. Its diagnosis is mainly based on the detection of elevated levels of serum antibodies.

第1节 螺旋体的一般特性
Section 1 General characteristics of spirochaetes

螺旋体是一类细长、柔软、弯曲呈螺旋状能活泼运动的原核单细胞微生物，在分类上介于细菌和原虫之间。因其基本结构、繁殖方式与细菌类似，故列入广义的细菌学范畴。螺旋体广泛分布于水生环境和动物体内，大部分无致病性，小部分可引起人和动物疾病。

形态： 螺旋状（**图16-1**）。

结构： 由螺旋柱状原生质体、轴鞭毛和外膜组成（**图16-1**）。

运动性： 轴鞭毛沿原生质柱缠绕，赋予螺旋体活跃的螺旋运动。

Spirochaetes are slender flexible and spiral prokaryotic monocellular microorganisms with active motility. In classification, they fall between bacteria and protozoa. Because their basic structure and reproductive pattern are similar to bacteria, spirochaetes are listed under the broad category of bacteriology. Spirochaetes are widely distributed in aquatic environment and animals, and most of them are non-pathogenic, with a few members causing human and animal diseases.

Morphology: spiral (**Figure 16-1**).

Structure: it is composed of spiral and columnar protoplasts, axial flagella and outer membrane (**Figure 16-1**).

Motility: the axial flagella wrap around the protoplasmic column, rendering spirochaetes active spiral movement.

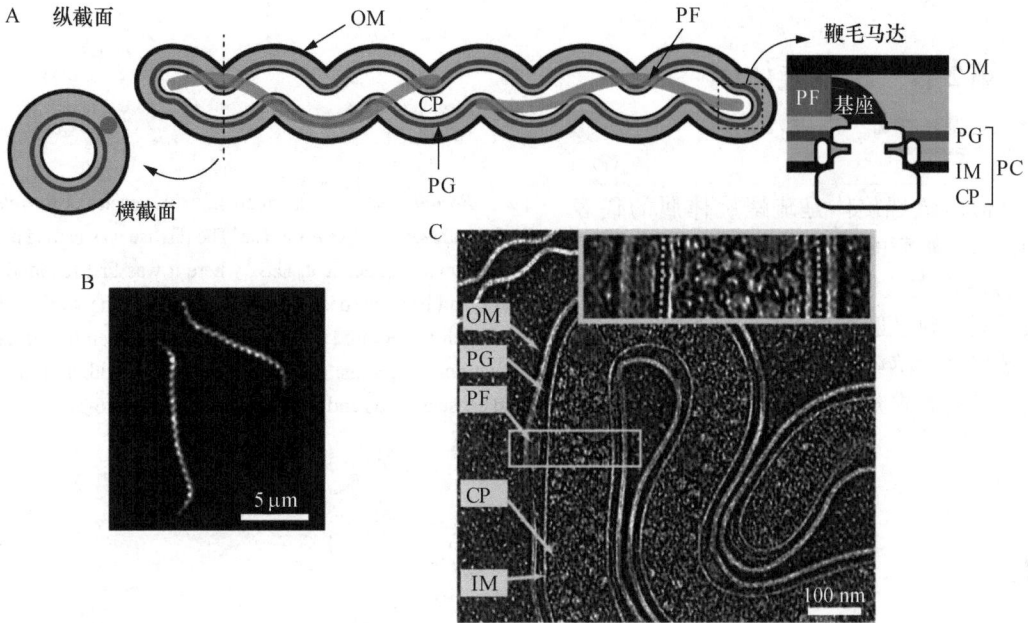

图16-1　螺旋体细胞结构

A. 螺旋体细胞结构和鞭毛马达的纵向和放大横截面图；外膜（OM）、外周鞭毛（PF）、肽聚糖层（PG）、内膜（IM）、细胞质（CP）和原生质柱（PC）；B. 暗视野下的钩端螺旋体显微照片；C.冷冻电镜断层扫描钩端螺旋体纵向切片照片。图像来源见英文。

Figure 16-1　Spirochaetal cell structure

A. Schematics of longitudinal and zoom-in cross-section views of the cell structure and the flagellar motor shared by spirochaete species; outer membrane (OM), periplasmic flagellum (PF), peptidoglycan layer (PG), inner membrane (IM), cytoplasm (CP), and protoplasmic cylinder (PC) are shown; B. Dark-field micrograph of *Leptospira biflexa*; C. Longitudinal slice image obtained by cryo-electron tomography of *L. biflexa*. Source: Nakamura S. 2020. Spirochete flagella and motility. Biomolecules, 10 (4): 550.

繁殖方式：以二分裂法繁殖。

核酸：核酸兼有 RNA 和 DNA。

染色特性：革兰氏染色呈阴性，但不易着色。吉姆萨染色、瑞氏染色和银染效果较好。

培养特性：多数需厌氧培养，钩端螺旋体可需氧培养。营养要求高，体外培养要求苛刻，有的螺旋体只能用易感动物进行增殖和保种。

分类：根据16S rRNA序列和螺旋的数量及排列方式进行分类。螺旋体中与兽医学有关的属主要有疏螺旋体属、密螺旋体属、短螺旋体属和钩端螺旋体属。

Propagation pattern: binary fission.

Nucleic acid: both RNA and DNA.

Staining characteristics: negative in Gram staining, but hard to stain; better to stain with Giemsa staining, Wright's staining and silver staining.

Cultivation characteristics: most spirochaetes require anaerobic culture, but *Leptospira* can be aerobically cultured. They are considered fastidious in culturing *in vitro* since they demand rich nutrition, and some can only be proliferated and passaged in susceptible animals.

Classification: spirochaetes are classified according to the 16S rRNA sequence and the number and arrangement of the helices. The genera of spirochaetes pertaining to veterinary medicine include *Borrelia*, *Treponema*, *Brachyspira* and *Leptospira*.

第2节 疏螺旋体属
Section 2 *Borrelia*

伯氏疏螺旋体是疏螺旋体属的代表种，为莱姆病的病原体。莱姆病的得名源于1975年在美国康涅狄格州的莱姆镇首次发现，现已遍布全世界。该病以硬蜱为传播媒介，感染人和犬、牛、马、猫等动物，可致关节炎等多种病症（**图16-2**）。

Borrelia burgdorferi, a representative species of *Borrelia*, is the pathogen for Lyme disease. The disease was named after Lyme town, Connecticut, U.S., where it was first reported in 1975, and has now been transmitted all over the world. The pathogen is vectored by *Ixodes* ticks for transmission, can infect humans, dogs, cattle, horses, cats and other animals, and cause arthritis and other diseases (**Figure 16-2**).

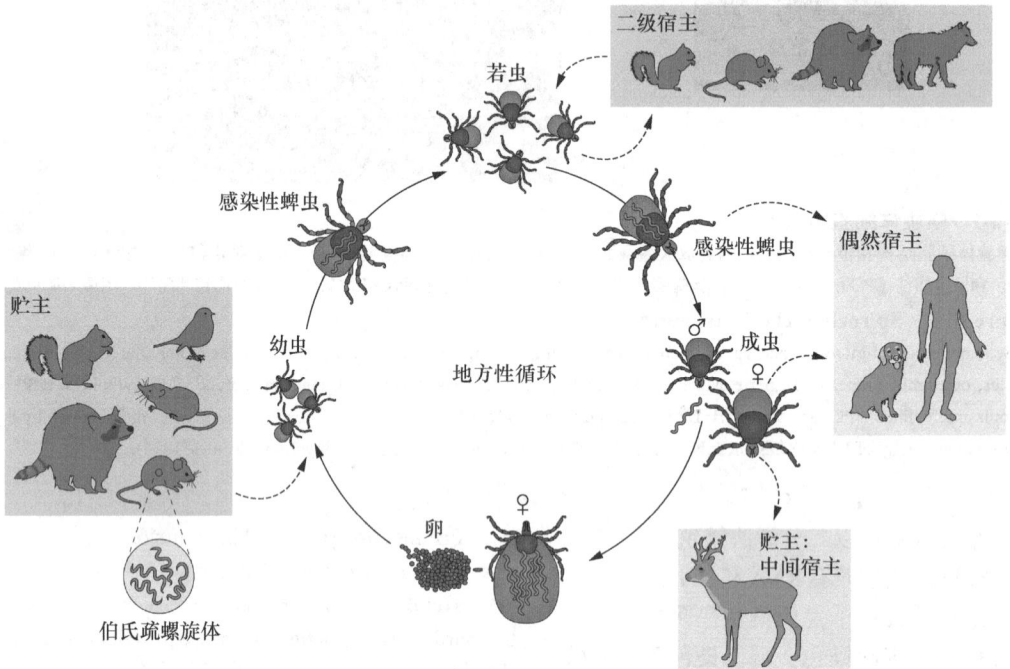

图16-2 硬蜱和伯氏疏螺旋体的生命周期

未感染螺旋体的蜱幼虫孵化寻找寄主，通常为小型哺乳动物或鸟类，也可能为大动物。由于伯氏疏螺旋体不经卵传播，所以该阶段是蜱通过进食感染螺旋体的主要机会。进食后，六条腿的幼虫蜕皮，形成八条腿的若虫，并在吸血过程中感染螺旋体。若虫寻找第二宿主，通常为小型或中型哺乳动物，通过吸血为螺旋体感染蜱提供第二次机会。若虫蜕皮至成年后，成年硬蜱寻找大型动物宿主，通常为白尾鹿。蜱交配后进行最后的血餐。交配后，饱腹的雌蜱从宿主身上释放并产卵，数量数百至数千不等。蜱只产一窝卵，然后死亡。图像来源见英文。

Figure 16-2 The life cycles of *Ixodes scapularis* and *Borrelia burgdorferi*

Uninfected larvae hatch and seek a host to feed on, which is typically a small mammal or bird, but may include larger animals. Because *Borrelia burgdorferi* is not transmitted transovarially, this life stage is the primary opportunity for spirochaetes to infect ticks that feed on an infected host. After feeding, the six-legged larvae moult and emerge as eight-legged nymphs, which may be infected with spirochaetes acquired during their initial bloodmeal. Nymphs seek a second host, typically a small or medium-sized mammal, and this bloodmeal may offer a second opportunity for spirochaetes to infect ticks. After fed nymphs have moulted to the adult stage, newly emerged adult *Ixodes scapularis* ticks search for a large animal host, typically white-tailed deer, for mating and a final bloodmeal. After mating, engorged females release themselves from hosts and eventually oviposit an egg mass, which may contain hundreds to thousands of eggs. *I. scapularis* ticks produce only a single clutch of eggs and then die. Source: Kurokawa C, Lynn GE, Pedra JHF, et al. 2020. Interactions between *Borrelia burgdorferi* and ticks. Nat Rev Microbiol, 18 (10): 587-600.

形态结构：伯氏疏螺旋体每个菌体有4～20个疏松螺旋。15～20根内鞭毛组成的轴丝位于外膜和细胞膜之间，缠绕着原生质柱，以螺旋方式活泼运动。

培养特性：伯氏疏螺旋体为微需氧菌，可用商品化BSK-H（Barbour Stonner Kelly-H）培养基培养，最适培养温度33～34 ℃，新分离的菌株菌落需2～5周才能在暗视野显微镜下观察到。

抗原结构：伯氏疏螺旋体外膜表面蛋白具有抗原性，其中OspA是研制疫苗的主要保护性抗原，OspF有可能成为新疫苗研究的重点。

微生物学诊断：可用暗视野显微镜和免疫荧光显微镜直接观察蜱叮咬部附近的皮肤、患病关节的关节液等组织或体液中的螺旋体；利用间接免疫荧光试验、酶联免疫吸附试验（ELISA）检测血清或体液抗体；应用属特异性引物对组织和体液样本进行PCR或实时荧光定量PCR检测。

Morphology: each *B. burgdorferi* cell has 4-20 loose spirals. Between the outer membrane and cell membrane there are the axial filaments that are composed of 15-20 internal flagella. The filaments wrap around the protoplasmic pillar to render the cell move actively in a spiral manner.

Cultivation characteristics: *B. burgdorferi* is a microaerophilic bacterium and can be cultured in the commercial medium BSK-H (Barbour Stonner Kelly-H). The optimum culture temperature is 33-34 ℃. Generally, the colonies of the primary isolates are not visible under the dark-field microscope until 2-5 weeks of culture.

Antigenic structure: the surface proteins on the outer membrane of *B. burgdorferi* are of good antigenicity. Of them, OspA is the main protective antigen for vaccine development, and OspF is likely to be a focus for new vaccine candidate.

Microbiological diagnosis: spirochaetes from tissues and body fluids of the skin near tick bites and in the synovial fluid of affected joints can be observed by dark-field microscope and immunofluorescence microscope. Indirect immunofluorescence assays and enzyme-linked immunosorbent assays (ELISA) can be used to detect the antibodies in the serum or other body fluids. Genus-specific primers for PCR or real-time fluorescence quantitative PCR can also be applied for the detection of tissue and body fluid samples.

第3节 短螺旋体属
Section 3 *Brachyspira*

猪痢短螺旋体是短螺旋体属的代表种，可引起8～14周龄幼猪黏膜出血性下痢和卡他性炎症（图16-3）。

形态结构：菌体多为2～4个弯曲，两端尖锐，形似双燕翅状。

培养特性：严格厌氧，使用含10%胎牛（或犊牛、兔）血清或血液的培养基，可形成扁平、半透明、针尖状、强β溶血菌落；有时向周围扩散呈云雾状表面生长而无可见菌落。

微生物学诊断：取发病猪血便或病变结肠黏膜刮取物及肠内容物直接做涂片，用吉姆萨或维多利亚蓝染色，也可用印度墨汁做负染或银染后镜检。活体检查可将待检样品制成压滴标本片，置于相差或暗

Brachyspira hyodysenteriae is a representative species of the genus *Brachyspira*. It can cause hemorrhagic dysentery and catarrhal inflammation in 8- to 14-week-old pigs (**Figure 16-3**).

Morphology: most of the cells have 2-4 curves, resembling two swallows' wings with sharp ends.

Cultural characteristics: it is strictly anaerobic, and forms conspicuous β-hemolytic, flat, needle-shaped, translucent colonies in the medium containing 10% fetal bovine (or calf, rabbit) serum or the blood culture medium. Sometimes it is in superficial and cloudy growth spreading to the periphery but forms no visible colonies.

Microbiological diagnosis: bloody stools, scrapes of pathological colonic mucosa or intestinal contents from the sick pig are taken directly for smears, which are then stained with Giemsa or Victoria blue. Indian ink for negative staining or silver staining can also be used, followed by microscopy. Biopsy samples can be made into a pressure drop specimen and observed under a phase contrast or dark-

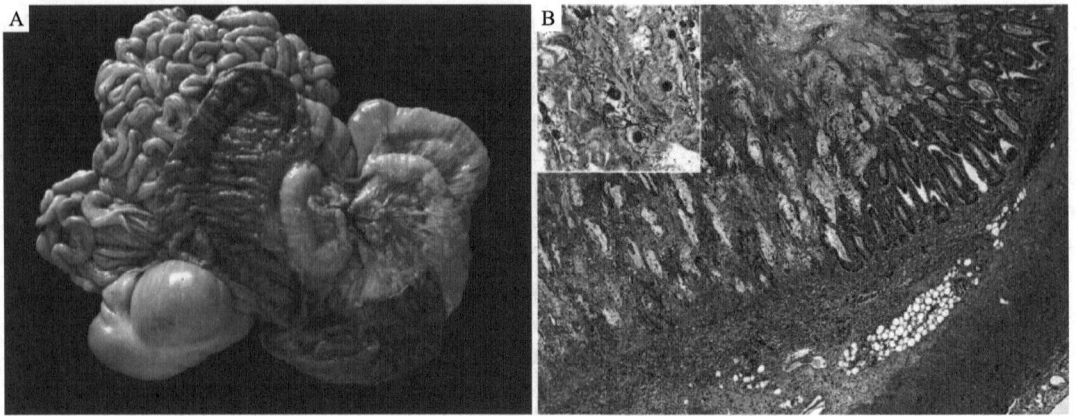

图16-3　共感染猪短螺旋体和波列基内阿米巴虫的猪大肠、盲肠宏观和微观病变
A. 弥漫性、重度、亚急性、纤维蛋白凝固性伤寒出血性内容物；B. 黏膜顶端穿透性坏死，产生大量黏液。插图：黏膜中含有大量呈银染阳性的螺旋体。图像来源见英文。

Figure 16-3　Macroscopic and microscopic lesions of colon and cecum of a pig coinfected with *Brachyspira hyodysenteriae* **and** *Entamoeba polecki*

A. Diffuse severe subacute fibrinonecrotizing typhlocolitis with hemorrhagic content; B. Apical-to-transmural necrosis of the mucosa with abundant production of mucus. Inset: numerous Warthin-Starry-positive, spiral-shaped bacteria are present in the mucosa. Source: Cuvertoret-Sanz M, Weissenbacher-Lang C, Lunardi M, et al. 2019. Coinfection with *Entamoeba polecki* and *Brachyspira hyodysenteriae* in a pig with severe diarrhea. J Vet Diagn Invest, 31 (2): 298-302.

视野显微镜下观察，视野中见有蛇样运动的较大螺旋体即可确诊。病料经镜检证实有可疑螺旋体存在后，可做分离培养。病料接种选择性培养基，42 ℃厌氧培养10 d，每48 h观察一次。凝集试验和ELISA具有较好的血清学诊断意义。

field microscope, and diagnosis can be confirmed when large spirochaetes with snake-like movement are observed. When the sample is suspected to have spirochaetes by microscopic examination, it can be subjected to primary isolation and pure culture. Selective medium is preferred for inoculation. The pathogen is then cultured anaerobically at 42 ℃ for 10 d, and observed every 48 h. Agglutination test and ELISA are of value as means of serological diagnosis.

第4节　钩端螺旋体属
Section 4　*Leptospira*

钩端螺旋体属是一大类菌体纤细、螺旋致密，一端或两端弯曲呈钩状的螺旋体，生存于水生环境中，可引起人和动物钩端螺旋体病，临床表现为发热、贫血、黄疸、血红蛋白尿、睫状体炎等（**图16-4**）。

形态及染色特性：纤细圆柱形，螺旋细密规则，暗视野显微镜下呈细长串珠样，菌体一端或两端呈弯钩（**图16-5**）。常用镀银染色或刚果红负染。

培养特性：微需氧，可选择柯氏培养基28～30 ℃培养。

Leptospira is a large class of slender and dense spirals with one or both ends bent in a hook shape. It lives in aquatic environment and can cause leptospirosis in humans and animals with the clinical symptoms of fever, anemia, jaundice, hemoglobinuria, cyclitis, *etc.* (**Figure 16-4**).

Morphology and staining characteristics: it's slender or cylindrical with dense and regular spirals. It has a slender bead-like morphology with hooks on one or both ends of the bacterial cells under a dark-field microscope (**Figure 16-5**). It is commonly stained by silver staining or Congo red negative staining.

Cultivation characteristics: it is microaerophilic and easy to grow at 28-30 ℃ in Korthof medium.

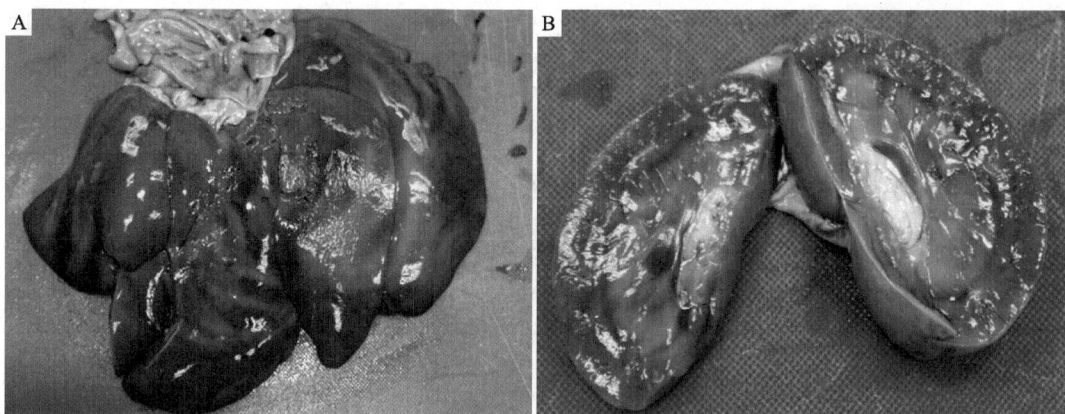

图16-4 感染钩端螺旋体的14周龄雄性流浪犬尸检照片

A. 肝脏呈弥漫性青铜色；B. 肾脏呈淡黄绿色。图像来源见英文。

Figure 16-4 Post-mortem examination of a 14 weeks old intact male mixed breed, stray puppy infected with *Leptospira*

A. Liver showing diffuse bronze discoloration; B. Kidney showing slight green yellow discoloration. Source: Larson CR, Dennis M, Nair RV, et al. 2017. Isolation and characterization of *Leptospira interrogans* serovar Copenhageni from a dog from Saint Kitts. JMM Case Rep, 4 (10): e005120.

彩图

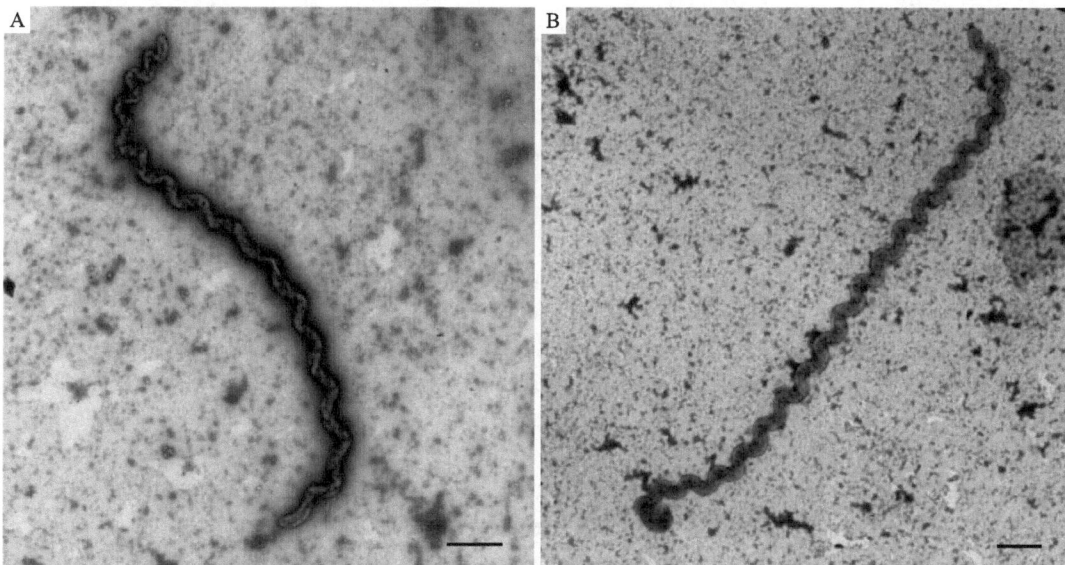

图16-5 钩端螺旋体 Eri-1ᵀ 株（A）和 Patoc I 株（B）的电镜照片

标尺：1 μm。图像来源见英文。

Figure 16-5 Electron micrograph of strain *Leptospira* Eri-1ᵀ (A) and *L. biflexa* strain Patoc I (B)

Scale bars: 1 μm. Source: Saito M, Villanueva SYAM, Kawamura Y, et al. 2013. *Leptospira idonii* sp. nov., isolated from environmental water. Int J Syst Evol Microbiol, 63 (Pt 7): 2457-2462.

　　抗原结构：有群和型特异性抗原之分，应用显微镜凝集实验（MAT）和凝集吸收实验（ATT）可进行分类。

　　Antigenic structure: *Leptospira* has group- and type-specific antigens. Microscopic agglutination test (MAT) and agglutination absorption test (ATT) can be applied for classification.

微生物学诊断：可用暗视野显微镜直接或镀银染色镜检发病早期血液和发病晚期尿液；分离培养可用无菌肝素抗凝血及尿样直接接种柯氏培养基；血清学检测常用MAT法，试验时用活菌体作为抗原，与被检血清作用后，用暗视野显微镜观察。若待检血清中有同型抗体，则见菌体相互凝集成"小蝌蚪状或小蜘蛛状"，继而膨胀并裂解。MAT效价≥1∶800为阳性；效价≤1∶400的，间隔10～14 d采样，抗体效价较上次增加4倍以上，判阳性；PCR技术可用于检测尿中的菌体DNA。

Microbiological diagnosis: a blood sample at the early stage or a urine sample at the late stage can be directly examined under microscope, or these samples are stained with silver staining and then observed under dark-field microscope. Sterile heparinized blood and urine samples can be directly inoculated in Korthof medium for isolation. Microscopic agglutination test (MAT) is usually used as serological test. Live bacteria are used as antigens in the test. After interacting with the sera tested, the mixtures are observed under dark-field microscope. If there are serum antibodies against the corresponding antigenic type of the bacterium, the bacteria will agglomerate into a "tadpole or spider" shape, and then swell and lyse. A titer of the MAT above 1∶800 is proposed to be positive. If the titer is less than 1∶400, samples are collected at intervals of 10-14 d. If the antibody level is increased by 4 times compared to that from the last testing, it is considered positive. PCR technology can be used to detect the specific fragment of the leptospiral DNA in urine.

思考题　Questions

1. 螺旋体具有哪些特点？

2. 如何对引起莱姆病的伯氏疏螺旋体进行微生物学诊断？

3. 为什么生活在我国南方的人群或动物比生活在北方的人群或动物患钩端螺旋体病的概率更高？

4. 如何对猪痢疾和马的"月盲症"进行微生物学诊断？

1. What are the characteristics of spirochaetes?

2. How to make microbiological diagnosis of *Borrelia burgdorferi* that causes Lyme disease?

3. Why are people or animals living in the south more likely to suffer from leptospirosis than those who live in the north of China?

4. How to make microbiological diagnosis of swine dysentery and horse "moon blindness"?

第17章 支　原　体
Chapter 17 / *Mycoplasma*

内容提要　支原体是一类无细胞壁、高度多形性、可通过细菌滤器、可用人工培养基培养增殖的最小原核微生物，大小为0.1～0.3 μm。支原体在分类上属柔膜体纲，其中支原体属及脲原体属是兽医领域的主要研究对象。

Introduction　*Mycoplasma* is a class of the most minuscule prokaryotic organisms with a size of 0.1-0.3 μm that are devoid of cell walls, highly polymorphic, and able to pass through bacterial filters and proliferate in artificial media. *Mycoplasma* belongs to the class *Mollicutes*. *Mycoplasma* and *Ureaplasma* are the main research subjects in veterinary science.

第1节　支原体的一般特性
Section 1　General characteristics of *Mycoplasma*

支原体是一类特性介于细菌与病毒之间的单细胞原核微生物，无细胞壁，与梭菌、链球菌及乳杆菌在基因水平关系较近。细胞柔软，高度多形性，能通过细菌滤器，能在无细胞的人工培养基中生长繁殖。基因组在细菌中是最小的，只有0.58～1.35 Mb。含有DNA和RNA，以二分裂或出芽方式繁殖。常污染实验室的细胞培养及生物制品，有30多种对人或畜禽有致病性。

支原体在分类上归为柔膜体纲，原在该纲的嗜热原体属现已划为古菌，兽医学研究的主要对象为支原体属及脲原体属的成员。

由于支原体细胞外围只有柔软的细胞膜，故具有多形性、可塑性和滤过性。常呈球状、两极状、环状、杆状，偶见分

Mycoplasma is a class of single-cell prokaryotic microorganisms that lie between bacteria and viruses. It has no cell walls, and is closely related to *Clostridium, Streptococcus*, and *Lactobacillus* at the gene level. Mycoplasmal cells are soft, highly polymorphic, and can pass through a bacterial filter, grow and reproduce in cell-free artificial media. The genomes of *Mycoplasma* spp. are the smallest among bacteria, only 0.58-1.35 Mb. They contain DNA and RNA and reproduce in binary fission or budding. They often contaminate the cell cultures in the laboratory and biological products. There are more than 30 *Mycoplasma* species that are pathogenic to humans or livestock and poultry.

Mycoplasma is classified as the *Mollicutes* class. The original genus *Thermoplasma* in this class is now classified as Archaea. The members of *Mycoplasma* and *Ureaplasma* are the main subjects of veterinary studies.

Mycoplasmal cells are polymorphic with plasticity and filterability because there is only a soft plasma membrane around the cells. They are often spherical, bipolar, ring or rod-shaped (**Figure 17-1**), with some in

枝丝状（**图17-1**）。球形细胞直径0.3～
0.8 μm，丝状细胞大小（0.3～0.4）μm×
（2～150）μm。无鞭毛，有些能滑动。革
兰氏染色呈阴性，通常着色不良，吉姆萨
或瑞氏染色良好，染色后呈淡紫色。

slender braches. The spherical cells have a diameter of
0.3-0.8 μm and the filamentous ones, (0.3-0.4) μm×(2-
150) μm in size. They have no flagella and some of
them can slide. Gram staining is negative, poorly stained
though. However, they can be well stained with Giemsa
or Wright's staining, being pale purple in color.

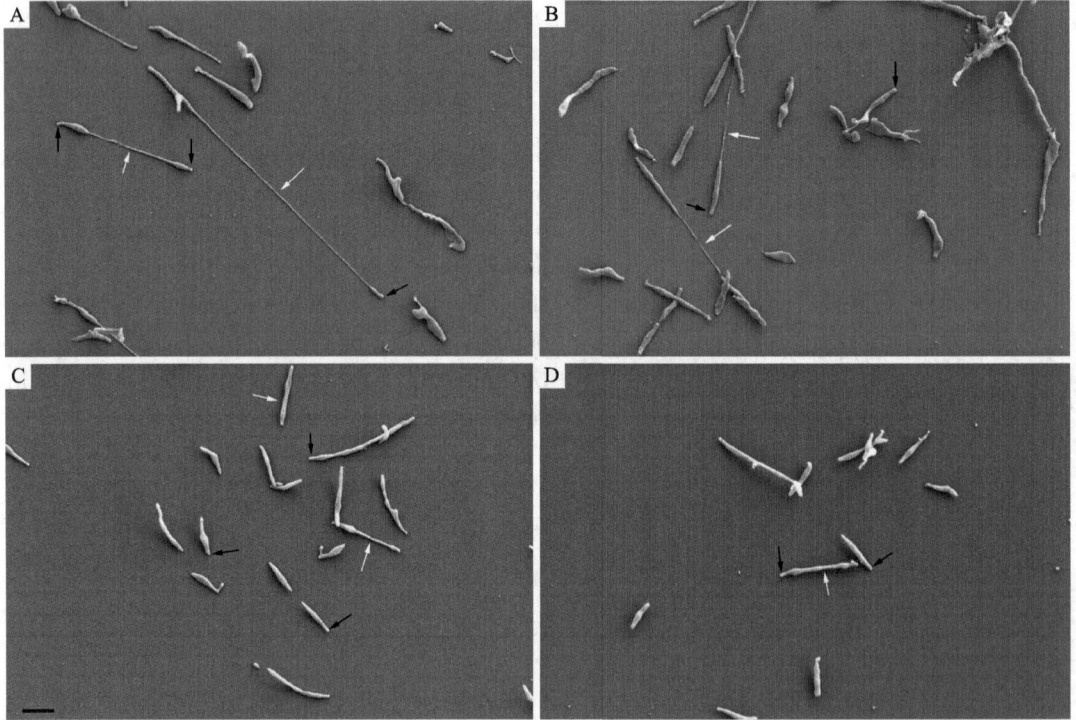

图17-1　扫描电镜观下的穿透支原体和衣阿华支原体

A. 穿透支原体GTU-54-6A1株；B. 穿透支原体HF-2；C. 衣阿华支原体K株；D. 衣阿华支原体N株。白色箭头为连接单个
支原体细胞的丝状结构；黑色箭头为相互黏附的支原体。标尺：1 μm。图像来源见英文。

**Figure 17-1　Scanning electron microscopy of *Mycoplasma penetrans* and *Mycoplasma iowae* cells attached
to glass**

A. *M. penetrans* strain GTU-54-6A1; B. *M. penetrans* strain HF-2; C. *M. iowae* serovar K; D. *M. iowae* serovar N. White arrows,
filamentous structures connecting individual cell bodies; black arrows, attachment organelles. Scale bar: 1 μm. Source: Jurkovic DA,
Newman JT, Balish MF. 2012. Conserved terminal organelle morphology and function in *Mycoplasma penetrans* and *Mycoplasma
iowae*. J Bacterio, 194 (11): 2877-2883.

　　支原体由于基因组小，编码基因少，
生物合成能力较弱，营养要求较高，培养
支原体的人工培养基需添加外源脂肪酸和
甾醇。
　　支原体生长缓慢，在琼脂培养基上孵
育2～6 d，才长出必须用低倍显微镜才能
观察到的微小菌落。支原体的典型菌落呈
"荷包蛋"状，菌落中心深入培养基中，致

Mycoplasma spp. have small genomes with relatively
lower numbers of coding genes, thus having weak
biosynthesis ability and high nutritional requirements.
Exogenous fatty acids and sterols are required to supplement
the artificial media for cultivation of *Mycoplasma*.

Mycoplasma spp. grow slowly and tiny colonies can
be observed with a microscope in a low magnification
after 2-6 d of incubation on an agar medium. Typical
colonies are "fried egg-shaped", in which the center

密、色暗；周围长在培养基表面，较透明
（**图17-2**）。猪肺炎支原体和肺支原体等的
菌落则不呈"荷包蛋"状，无中心生长点。

of the growth is indented into the medium with dense
and dark appearance, while the peripheral part is
superficial and translucent (**Figure 17-2**). The colonies of
M. hyopneumoniae and *M. pneumoniae* are not "fried
egg-shaped" and have no central indentation.

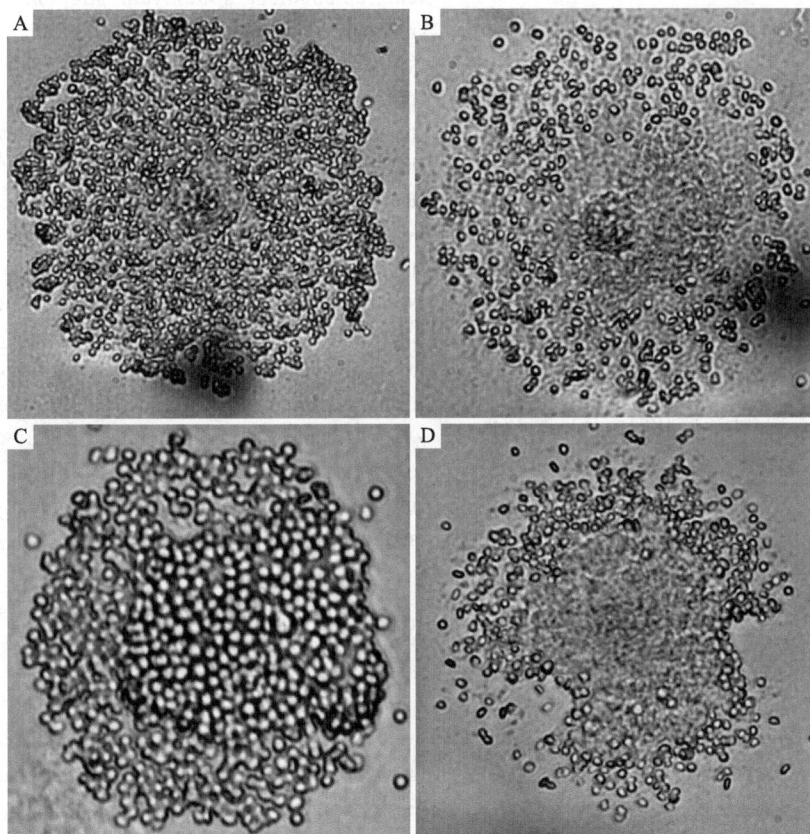

图17-2　吸附红细胞后的穿透支原体和衣阿华支原体菌落
A. 穿透支原体GTU-54-6A1株；B. 穿透支原体HF-2；C. 衣阿华支原体K株；D. 衣阿华支原体N株。图像来源见英文。
Figure 17-2　Colony hemadsorption of *M. penetrans* and *M. iowae*
A. *M. penetrans* strain GTU-54-6A1; B. *M. penetrans* strain HF-2; C. *M. iowae* serovar K; D. *M. iowae* serovar N. Source: Jurkovic
DA, Newman JT, Balish MF. 2021. Conserved terminal organelle morphology and function in *Mycoplasma penetrans* and *Mycoplasma
iowae*. J Bacteriol, 194 (11): 2877-2883.

支原体与细菌L型极相似，均无细胞
壁，形体柔软、形态多样，均具滤过性，
对作用于细胞壁的抗生素有抵抗作用，细
菌L型菌落也似"荷包蛋"状。

支原体的抗原由细胞膜上的蛋白质和
类脂组成。各种支原体的抗原结构不同，
交叉很少，有鉴定意义。可用相应抗血清
做生长抑制试验、代谢抑制试验、免疫荧
光试验及ELISA等，进行支原体的血清学

Mycoplasma is very similar to L-form bacteria,
in that both are devoid of cell walls, flexible and
morphologically diverse, filterable, and resistant to the
antibacterial agents targeting the cell wall. The L-form
bacterial colonies are also "fried egg-shaped".

The antigens of *Mycoplasma* are composed of proteins
and lipids on the cell membrane. The antigenic structures of
various *Mycoplasma* spp. are different, with few cross-reactions,
and have significance in identification. Corresponding antiserum
against particular *Mycoplasma* sp. can be used for tests of
growth inhibition, or metabolic inhibition, immunofluorescent

鉴定或分型。

支原体对重金属盐类、石炭酸、来苏尔等消毒剂均比细菌敏感；对放线菌素D和丝裂霉素C最为敏感，对四环素族、强力霉素、红霉素、氯霉素、螺旋霉素、链霉素等敏感，对青霉素、先锋霉素有抵抗作用。

病原性支原体常定居于多种动物呼吸道、泌尿生殖道、消化道黏膜表面、乳腺及眼等，对胸腺、腹膜和关节滑液囊的间质细胞，以及中枢神经系统有较强亲和力。

支原体感染潜伏期长，呈慢性经过，地方性流行，多具有种的特性。动物自然发生支原体病后具有免疫力，很少发生再次感染，机体感染支原体主要引起体液免疫应答，可产生特异性IgM、IgG和IgA。

微生物学诊断： 应取病料进行分离培养和形态学、生化及血清学鉴定。

assays and ELISA for serological identification or typing.

They are more sensitive to heavy metal salts, carbonic acid, lysol and other disinfectants than bacteria. They are most sensitive to actinomycin D and mitomycin C, and sensitive to tetracycline, doxycycline, erythromycin, chloramphenicol, spiramycin, streptomycin, *etc*., and resistant to penicillin and cephalexin.

Pathogenic *Mycoplasma* spp. often reside in the mucosal surface of respiratory tract and urogenital tract, digestive tract, mammary glands and eyes in a variety of animals species, and have high affinity for the thymus, peritoneum, interstitial cells of the joint synovial bursa and central nervous system.

Mycoplasmosis has a long latent period and often takes a chronic course. Its occurrence is usually species-specific and endemic in certain regions. Animals develop immunity after natural infections and rarely get reinfected. *Mycoplasma* infections mainly elicit humoral immune responses and produce specific IgM, IgG and IgA.

Microbiological diagnosis: isolation and culture of the target organism in the clinical samples for morphological examination, biochemical profiling and serological identification.

第2节 猪的支原体
Section 2 *Mycoplasma* in pigs

以猪为宿主的支原体有猪肺炎支原体、猪鼻支原体、猪滑液支原体、絮状支原体等，前3种对猪具有致病性。

猪肺炎支原体

该菌是猪地方流行性肺炎（猪喘气病）的病原，曾被认为是病毒，直至1965年在无细胞的人工培养基上培养成功，才被命名为猪肺炎支原体。1973年我国首次分离到猪肺炎支原体。

形态多样，大小不等。在液体培养物和肺触片中，以环形为主，也见球状、两极杆状、新月状或丝状。可通过0.3 μm孔径滤膜，革兰氏染色阴性，着色不佳，吉姆萨或瑞氏染色良好。

常用化学消毒剂、1%氢氧化钠、20%草木灰等均可数分钟内将其灭活；对放线菌

Mycoplasma spp. in pigs include *M. hyopneumoniae*, *M. hyorhinis*, *M. hyosynoviae*, *M. flocculare*, *etc*. The first 3 species are confirmed to be pathogenic to pigs.

Mycoplasma hyopneumoniae

This bacterium is the causing pathogen of endemic pneumonia in pigs. It was once thought to be a virus. It was not named until 1965 when it was successfully cultured on cell-free artificial media. In 1973, it was isolated for the first time in China.

It has various forms and sizes. In the liquid culture or lung contact smears, it is mainly ring-shaped, but can also be spherical, bipolarly rod, crescent or filamentous. It can be filtered through a membrane with a pore size of 0.3 μm in diameter. It is Gram-negative, but poorly stained. Good staining could be achieved with the Giemsa or Wright's staining.

Common chemical disinfectants, 1% sodium hydroxide or 20% plant ash can inactivate the bacterium within a few minutes. It is most sensitive to actinomycin D and mitomycin

素D、丝裂霉素C最敏感，对四环素、土霉素、泰乐菌素、螺旋霉素、林可霉素敏感，青霉素、链霉素、红霉素和磺胺对其无效。

　　自然感染仅见于猪，引起猪地方流行性肺炎。猪肺炎支原体黏附于猪呼吸道纤毛上皮细胞，启动在宿主体内的定植。该黏附具有动态性且过程复杂，涉及病原体和宿主细胞表面分子的相互作用（**图17-3**）。

C, and fairly sensitive to tetracycline, oxytetracycline, tylosin, spiramycin and lincomycin. Penicillin, streptomycin, erythromycin and sulfonamides have no effects on it.

　　Natural infections are found only in pigs, causing enzootic pneumonia. Adhesion of *M. hyopneumoniae* to ciliated epithelial cells of the respiratory tract is the initial event in host colonization. This adhesion process is complex and dynamic, and is known to involve several surface-displayed molecules of both the pathogen and the host cells (**Figure 17-3**).

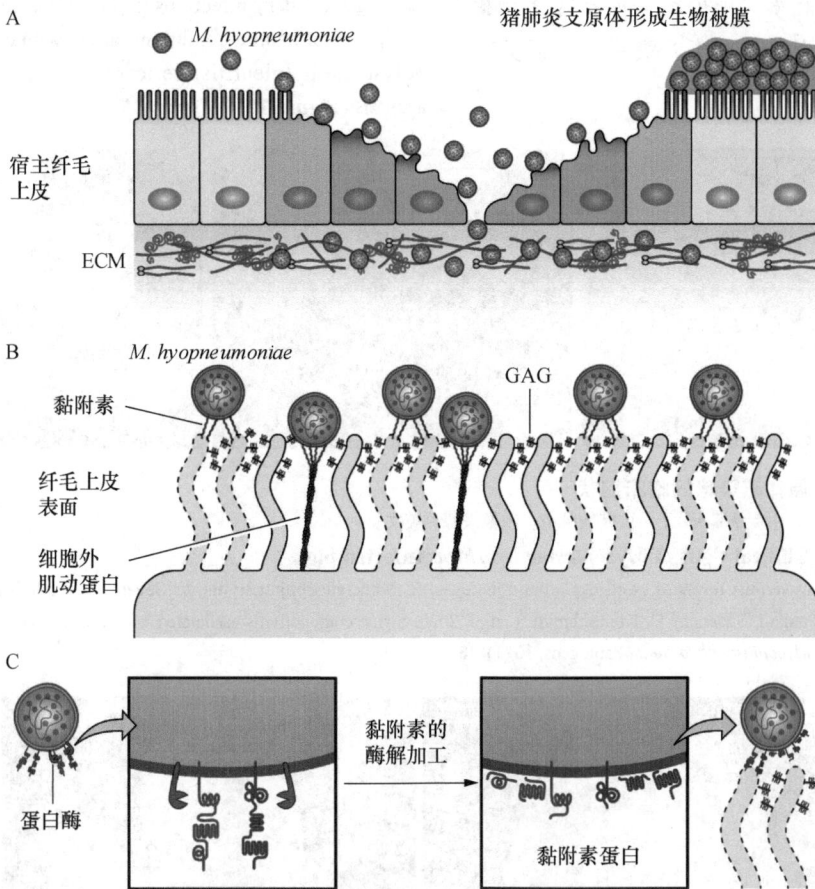

图17-3　猪肺炎支原体与猪呼吸道纤毛上皮黏附示意图

A．猪肺炎支原体附着于呼吸道纤毛上皮细胞后，引起纤毛退化、脱落，随后导致上皮细胞死亡。猪肺炎支原体可在纤毛上皮表面形成生物被膜，也与细胞外基质（ECM）互作，如纤连蛋白和纤溶酶原；B．猪肺炎支原体对纤毛细胞的黏附由可与宿主配体互作的黏附素介导，如位于纤毛表面的GAG和细胞外肌动蛋白；C．猪肺炎支原体黏附素被位于表面的蛋白酶水解后，生成大量黏附素蛋白，暴露于支原体表面。虚线代表受损的纤毛上皮细胞。图像来源见英文。

Figure 17-3　Schematic representation of *M. hyopneumoniae* adhesion to swine ciliated respiratory epithelium

A. *M. hyopneumoniae* cells attach to the cilia of respiratory epithelium, causing ciliostasis, cilium loss, and subsequent epithelial cell death. *M. hyopneumoniae* cells may form biofilms on the ciliated epithelial surface. *M. hyopneumoniae* also interacts with molecules from the extracellular matrix (ECM), such as fibronectin and plasminogen; B. *M. hyopneumoniae* adhesion to ciliated cells is mediated by adhesins that interact with host ligands, as GAGs (displayed on cilia surface), and extracellular actin; C. Adhesins of *M. hyopneumoniae* are endoproteolytically processed by surface-displayed proteases, generating a combinatorial library of adhesin proteoforms exposed on the surface. Dashed lines represent damaged ciliated epithelial cells. Source: Leal Zimmer FMA, Paes JA, Zaha A, et al. 2020. Pathogenicity & virulence of *Mycoplasma hyopneumoniae*. Virulence, 11 (1): 1600-1622.

不同年龄、性别、品种的猪均可感染，但以哺乳仔猪和幼猪最为易感。表现为咳嗽和气喘，发病率高，死亡率低。

Pigs of different ages, genders and breeds can all be infected, but suckling piglets and juveniles are most susceptible. Infected animals cough but wheeze with a high morbidity but a low mortality.

猪鼻支原体

猪鼻支原体是猪鼻腔的常在病原，猪喘气病最常见的继发病原，在自然条件下可引起猪的肺炎，伴发多发性结膜炎、浆膜炎（胸膜炎、心包炎、腹膜炎）或关节炎（图17-4，图17-5）。

Mycoplasma hyorhinis

M. hyorhinis is a commensal agent in the pig nasal cavity that can become the most common pathogen causing secondary infections in pigs. It causes pneumonia in pigs under natural conditions, along with conjunctivitis, polyserositis (pleuritis, pericarditis and peritonitis) or arthritis (**Figure 17-4, Figure 17-5**).

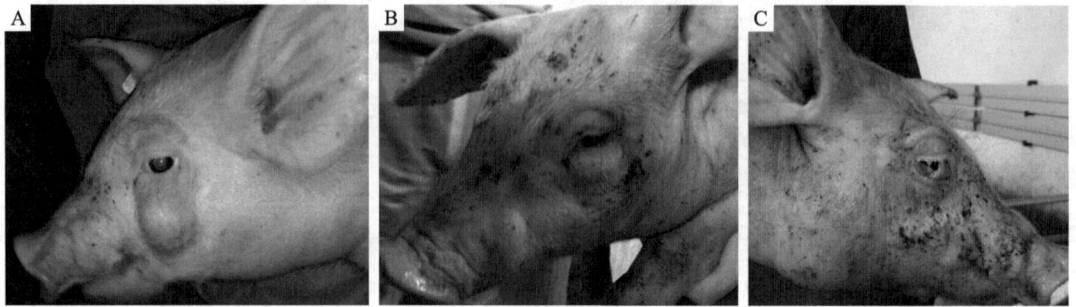

图17-4　猪支原体感染猪的临床症状
A. 眼周肿胀；B. 中度结膜炎；C. 重度结膜炎。图像来源见英文。
Figure 17-4　Clinical signs of *Mycoplasma hyorhinis* infected pigs
A. Pigs displaying various forms of swellings around the eyes; B. Moderate conjunctivitis; C. Severe conjunctivitis.
Source: Hennig-Pauka I, Sudendey C, Kleinschmidt S, et al. 2020. Swine conjunctivitis associated with a novel *Mycoplasma* species closely related to *Mycoplasma hyorhinis*. Pathogens, 10 (1): 13.

彩图

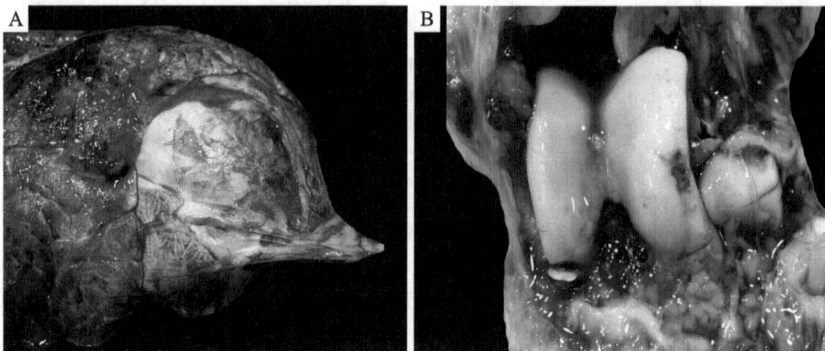

图17-5　猪鼻支原体病变
A. 纤维性心外膜炎，心包明显增厚、不透明，心包附着于心外膜；B. 增生性关节炎，伴有滑膜绒毛增生、出血性关节液和纤维蛋白。图像来源见英文。
Figure 17-5　*Mycoplasma hyorhinis* pathologic lesions
A. Fibrosing epicarditis with marked thickening and opacity of the pericardium and attachment of the pericardium to the epicardium; B. Proliferative arthritis with hyperplasia of synovial villi, abundant serohemorrhagic joint fluid, and fibrin. Source: Giménez-Lirola LG, Meiroz-De-Souza-Almeida H, Magtoto RL, et al. 2019. Early detection and differential serodiagnosis of *Mycoplasma hyorhinis* and *Mycoplasma hyosynoviae* infections under experimental conditions. PLoS One, 14 (10): e0223459.

彩图

猪滑液支原体

该菌可引起猪急性滑膜炎和关节炎，3～6月龄猪最易感。发病率低、散发，病猪以急性跛行为特征，可从关节、黏膜分泌物等分离到病原。

第3节 禽的支原体
Section 3 *Mycoplasma* in poultries

禽的支原体有多种，对禽类具明显致病性的是鸡毒支原体、滑液支原体和火鸡支原体。

鸡毒支原体

鸡毒支原体又名禽败血支原体，是引起鸡、火鸡等多种禽类慢性呼吸道病（CRD）或火鸡传染性鼻窦炎的病原，从鸡、火鸡、雉、珍珠鸡、鹌鹑、鹧鸪、鸭、鸽、孔雀、麻雀等多种禽类均可分离到。

菌体通常为球形或卵圆形，直径0.2～0.5 μm，细胞的一端或两端具有"小泡"极体，该结构与菌体的吸附性有关。吉姆萨或瑞氏染料着色良好，革兰氏染色为弱阴性。

该菌主要感染鸡和火鸡，引起CRD，亦可感染珍珠鸡、鸽、鹧鸪、鹌鹑、野鸡等。病原体主要存在于病鸡和带菌鸡的呼吸道、卵巢、输卵管和精液中，带菌鸡胚可垂直传递给后代，公鸡可通过交配将病传遍全群，鸡群一旦染病即难以彻底根除。

血清学诊断一般常用的方法有平板凝集试验、试管凝集试验、血凝抑制试验等。多采用抽样检查法，一旦检出抗体阳性鸡，即可作为整个鸡群感染的定性指标，判为阳性鸡群。

Mycoplasma hyosynoviae

The bacterium can cause acute synovitis and arthritis in pigs, particularly those at 3-6 months old. It causes sporadic infections with a low incidence and is clinically characterized by acute lameness. The bacterium can be isolated from the joint, mucosal secretions, *etc.*

There are many *Mycoplasma* species of birds. *M. gallisepticum*, *M. synoviae* and *M. meleagridis* are found to be pathogenic to birds.

Mycoplasma gallisepticum

M. gallisepticum is also known as poultry sepsis *Mycoplasma*. It causes chronic respiratory disease (CRD) or turkey infectious sinusitis. It can be isolated from chicken, turkey, pheasant, guinea fowl, quail, partridge, duck, pigeon, peacock, sparrow and other birds.

The organism is usually spherical or ovoid, with a diameter of 0.2-0.5 μm. A "vesicule" polar body, a structure associated with adsorption, is present at either or both ends of the bacterial cell. It is well stained in Giemsa and Wright's staining, and weakly negative in Gram staining.

The bacterium mainly infects chickens and turkeys, causing CRD. It can also infect guinea fowls, pigeons, partridges, quails and pheasants. The pathogen exists in the respiratory tract, ovary, fallopian tubes and semen of the sick chickens or carrier chickens. The bacterium can be vertically transmitted to the offsprings, and the rooster can spread the disease to the whole flock through mating. Once the infection is established in the chicken population, it is difficult to eradicate.

Commonly-used methods for its serological diagnosis include plate coagulation test, tube coagulation test, hemagglutination inhibition test, *etc.* Random samples are often used. Once the specific antibodies to the bacterium are detected in the blood, infection of the whole chicken flock is qualitatively confirmed and thus considered as a *M. gallisepticum* positive flock.

滑液支原体

滑液支原体引致鸡和火鸡传染性滑液囊炎（**图17-6**），可通过蛋传播。人工接种雉和鹅也可感染，兔、豚鼠、小鼠、猪、羔羊等不感染。菌体多呈球状或球杆状，比鸡毒支原体稍小，直径0.2～0.4 μm。营养要求比鸡毒支原体更高。

Mycoplasma synoviae

M. synoviae leads to infectious synovial bursitis in chickens and turkeys (**Figure 17-6**). It can be transmitted through eggs. Pheasants and geese can also be infected by artificial inoculation, while rabbits, guinea pigs, mice, pigs, lambs and others are not susceptible. The bacterium is mostly spherical or club-shaped and about 0.2-0.4 μm in diameter which is slightly smaller than *M. gallisepticum*, but more fastidious than the latter.

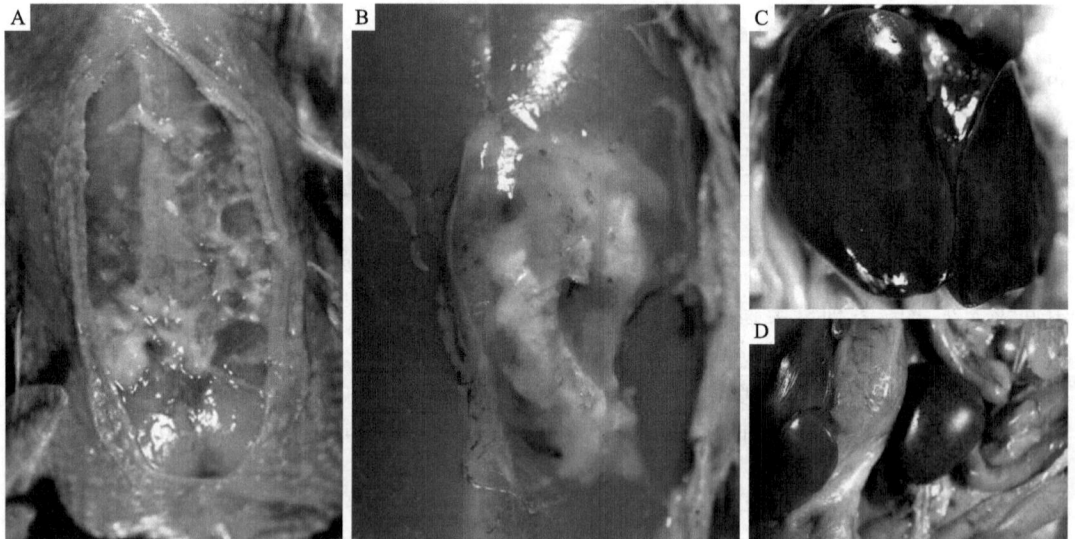

图17-6　滑液支原体引起的大体病变
A. 感染滑液支原体的肉鸡龙骨囊内及其周围皮下组织弥漫性纤维蛋白渗出；B. 龙骨囊局部积累干酪样渗出物；C. 肝脏肿大，略呈绿色和散在淡黄色网状；D. 脾脏斑驳肿大。图像来源见英文。

Figure 17-6　*Mycoplasma synoviae* pathologic lesions
A. Subcutaneous, diffuse accumulation of fibrinous exudate in and around the keel bursa of a broiler chicken infected with *M. synoviae*; B. Extensive localized accumulation of caseous exudate in the keel bursa; C. Enlarged liver with greenish tinge and scattered pale reticular pattern; D. Enlarged mottled spleen. Source: Senties-Cué G, Shivaprasad HL, Chin RP. 2005. Systemic *Mycoplasma synoviae* infection in broiler chickens. Avian Patho, 34 (2): 137-142.

彩图

火鸡支原体

火鸡支原体会引起火鸡孵化率下降、雏火鸡气囊炎、骨畸形等症状，经蛋传递。液体培养的菌体呈球状，直径约0.4 μm，有时单在，双球或成小丛排列。

Mycoplasma meleagridis

M. meleagridis causes a lower hatching rate in turkeys, air sacculitis in juvenile turkeys, and bone deformity. It is transmitted through eggs. Liquid-cultured bacterial cells are globular, about 0.4 μm in diameter. They are sometimes present singly, but other times arranged in

在含2%马红细胞的火鸡肉汤琼脂上，能溶解马红细胞。不吸附和凝集禽类红细胞。

pairs or small clumps. It disintegrates equine red blood cells (RBCs) on turkey broth agar containing 2% equine RBCs, and does not adsorb and coagulate avian RBCs.

第4节 牛羊的支原体
Section 4 *Mycoplasma* in bovine and caprine

丝状支原体群共有5个成员，均与牛、绵羊及山羊的呼吸道疾病有关，分别为：丝状支原体丝状亚种、丝状支原体山羊亚种、山羊支原体山羊肺炎亚种、山羊支原体山羊亚种和利氏支原体。丝状支原体丝状亚种及山羊支原体肺炎亚种在临床上较为重要。

M. mycoides cluster has 5 members, all being associated with respiratory diseases in cattle, sheep and goats: *M. mycoides* ssp. *mycoides* (Mmm) and ssp. *capri* (Mmc), *M. capricolum* ssp. *capripneumoniae* (Mccp) and ssp. *capricolum* (Mcc), and *M. leachii*. Among them, *M. mycoides* ssp. *mycoides* and *M. capricolum* ssp. *capripneumoniae* are of more clinical significance.

丝状支原体丝状亚种

丝状支原体丝状亚种是最早确认与动物致病有关的支原体，1898年即从患牛肺疫（牛传染性胸膜肺炎，CBPP）的病牛中成功分离。牛肺疫在历史上曾是最严重的牛病之一。

Mycoplasma mycoides ssp. *mycoides*

M. mycoides ssp. *mycoides* was the earliest *Mycoplasma* confirmed to be associated with animal diseases, successfully isolated in 1898 from sick cattle with contagious bovine pleuropneumonia (CBPP), also called bovine pulmonary pestilence, one of the most serious diseases in cattle in history.

丝状支原体山羊亚种

丝状支原体山羊亚种引致山羊呼吸系统广泛的病变。细胞呈多形性，可形成丝状，专性需氧，营养要求不严，供给少量甾醇即能生长。

Mycoplasma mycoides ssp. *capri*

M. mycoides ssp. *capri* causes extensive lesions in the goat's respiratory system. The Mycoplasmal cells are polymorphic and can form filaments. They are strictly aerobic and don't have high nutritional requirement for growth, but a small amount of sterol.

山羊支原体

山羊支原体有两个亚种：肺炎亚种（图17-7）和山羊亚种。

Mycoplasma capricolum

There are two subspecies of *Mycoplasma capricolum*: ssp. *capripneumoniae* (**Figure 17-7**) and ssp. *capricolum*.

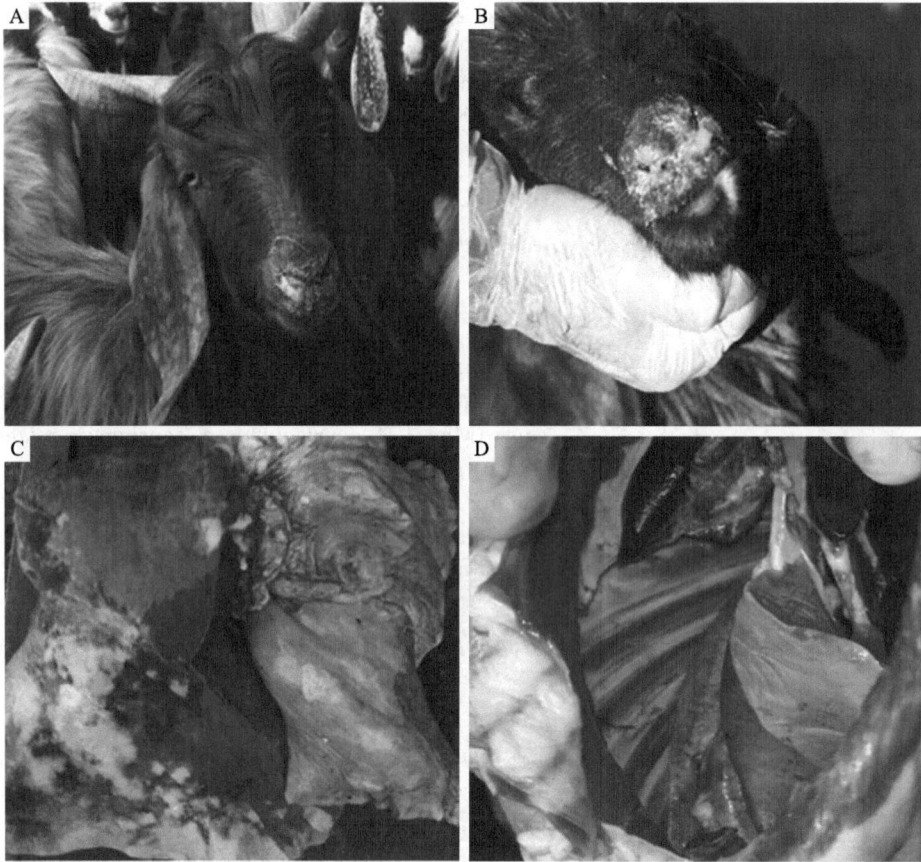

图 17-7　山羊支原体肺炎亚种引起的病变

A、B．山羊表现疼痛咳嗽，流出重度黏液脓性鼻涕；C、D．山羊剖检后肺部表现为典型的单侧红色肝变和纤维蛋白性胸膜肺炎。图像来源见英文。

Figure 17-7　*Mycoplasma capricolum* subsp. *capripneumoniae* pathologic lesions

A, B. Goat exhibiting painful cough, and severe mucopurulent nasal discharges; C, D. Post-mortem examination of goat; the lungs showed typical unilateral red hepatisation, and fibrinous pleuropneumonia. Source: Ahmad F, Khan H, Khan FA, et al. 2021. The first isolation and molecular characterization of *Mycoplasma capricolum* subsp. *capripneumoniae* Pakistan strain: A causative agent of contagious caprine pleuropneumonia. J Microbiol Immunol Infect, 54 (4): 710-717.

第5节　嗜血支原体
Section 5　*Mycoplasma haemophilus*

嗜血支原体是一种黏附于宿主红细胞表面的病原体，可引起以贫血、高热、黄疸、消瘦和流产为主要临床症状的嗜血支原体病。嗜血支原体病在全球范围内普遍存在，主要感染羊、牛、猪、鹿等动物及人。

嗜血支原体寄生于红细胞表面，大多球状，直径小于0.9 μm。对青霉素及其类

Mycoplasma haemophilus resides on the surface of host erythrocytes and can cause hemophilic mycoplasmosis with major clinical symptoms of anemia, high fever, jaundice, emaciation, and miscarriage. The disease is widespread worldwide in animals like sheep, cattle, pigs, deer, *etc*., and humans.

M. haemophilus lives on the red blood cell surface, mostly globular and less than 0.9 μm in diameter. It is resistant

同物有抗性，体外不能培养。

感染多无临床症状，在隐性感染的动物体内往往可存在数年之久。传播媒介有蚤、虱、蜱、蚊及吸血蝇类。

嗜血支原体具有黏附素，由于菌体的黏附导致红细胞凹陷，进而破坏红细胞的细胞骨架，增加了细胞的脆性，因此其感染的主要特点是引致贫血，本质是自身免疫病。

附红细胞体与嗜血巴通体均无细胞壁，黏附并生长于红细胞表面（**图17-8**）。根据

to penicillin and its derivatives, and cannot be cultured *in vitro*.

Animals infected are mostly asymptomatic and the bacterium can often persist in subclinically infected animals for years. Vectors for its transmission include fleas, lice, ticks, mosquitoes and blood sucking flies.

M. haemophilus has adhesion proteins which could depress the surface of the red blood cells and destroy their cytoskeleton, thus increasing the brittleness of the cells and clinically characterized as anemia, which is an autoimmune disease in nature.

Neither *Eperythrozoon* nor *Haemobartonella* has a cell wall. They adhere to and grow on the surface of host red blood cells (**Figure 17-8**). Based on their staining

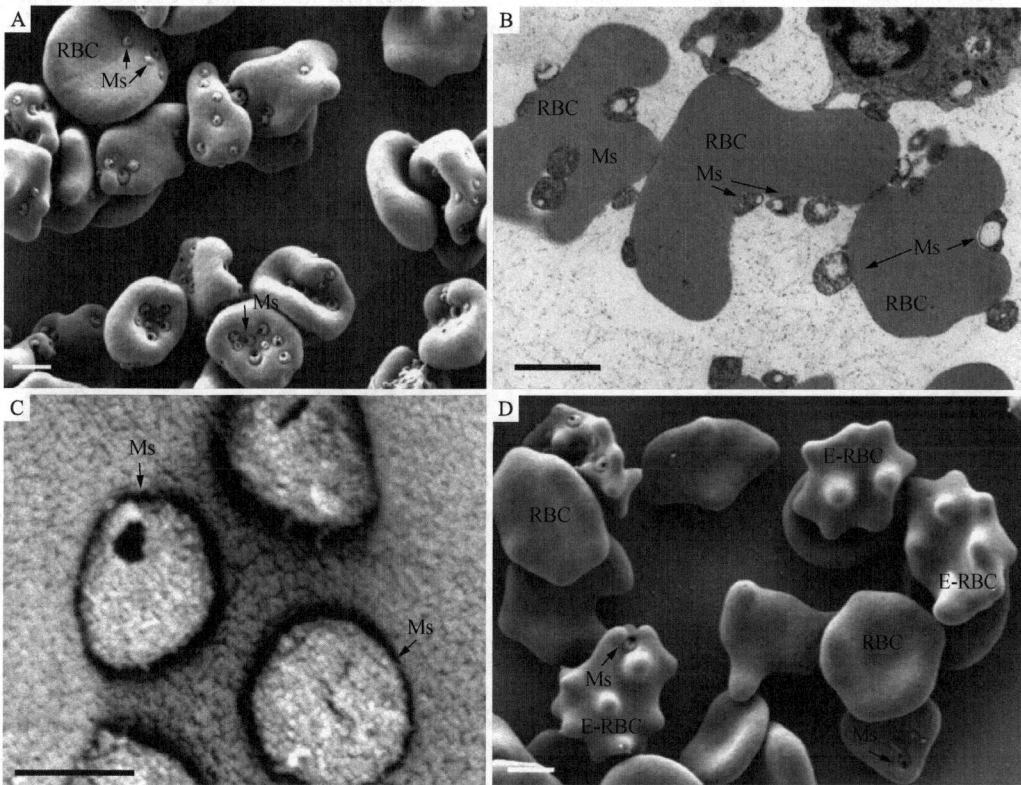

图17-8 猪支原体（Ms）感染红细胞（RBC）的扫描显镜（A、C、D）和透射电镜（B）图像

A. 在感染的急性期，红细胞表面可见大量猪支原体（标尺：2 μm）；B. 红细胞在猪支原体黏附处形成凹陷（标尺：2 μm）；C. 猪支原体与红细胞表面之间的接触紧密，由纤维丝介导（标尺：200 nm）；D. 在猪支原体感染过程中的红细胞（红斑红细胞，E-RBC）（标尺：2 μm）。图像来源见英文。

Figure 17-8 Scanning electron microscope (A, C, D) and transmission electron microscope (B) images of *Mycoplasma suis* (Ms) infected red blood cells

A. During the acute phase of infection, numerous *M. suis* are visible on the surface of the RBCs (scale bar: 2 μm); B. The RBCs form invaginations upon the adhesion of *M. suis* (scale bar: 2 μm); C. The contact between *M. suis* and the RBC surface is intimate and appears to be mediated by fibrils (scale bar: 200 nm); D. During the course of *M. suis* infection, eryptotic RBCs (E-RBCs) are observed (scale bar: 2 μm). Source: Hoelzle LE, Zeder M, Felder KM, et al. 2014. Pathobiology of *Mycoplasma suis*. Vet J, 202 (1): 20-25.

其染色特点、对四环素敏感及需要昆虫媒介等特征，其属于支原体属的新成员。

猪支原体曾被称为猪附红细胞体及猪血支原体，引致猪附红细胞体或贫血性黄疸（**图17-8**）。

犬血支原体导致犬血巴通体病（**图17-9**）。血红扇头蜱可作为媒介和贮主，既可经卵又可机械性传递；输血可致医源性感染。预防的主要措施为杀灭吸血昆虫，急性发病犬可用四环素类药物治疗。

猫血支原体曾称为猫血巴通体大型株；另一种所谓猫血巴通体小型株，现改称鼠血支原体。二者均引致猫传染性贫血或血巴通体病。

characteristics, sensitivity to tetracycline, and vector-borne transmission via insects, they are recognized as new members of the *Mycoplasma* genus.

M. suis, once known as porcine *Eperythrozoon* and *M. haemosuis*, causes eperythrozoonosis featured by icterus and anemia in pigs (**Figure 17-8**).

M. haemocanis causes haemobartonellosis in dogs (**Figure 17-9**). The tick species *Rhipicephalus sanguineus* can act as a vector and reservoir, which can transmit the pathogen to the next generations vertically via eggs or mechanically. Blood transfusion can cause an iatrogenic infection. The main measure of prevention is to use insecticides to kill blood-sucking ticks. Acute cases can be treated with tetracyclines.

M. haemofelis was formerly known as the large form of *Haemobartonella felis*. The small form of *H. felis* is now called as *M. haemomuris*. Both can cause feline infectious anemia or haemobartonellosis.

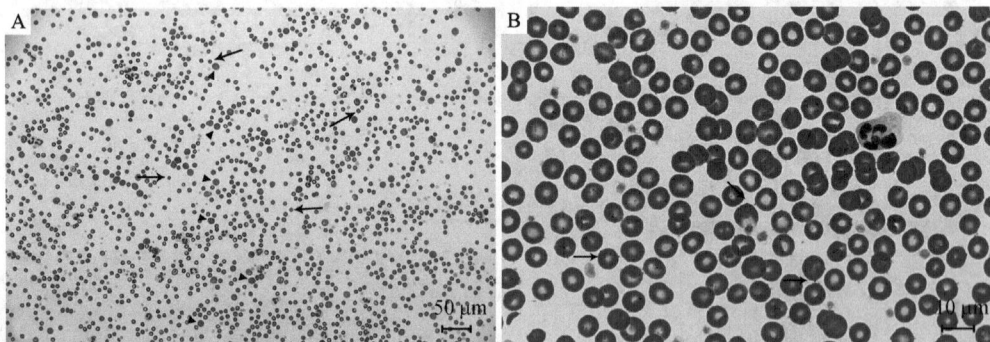

图17-9 犬血支原体感染的血液检查
A. 出现红细胞（箭头）和多染红细胞（三角箭头）聚集，说明出现了免疫溶血性贫血；B. 在红细胞表面出现了嗜碱性染色的犬血支原体（箭头）。一些红细胞表面凹陷处出现许多形成丝状的微生物，即犬血支原体。迪夫快速染色。图像来源见英文。

彩图

Figure 17-9　Blood film examination of *Mycoplasma haemocanis* infection
A. Spherocytosis (arrows), polychromasia (arrowheads), and agglutination of the red blood cells indicate immune-mediated hemolytic anemia; B. Basophilic organisms located on the surface of erythrocytes (arrows). Some erythrocytes have many organisms forming filamentous chains in the deep grooves on their surfaces. Diff-Quik staining. Source: Kim J, Lee D, Yoon E, et al. 2020. Clinical case of a transfusion-associated canine *Mycoplasma haemocanis* infection in the Republic of Korea: a case report. Korean J Parasitol, 58 (5): 565-569.

思考题　Questions

1. 支原体与其他微生物比较有何特点？

1. What are the characteristics of *Mycoplasma* compared with other microorganisms?

2. 试述支原体与细菌L型的相似之处与区别。

2. What are the similarities and differences between *Mycoplasma* and bacterial L-form?

第18章 立克次体和衣原体

Chapter 18 / *Rickettsia* and *Chlamydia*

内容提要 立克次体和衣原体都是专性细胞内寄生的原核微生物，结构和繁殖方式与细菌类似，生长要求类似病毒。有些立克次体对动物和人具有致病性，以节肢动物为传播媒介。衣原体具有独特的发育周期，可形成元体和网状体两种形式，可引起畜禽肺炎、流产和关节炎等。

Introduction Both *Rickettsia* and *Chlamydia* are obligate intracellular prokaryotic microorganisms. They are similar to bacteria in morphology and replication, but are like viruses in terms of growth requirements. Some *Rickettsia* can cause rickettsiosis of humans and animals. Pathogenic *Rickettsia* is spread by arthropod vectors. *Chlamydia* has a unique developmental cycle, and forms elementary bodies and reticular bodies. They can cause pneumonia, abortion and arthritis in livestock and poultry.

第1节 立 克 次 体
Section 1 *Rickettsia*

立克次体是一类专性细胞内寄生的革兰氏阴性原核单细胞微生物。因纪念发现落基山斑点热病原体的美国医生霍华德·泰勒·立克次而命名。立克次体在形态结构和繁殖方式等特性上与细菌相似，而在生长要求上又酷似病毒，是一类介于细菌与病毒之间的微生物。某些成员可引致人和动物立克次体病、斑点热或斑疹伤寒等。

生物学特性： 立克次体具有多形性，主要是球杆状，无鞭毛。大小介于细菌和病毒之间，球状菌直径为0.2～0.7 μm，杆状菌大小为（0.3～0.6）μm×（0.8～2）μm，吉姆萨染色呈紫色或蓝色（**图18-1**）。立克次体不能通过细菌滤器。革兰氏染色阴性，有类似于革兰氏阴性细菌的细胞壁结构和化学组成，细胞壁中含有肽聚糖、脂

Rickettsia is an obligate intracellular Gram-negative single-cell prokaryotic microorganism. It is named in honor of Howard Taylor Ricketts, an American doctor who discovered the pathogen of Rocky Mountain Spotted Fever. It is similar to bacteria in morphology and replication, but it appears more like a virus in terms of growth requirements. *Rickettsia* is a kind of microorganism between virus and bacteria. Some members can cause rickettsiosis in humans and animals, which presents clinically as spotted fever or typhus.

Biological characteristics: *Rickettsia* is pleomorphic under microscope. It has a spherical rhabdoid form with no flagella. The size of the organism is between bacteria and viruses. The diameter of coccal cells ranges from 0.2 μm to 0.7 μm, and the size of short rod-shaped bacteria is (0.3-0.6) μm×(0.8-2) μm. It is in purple or blue in Giemsa staining (**Figure 18-1**). *Rickettsia* could not pass through bacterial filter. It is Gram-negative.

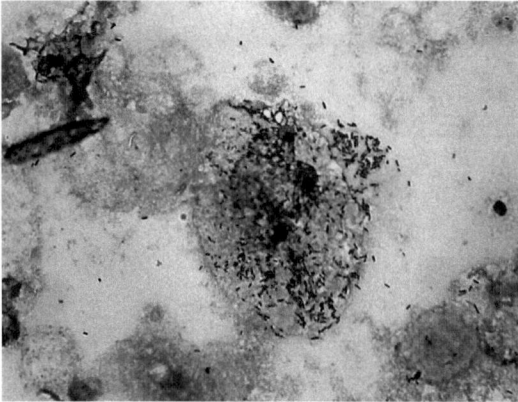

图18-1　吉姆萨染色的细胞离心涂片（放大100倍）
人脐静脉内皮细胞接种卡延钝眼蜱匀浆48 d后感染立克次体。
图像来源见英文。

Figure 18-1　Giemsa-stained cytocentrifuge smear (100×magnification)

Showing infection of human umbilical vein endothelial cell with *Rickettsia amblyommatis* at day 48 after inoculation with *Amblyomma cajennense* tick homogenate. Source: Santibáñez S, Portillo A, Palomar AM, et al. 2017. Isolation of *Rickettsia amblyommatis* in HUVEC line. New Microbes New Infect, 21: 117-121.

彩图

多糖和蛋白质，细胞质内有DNA、RNA及核糖体。

立克次体为专性细胞内寄生，以二分裂方式繁殖，不能在人工培养基上生长，可接种鸡胚卵黄囊或采用脊椎动物或节肢动物的巨噬细胞、上皮细胞等真核细胞进行培养。立克次体主要寄生于节肢动物，通过蜱、跳蚤、虱子、螨虫叮咬感染人和动物使其全身皮肤出现红色斑点或斑疹并伴随发热，因此立克次体病又称斑点热或斑疹伤寒。

分类：立克次体归类于立克次体目。立克次体目现分为立克次体科和艾立希体科。根据16S rRNA基因序列，立克次体科目前包括立克次体属、东方体属及鱼立克次体属3个属。立克次体属有两个生物群，分别是斑疹伤寒群和斑点热群。引起Q热的柯克斯体，曾属于立克次体目，现已列入军团菌目的柯克斯体科。

魏斐二氏反应：其原理是由于变形杆菌可与立克次体多糖抗原的抗体发生交叉反应，故在医学上，用变形杆菌菌株制作凝集原，通过凝集试验诊断某些立克次体病，该凝集试验称魏斐二氏反应。

立氏立克次体可导致犬和人的落基山斑点热，经蜱虫叮咬而感染，落基山斑点热常发于北美和南美（**图18-2**），无有效疫苗进行预防。

斑疹伤寒立克次体引起人的鼠斑疹伤寒，由寄生于鼠、猫、犬的跳蚤传给人，

The cell wall structure and chemical composition are similar to those of Gram-negative bacteria. It contains peptidoglycans, lipopolysaccharides (LPS) and proteins. There are DNA, RNA and ribosomes in the cytoplasm.

Rickettsia is a specific intracellular microbe. It reproduces in binary fission. It does not grow in standard artificial media but can be cultivated by inoculation in the yolk sacs of chicken embryos or eukaryotic cells, such as macrophages, and epithelial cells of vertebrates or arthropods. *Rickettsia* mainly lives in arthropods and infects humans and animals through biting of ticks, fleas, lice and mites, causing red spots or maculae all over the skin accompanied by fever. Rickettsiosis is also known as spotted fever or typhus.

Taxonomy: *Rickettsia* is classified as a member in the order *Rickettsiales*, which is divided into *Rickettsiaceae* and *Ehrlichiaceae*. Currently *Rickettsiaceae* contains *Rickettsia*, *Orientia* and *Piscirickettsia* based on 16S rRNA sequences. *Rickettsia* has two biotic groups: typhus group and spotted fever group. *Coxiella* which causes Q fever once belonged to *Rickettsiales*. Now it has been listed in the family *Coxiellaceae* of the order *Legionellales*.

Wei-Felix reaction: the principle is that *Proteus* can cross-react with the antibodies against the *Rickettsia* polysaccharide antigen. Therefore, in medicine, *Proteus* strains have been used as agglutinogens in the agglutination test to detect antibodies to *Rickettsia* for diagnosis of rickettsiosis. This agglutination test is called as Wei-Felix reaction.

Rickettsia rickettsii is responsible for a century-old disease, Rocky mountain spotted fever (RMSF), in dogs and humans. Canines and humans are infected following biting by infected ticks. The disease is commonly seen in both North and South America (**Figure18-2**). No vaccines are available.

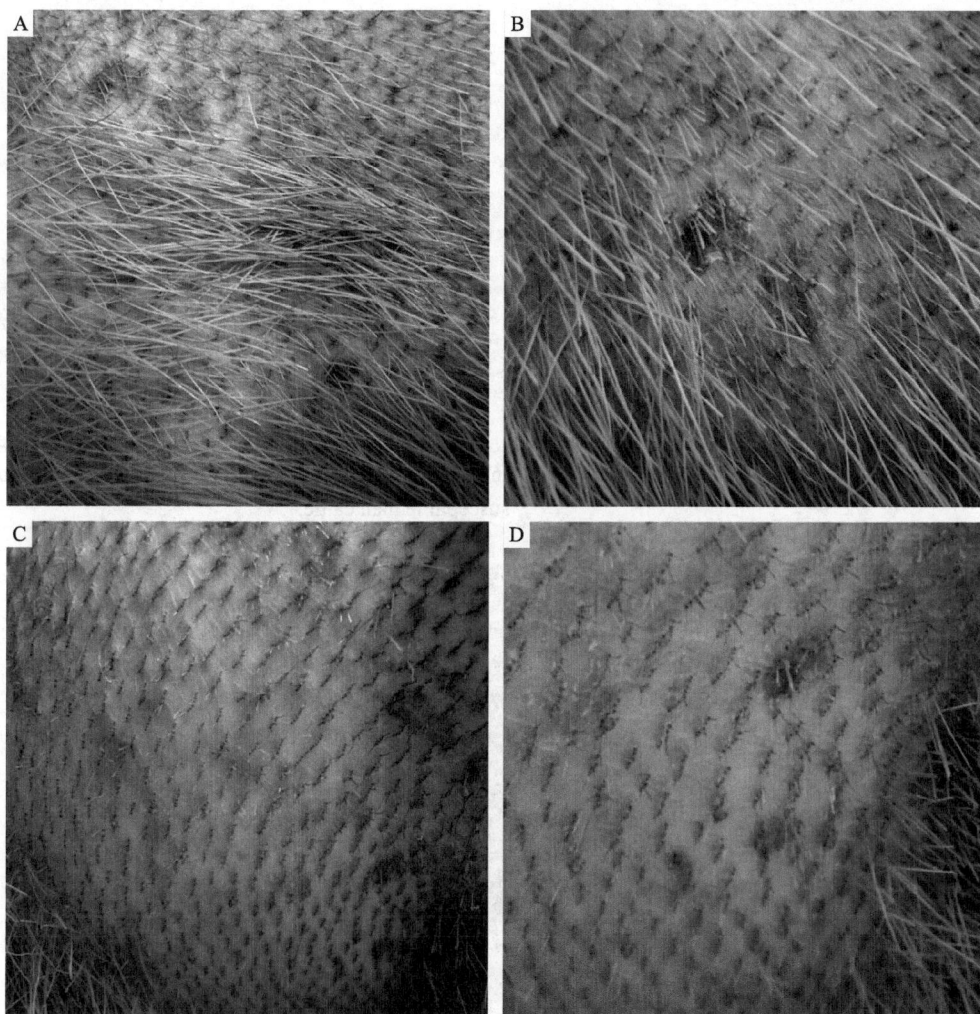

图18-2 蜱传立氏立克次体感染（Ⅱ期）的水豚皮肤

A、B. 水豚皮肤紫斑（感染后第18天）；C、D. 水豚腹部皮疹（感染后第10天）。图像来源见英文。

Figure 18-2 Skin of capybaras during primary infection (phase Ⅱ) with *Rickettsia rickettsii* via tick exposure

A, B. Purplish macules in capybara (18 dpi); C, D. Abdominal rash in capybara (10 dpi). Source: Ramírez-Hernández A, Uchoa F, de Azevedo Serpa MC, et al. 2020. Clinical and serological evaluation of capybaras (*Hydrochoerus hydrochaeris*) successively exposed to an Amblyomma sculptum-derived strain of *Rickettsia rickettsii*. Sci Rep, 10 (1): 924.

可通过人虱在人群中传播。

犬艾立希体是引起犬单核细胞增多艾立希体病的主要病原（**图18-3**），可用犬巨噬细胞进行培养。

Rickettsia typhi causes murine typhus in humans. The pathogen is transmitted from rats, cats and dogs to humans by fleas and can spread among populations by human lice.

Ehrlichia canis is the major pathogen responsible for canine monocytic ehrlichiosis (**Figure18-3**) and can be cultivated *in vitro* using a canine macrophage cell line.

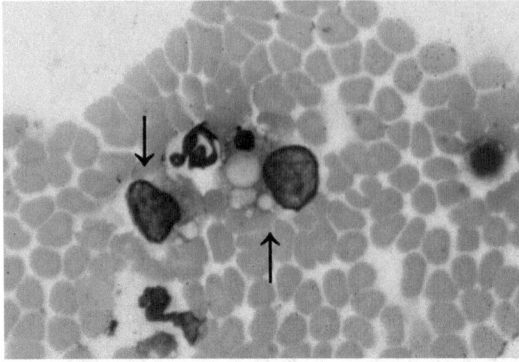

图18-3 感染犬艾立希体犬的血液涂片

注意反应性单核细胞（箭头）、红细胞吞噬、核和血小板吞噬。图像来源见英文。

Figure 18-3 Blood smear of a dog infected with *Ehrlichia canis*

Note the reactive monocytes (arrows), erythrophagocytosis, phagocytosis of nuclear material and megaplatelets. Source: Harrus S, Waner T. 2011. Diagnosis of canine monocytotropic ehrlichiosis (*Ehrlichia canis*): an overview. Vet J, 187 (3): 292-296.

彩图

反刍动物艾立希体引起反刍动物的心水病（**图18-4**），山羊和绵羊比牛更易感。

Ehrlichia ruminantium causes heartwater disease in ruminants (**Figure18-4**). Goats and sheep are more susceptible than cattle.

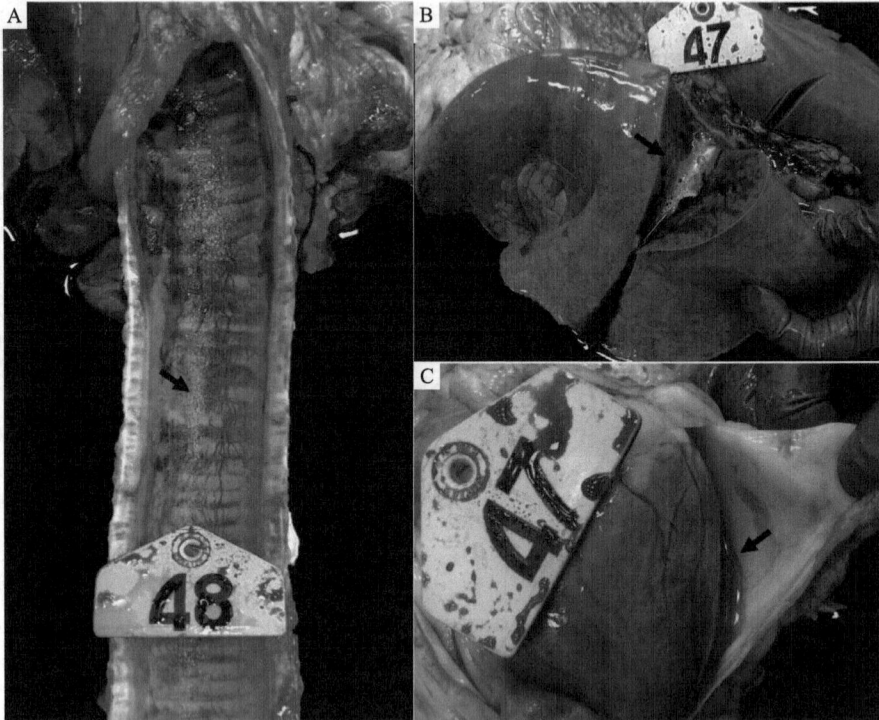

图18-4 感染反刍动物艾立希体杜泊羊的大体病变

感染反刍动物艾立希体第28天后采集的绵羊组织样本。A. 气管：黏膜下充血和出血（箭头）；B. 肺：弥漫性严重充血和水肿。注意从切部渗出的大量泡沫液体（箭头）；C. 心脏：心包积液（箭头）。图像来源见英文。

Figure 18-4 Gross pathological lesions observed in White Dorper breed sheep infected with *E. ruminantium*

Tissue samples collected on day 28 post infection were assessed in sheep infected with *E. ruminantium*. A. Trachea, submucosal congestion and hemorrhage (arrow); B. Lung, diffuse and severe congestion and edema. Note the copious amounts of foamy fluid oozing from the cut section (arrow); C. Heart, pericardial serosanguineous fluid (arrow). Source: Nair A, Hove P, Liu H, et al. 2021. Experimental infection of North American sheep with *Ehrlichia ruminantium*. Pathogens, 10 (4): 451.

第2节 衣 原 体
Section 2 *Chlamydia*

衣原体是一类严格细胞内寄生，具有独特发育周期、形成包涵体样结构的革兰氏阴性原核细胞型微生物，以二分裂繁殖，可通过0.45 μm孔径的细菌滤器，是一类介于立克次体和病毒之间的微生物，能引起人和家畜的衣原体病。

生物学特性：衣原体呈圆形或椭圆形，大小0.2～1.0 μm，只能在细胞内生长和繁殖，类似立克次体。它们具有像其他革兰氏阴性菌一样的细胞壁，并具有核糖体、DNA和RNA。细胞壁含有肽聚糖、LPS和主要外膜蛋白。常用吉姆萨染色法对衣原体进行染色。衣原体主要通过尘埃、飞沫及接触传播，感染宿主的上皮细胞和黏膜。

分类：衣原体目，有4个科：衣原体科、副衣原体科、西氏衣原体科和沃氏衣原体科。衣原体科有2个属，衣原体属和嗜衣原体属。衣原体属包括沙眼衣原体、小鼠衣原体和猪衣原体。嗜衣原体属包括山鸡衣原体、肺炎衣原体、鹦鹉热衣原体、流产衣原体等。

发育周期：衣原体具有独特的发育周期，其独特之处在于在发育过程中可以形成元体和网状体两种形式，分别缩写为EB和RB（**图18-5**）。

元体：呈卵圆形，小而致密，大小0.2～0.4 μm，有细胞壁。元体是发育成熟的衣原体，具有感染性，通过受体介导的吞饮作用进入易感细胞，宿主细胞膜包围元体形成空泡，感染宿主细胞后6～36 h，空泡内的元体增大，发育成为大而疏松无细胞壁的结构，称为网状体。

网状体：又称始体，呈圆形或椭圆形，比元体大，为0.6～1.0 μm。网状体无细胞壁，在细胞空泡中以二分裂方式进行

Chlamydia is an obligate intracellular Gram-negative single-cell prokaryotic microorganism that has a unique developmental cycle, and forms inclusion body-like structures. It reproduces in binary fission, and can pass through the bacterial filter with a pore size of 0.45 μm. It is a microorganism similar to both *Rickettsia* and viruses, and is able to cause chlamydiosis in humans and livestock.

Biological characteristics: members of the order Chlamydiales are round or oval, with a size ranging from 0.2 μm to 1.0 μm. Similar to *Rickettsia*, they can grow and propagate only inside host cells. They all have a cell wall like that of the most other Gram-negative organisms and possess ribosomes, DNA and RNA. The cell wall consists of peptidoglycan, lipopolysaccharides and major outer membrane proteins. Giemsa staining is used for these organisms. They are transmitted primarily by dust particles and droplets, as well as by contact, and infect the host epithelial cells and mucous membranes.

Taxonomy: the order *Chlamydiales* contains 4 families of *Chlamydiaceae*, *Parachlamydiaceae*, *Simkaniaceae* and *Waddliaceae*. The family *Chlamydiaceae* has two genera: *Chlamydia* and *Chlamydophila*. The *Chlamydia* species include *C. trachomatis*, *C. muridarum* and *C. suis*. The *Chlamydophila* species are composed of *C. pecorum*, *C. pneumoniae*, *C. psittaci*, *C. abortus*, etc.

Life cycle: *Chlamydia* has a unique developmental cycle. In the process of its development, it forms elementary bodies and reticular bodies, abbreviated as EB and RB, respectively (**Figure 18-5**).

Elementary bodies are oval, small and dense with a diameter of 0.2-0.4 μm and have cell walls. In fact, they are fully developed *Chlamydia* that are infectious and can enter susceptible cells through receptor-mediated endocytosis. The host cell membrane surrounds EBs to form vacuoles. About 6-36 h after infection, the EBs will become a large, loose structure without cell wall. This structure is called as reticular body.

Reticular bodies are also called as initial bodies. They are round or oval, larger in size (0.6-1.0 μm) compared to EBs. The RBs replicate by binary fission in the vacuoles, which later transform into a large number

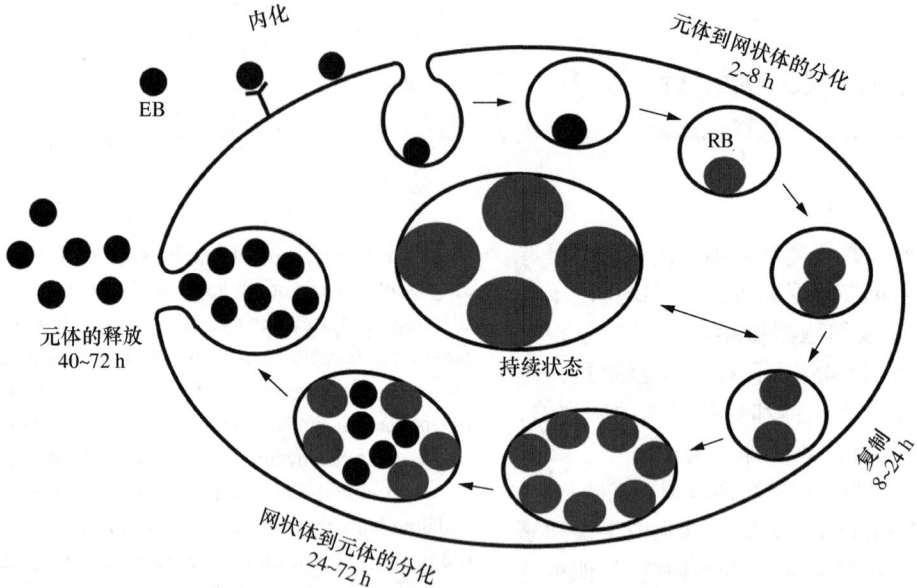

图18-5　衣原体的发育周期

当感染性元体（EB）吸附宿主细胞表面的硫酸乙酰肝素蛋白多糖和宿主细胞受体时，感染开始。EB被宿主膜衍生的包涵体内化。在包涵体中，EB分化为网状体（RB），并通过二分裂法复制。在发育后期RB以非同步方式分化为EB并通过裂解宿主细胞或包涵体释放新的感染后代。此外，为应对细胞应激，RB可进入可逆的持续状态。图像来源见英文。

Figure 18-5　*Chlamydia* developmental cycle

Infection is initiated when the infectious EB attaches to host cell heparan sulfate proteoglycan and host cell receptors at the host cell surface. The EB is internalized into a host membrane-derived inclusion. Within the inclusion the EB differentiates into RB that replicates via binary fission. Late in the developmental cycle, RBs differentiate to EBs in an asynchronous manner and new infectious progeny are released via host cell lysis or extrusion of the inclusion. Alternatively, in response to cellular stress the RB can enter a reversible state of persistence. Source: Christensen S, McMahon RM, Martin JL, et al. 2019. Life inside and out: making and breaking protein disulfide bonds in *Chlamydia*. Crit Rev Microbiol, 45 (1): 33-50.

繁殖，形成大量子代元体，从而形成各种形态的包涵体。子代元体通过宿主细胞裂解或胞吐释出，再感染新的易感细胞，开始新的发育周期。网状体是衣原体的一种繁殖型个体形态，无感染性。

　　包涵体样结构是衣原体在细胞空泡内所形成的集团形态，内含无数子代元体和正在分裂的网状体。繁殖过程中形成的包涵体不断成熟增大，最终导致宿主细胞裂解。

　　鹦鹉热衣原体在鹦鹉类禽鸟中引起鹦鹉热，在其他鸟类则称为鸟疫。鸟类和哺乳动物为其天然宿主，有8个血清型，可致畜禽肺炎、流产、关节炎等，偶致人的肺炎。

　　牛羊衣原体引起牛和绵羊的多发性关节炎、脑脊髓炎和腹泻（**图18-6，图18-7**）。

of EBs, forming various types of inclusion bodies. The progeny EBs are released after complete lysis of the infected host cells or by exocytosis, then infect new susceptible cells and start a new development cycle. The RBs are the replicating form of *Chlamydia* and do not have infectivity.

　　Inclusion body-like structure is the group form of *Chlamydia* formed in cell vacuoles, which contains numerous elementary bodies and dividing reticular bodies. The inclusion bodies formed during the reproduction process continue to mature and increase in size, eventually leading to lysis of the host cells.

　　Chlamydia psittaci causes infections in birds, referred to as parrot fever in parrots, and ornithosis in all other bird species. Birds and mammals are their natural hosts. There are 8 serotypes of *C. psittaci*, which cause pneumonia, miscarriage, arthritis in livestock and poultry, and occasionally cause pneumonia in humans.

　　Chlamydia pecorum causes polyarthritis, encephalomyelitis and diarrhea in cattle and sheep (**Figure 18-6, Figure 18-7**).

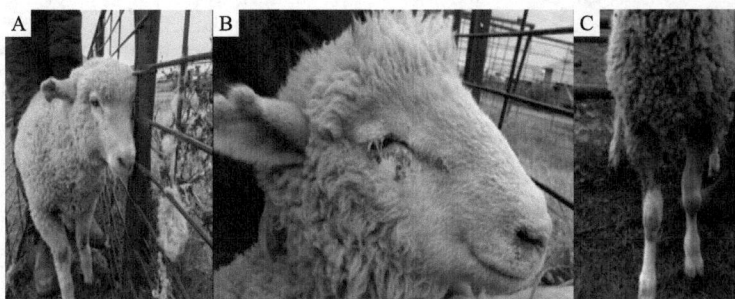

图18-6　衣原体相关性关节炎病例的临床特征

A. 羔羊严重抑郁，站立需要辅助；B. 并发双侧结膜炎；C. 膝关节肿胀。图像来源见英文。

Figure 18-6　Clinical features of *Chlamydia pecorum* associated arthritis

A. Affected lambs demonstrating severe depression and requiring assistance to stand; B. Concurrent bilateral conjunctivitis; C. Carpal swelling. Source: Walker E, Moore C, Shearer P, et al. 2016. Clinical, diagnostic and pathologic features of presumptive cases of *Chlamydia pecorum*-associated arthritis in Australian sheep flocks. BMC Vet Res, 12 (1): 193.

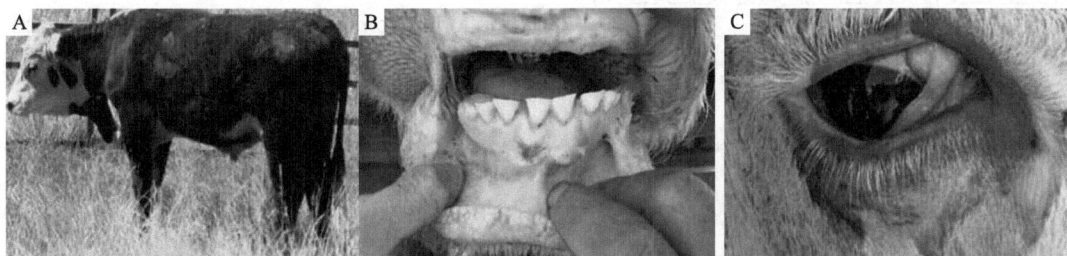

图18-7　确诊肉牛脑脊髓炎病例临床症状

A. 精神沉郁，外表僵硬和蜷缩，下颌骨、四肢和关节水肿；B. 口腔黏膜苍白；C. 眼结膜苍白，有泪痕溢出。图像来源见英文。

Figure 18-7　Clinical signs observed in confirmed cases of bovine encephalomyelitis in beef cattle

A. Depression, stiffness and tucked-up appearance. Dependent oedema of mandible, limbs and joints; B. Pale mucus membranes; C. Pale conjunctiva and epiphora. Source: Walker E, Lee EJ, Timms P, et al. 2015. *Chlamydia pecorum* infections in sheep and cattle: a common and under-recognised infectious disease with significant impact on animal health. Vet J, 206 (3): 252-260.

思考题　Questions

1. 立克次体有哪些特点？对人或动物致病的立克次体主要有哪些？

1. What are the characteristics of *Rickettsia*? Which *Rickettsiae* spp. cause diseases in humans or animals?

2. 衣原体是怎样完成其发育周期的？

2. How does *Chlamydia* complete its life cycle?

3. 简述鹦鹉热的病原及其主要特点。

3. Please describe the causal pathogen and main characteristics of psittacosis.

第19章 病毒的结构和分类
Chapter 19 / Structure and taxonomy of viruses

内容提要 病毒是一类结构简单，只能在人、动物、植物或细菌等生物活细胞内复制并具有侵染性的非细胞生物体。病毒核酸类型、复制方式和囊膜存在与否是病毒分类的主要依据。国际病毒分类委员会负责病毒的分类和命名。

Introduction Viruses are a class of acellular infectious agents that are small in size and simple in structure, and can multiply only in living cells of humans, animals, plants and bacteria. Genome type, replication strategy and presence of an envelope are used as the principal basis of virus classification. The International Committee on Taxonomy of Viruses (ICTV) authorizes and organizes the taxonomic classification and the nomenclatures for viruses.

第1节 病毒的结构特征
Section 1　Structural characteristics of viruses

病毒一般以病毒颗粒或病毒粒子的形式存在，具有一定的形态、结构和功能。病毒直径的测量单位为纳米，范围在20～400 nm之间，大多小于150 nm，必须借助电子显微镜才能观察到。最大的脊椎动物病毒为痘病毒，约300 nm，最小的为圆环病毒，仅17 nm。病毒的形态有多种，动物病毒多呈球状或近似球形，植物病毒多呈杆状，少数为杆状丝状（丝状病毒）、子弹状（弹状病毒）、蝌蚪状（噬菌体）或砖块状（痘病毒）。有些病毒表现为多形性，如副黏病毒和冠状病毒（**图19-1**）。

病毒是一类个体微小、结构简单、严格细胞内寄生、只含一种核酸（DNA或RNA）、能够自我复制并具有侵染性的非细胞生物体。

Viruses generally exist as viral particles or virions and possess certain morphology, structure, and functions. The diameter of viruses is in nanometers ranging from 20 nm to 400 nm, mostly less than 150 nm. Because of the small size, they can be visualized only under electron microscope. The known largest vertebrate virus is poxvirus of about 300 nm, whereas the smallest is circovirus of some 17 nm. Viruses have diverse morphology. Most animal viruses appear spherical or nearly spherical, plant viruses appear to be rod-shaped, and a few are rod-shaped (filovirus), bullet-shaped (rhabdovirus), tadpole-shaped (bacteriophage), or brick-shaped (poxvirus). Some viruses can exist as pleomorphic, such as paramyxoviruses and coronaviruses (**Figure 19-1**).

Viruses are acellular infectious agents that are tiny in size and simple in structure, contain only one type of nucleic acids (DNA or RNA), and rely exclusively on intracellular infection for their productive replication.

病毒的天然形态和分类

病毒大小为20~400 nm，由核酸和蛋白衣壳组成；部分病毒有囊膜。病毒只有一种核酸，DNA或RNA。

· 病毒复制依赖宿主细胞。
· 某些病毒对宿主细胞具有专嗜性。
· 某些病毒在环境中稳定，也有些病毒对热、干燥、去污剂和消毒剂敏感。

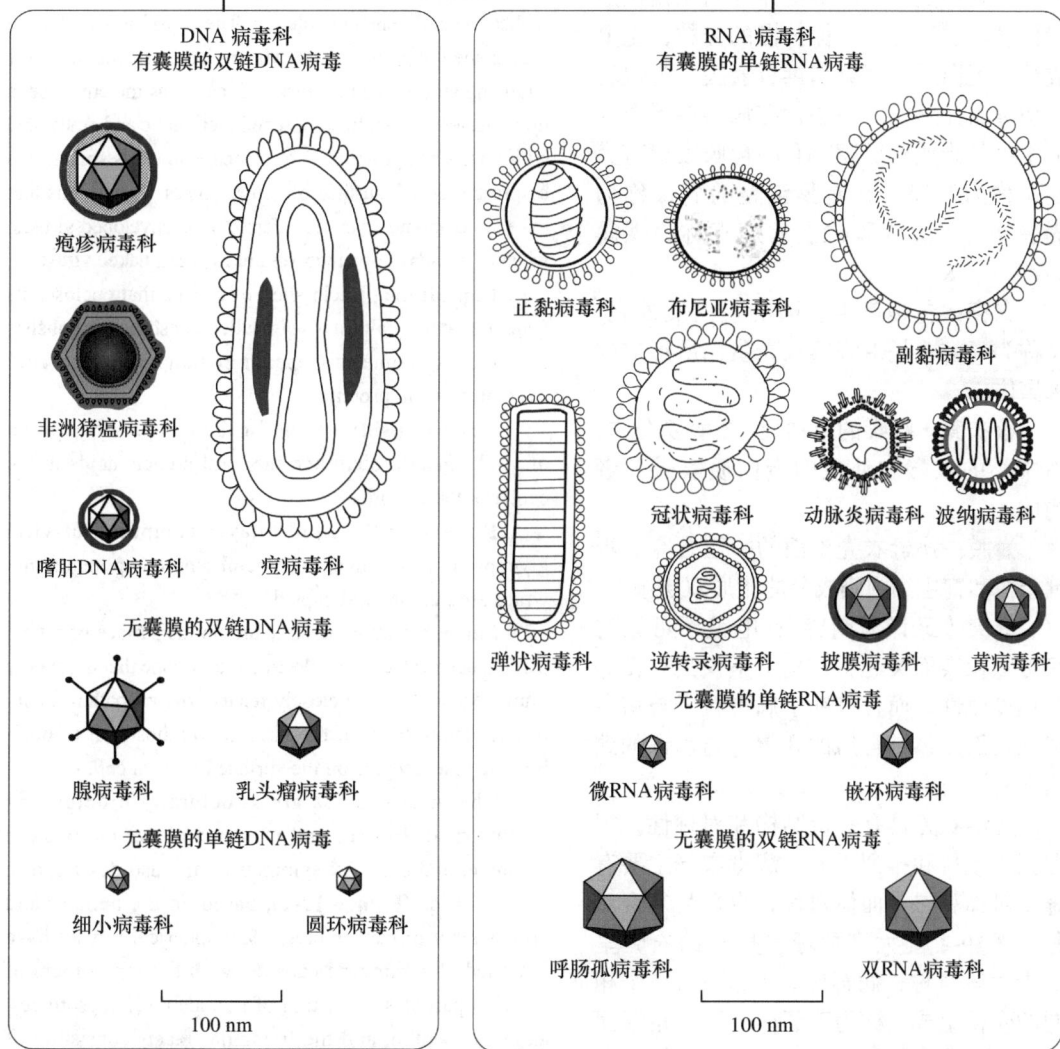

DNA 病毒科
有囊膜的双链DNA病毒

疱疹病毒科

非洲猪瘟病毒科

嗜肝DNA病毒科

痘病毒科

无囊膜的双链DNA病毒

腺病毒科

乳头瘤病毒科

无囊膜的单链DNA病毒

细小病毒科

圆环病毒科

100 nm

RNA 病毒科
有囊膜的单链RNA病毒

正黏病毒科

布尼亚病毒科

副黏病毒科

冠状病毒科

动脉炎病毒科

波纳病毒科

弹状病毒科

逆转录病毒科

披膜病毒科

黄病毒科

无囊膜的单链RNA病毒

微RNA病毒科

嵌杯病毒科

无囊膜的双链RNA病毒

呼肠孤病毒科

双RNA病毒科

100 nm

图19-1 常见的病毒及其相对大小

注：病毒核酸未按比例绘制。图像来源见英文。

Figure 19-1 Most common viruses with their relative size

The nucleic acids are not to scale. Source: Quinn PJ, Markey BK, Leonard FC, et al. 2016. Concise Review of Veterinary Microbiology, 2nd ed. New York: John Wiley & Sons.

病毒粒子是指一个形态和结构上完整的成熟病毒个体，又称病毒颗粒。无囊膜的病毒，核衣壳就是病毒完整的成熟个体；有囊膜的病毒，在核衣壳外包裹囊膜后才成为病毒粒子。

病毒粒子的基本构造为核衣壳，由核酸和衣壳组成。核酸构成病毒粒子的芯髓，为病毒复制、遗传变异等功能提供遗传信息。衣壳是一层保护性的壳粒，主要成分是蛋白质。部分病毒有囊膜，为包裹在病毒衣壳外面的一层含有病毒糖蛋白和宿主细胞脂质双层膜。有的囊膜表面有突起，称为纤突。有囊膜包裹的病毒称为囊膜病毒，无囊膜包裹的病毒称为裸露病毒。

衣壳： 包裹病毒核酸的蛋白质外壳，又称为壳粒，是由病毒衣壳蛋白所形成的寡聚体。

核衣壳： 病毒的蛋白质-核酸复合体，由中心的核酸和外包的蛋白质衣壳两者构成。

囊膜： 病毒衣壳外面的一层含有病毒糖蛋白和宿主细胞膜成分的脂质双层膜。

纤突（又译为刺突）： 电镜下可见的病毒颗粒表面突起状结构，不仅具有良好的免疫原性，而且有助于病毒粒子与宿主细胞表面受体结合后的入侵，与病毒的致病性密切相关。

病毒衣壳具有一定结构和对称性。根据壳粒数目和排列不同，病毒衣壳主要有螺旋对称和二十面体对称，少数为复合对称（**图19-2**）。大多数球状病毒的衣壳呈二十面体对称，病毒颗粒的顶角由5个相同的壳粒组成，称为五邻体；而三角形面由6个相同的壳粒组成，称为六邻体。复合对称的壳粒排列既有螺旋对称又有立体对称的形式，如噬菌体和痘病毒。

Virion is a morphologically and structurally mature individual virus, also called as viral particle. For non-enveloped viruses, the nucleocapsid is a mature virion; for enveloped viruses, the nucleocapsid is further enveloped by membrane to become a mature virion.

Nucleocapsid is the basic structure of viruses and is composed of nucleic acids and a capsid protein. The nucleic acids constitute the core of a virion and carry genetic information in the form of genomes important for virus replication and genetic evolution. The capsid is a protective coat or shell of protein. Some viruses have a membrane layer wrapping around the nucleocapsid, called as the envelope, a lipid bilayer derived from host cell membrane and embedded with viral glycoproteins. Protrusions, named as spikes, can be present on the surface of some viruses. The virions that have a membrane layer are referred to as enveloped viruses, whereas non-enveloped viruses are called as naked viruses.

Capsid: the protein shell of a virus that encloses its genetic material, alternatively called capsomer. It consists of several oligomeric (repeating) subunits of the viral structural capsid protein.

Nucleocapsid: a term for the complex composed of viral capsid protein and the viral nucleic acids in the center of the capsid.

Envelope: the lipid bilayer composed of viral glycoproteins and host cell membrane components that wraps around the viral capsid.

Spike: a protruding structure on the surface of a viral particle observable under electron microscope that is not only immunogenic but also closely related to viral pathogenicity by facilitating entry of the virion into a host cell through binding to a receptor on the surface of a host cell.

The viral capsids are structurally ordered and symmetrical. Helical and icosahedral symmetries are common, and complex symmetries are also described in some viruses (**Figure 19-2**), based on the number and arrangement of capsomeres. Most spherical viruses have icosahedral symmetric capsids, with the top corners of the viral particles consisting of five identical capsomeres, called as penton, and the triangular facets consisting of six identical capsomeres, called as hexon. The complex symmetric shell arrangement has both helical and cubical symmetries, such as phages and poxviruses.

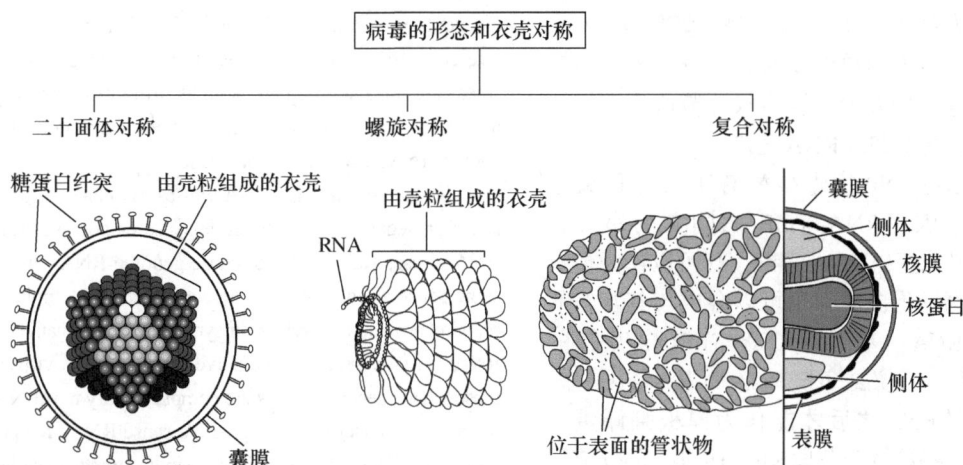

图19-2 病毒衣壳的对称性

完整的病毒的粒子具有二十面体对称性，这种对称结构为特定表面积的病毒提供了最大承载力和强度。一些RNA病毒具有保护性的螺旋对称衣壳，是通过在螺旋状核酸的每一圈之间插入蛋白质亚单位形成。核衣壳由脂双层和糖蛋白组成的囊膜包裹，当核衣壳通过细胞膜出芽时获得囊膜。病毒蛋白由核酸编码，并通过宿主细胞的分区机制以糖蛋白的形式整合到膜中，是病毒囊膜的组成部分；膜粒或纤突是位于病毒囊膜上的球状蛋白突起。图像来源见英文。

Figure 19-2 The capsid symmetry of viruses

Closed-shell, isometric viruses have a structure based on icosahedral symmetry. This structural form offers the maximum capacity and greatest strength for a given surface area. The protective, helical capsid of many RNA viruses is formed by the insertion of protein subunits between each turn of the nucleic acid helix. In many types of viruses, the nucleocapsid is covered by an envelope composed of a lipid bilayer and associated glycoproteins. The envelope is acquired when the nucleocapsid buds through a membrane of the cell. Proteins, encoded by viral nucleic acid and integrated as glycoprotein into the appropriate membrane by the compartmentalization mechanisms of the host cell, are an integral part of the viral envelope. Peplomers or spikes are knob-like projections on the envelope of certain viruses. Source: Quinn PJ, Markey BK, Leonard FC, et al. 2016. Concise Review of Veterinary Microbiology, 2nd ed. New York: John Wiley & Sons.

第2节 病毒的化学组成
Section 2 Chemical composition of viruses

病毒的化学组成有核酸、蛋白质、脂类和糖，前两种为最主要成分。脂类主要存在于病毒囊膜，少数无囊膜的病毒也存在脂类，主要成分是磷脂，其次是胆固醇。常使用乙醚或氯仿溶解囊膜中的脂质，使部分病毒失活。糖类一般以糖蛋白的形式存在，是某些病毒囊膜或纤突的成分，与病毒吸附细胞受体有关。

病毒携带的核酸只有一种，DNA或RNA。核酸可分为单链或双链、线状或环状、分节段或不分节段。DNA病毒的核酸

The major chemical compositions of viruses include nucleic acids, proteins, lipids, and sugars. The lipids are mainly present in the virus envelope. In some non-enveloped viruses, there are also lipid components, mainly phospholipids, followed by cholesterol. As a result, some chemicals such as chloroform and ethyl ether are often used to decompose lipids, leading to inactivation of the viruses. Sugars, generally in the form of glycoproteins, are components of the envelope proteins or spikes of some viruses that are associated with viral adhesion to cellular receptors.

The nucleic acids of viruses can be DNA or RNA, single-or double-stranded, linear or circular, segmented or non-segmented. Most of the nucleic acids of DNA viruses

大多为双链、线状，如疱疹病毒。RNA病毒的核酸多为单链线状、不分节段，如正黏病毒。少数为双链线状分节段病毒，如呼肠孤病毒和双RNA病毒。

单链RNA病毒核酸有正义和负义之分，基因组RNA与mRNA同义的病毒称为正义或正链RNA病毒，既可以作为模板直接翻译蛋白质，也可以合成与其互补的负链RNA；基因组RNA与mRNA互补的病毒称为反义或负链RNA病毒，需要先合成正链RNA之后才能作为模板翻译蛋白质。有些单链分节段RNA病毒，部分节段为负链，另一部分节段为正链，称之为双义RNA病毒。

结构蛋白是参与病毒粒子组装和形成的蛋白质，包括病毒衣壳蛋白、基质蛋白和囊膜蛋白，通常在病毒复制后期才大量合成，具有保护核酸、介导病毒黏附和入侵宿主细胞等功能。

非结构蛋白是病毒壳粒或囊膜蛋白组分之外的蛋白质，如DNA或RNA聚合酶、解旋酶等，通常参与病毒基因组复制。近年来发现一些非结构蛋白具有拮抗宿主免疫系统的功能，有利于病毒感染，称为免疫逃逸。

基质蛋白是一类连接病毒囊膜和核衣壳的病毒结构蛋白，主要参与病毒粒子的装配。

囊膜蛋白是位于病毒脂质双层的病毒膜蛋白，具有与宿主细胞表面受体结合和膜融合功能。

病毒样颗粒是只含有蛋白质不含核酸的一种特殊形式的病毒颗粒，不具有感染性。

are double-stranded and linear, such as herpes viruses. The nucleic acids of RNA viruses are mostly single-stranded, linear and non-segmented, such as orthomyxoviruses, and a few are double-stranded, linear and segmented viruses, such as reoviruses and birnaviruses.

The single-stranded RNA viruses can either be positive-sense or negative-sense. Positive-sense or positive-stranded RNA viruses refer to those whose genome RNAs are in the same sense as mRNA that can act directly as templates for protein translation as well as for synthesis of the negative-strand RNA. Negative-sense or negative-stranded RNA viruses are those whose genome RNAs are complementary to mRNA and have to be transcribed into positive-sense RNAs for protein translation. The genomes of some single-stranded segmented RNA viruses are ambisense, containing negative-sense RNA in some segments, yet positive-sense RNA in the others.

Structural proteins refer to the proteins involved in virus assembly and morphogenesis. They are often components of viral particles such as capsid proteins, matrix proteins, and membrane proteins. Typically, structural proteins are synthesized at a large scale later in viral replication and play multiple roles in protection of nucleic acids, attachment to and entry into host cells, *etc.*

Non-structural proteins are the proteins that are not involved in forming the viral capsomer or envelope but related to viral genome replication in infected cells in the form of DNA or RNA polymerases, helicases, *etc.* In recent years, some non-structural proteins have been found to possess antagonistic activity against host immune systems in favor of viral infection that is generally referred to as evasion of host immunity.

Matrix proteins are a class of viral proteins that mediate interactions of the nucleocapsids with the viral envelope important for assembly of virus particles.

Envelope proteins are the viral membrane proteins located in the lipid bilayer that are involved in binding to host cell receptors and membrane fusion.

Virus-like particles (VLPs) are multiprotein structures that mimic the organization and conformation of native viruses and are non-infectious due to lack of the viral genome.

第3节 病毒的分类
Section 3 Taxonomy of viruses

病毒分类指的是将自然界中的病毒种群，按照性质相似性和亲缘关系进行归

Virus classification is to name viruses and place them into a taxonomic system according to their shared

纳，对病毒进行命名，并将它们放在同一系统中，以便更好地了解病毒的共性和个性特点。国际病毒分类委员会（ICTV）是国际公认的病毒分类与命名的权威机构。

国际病毒分类委员会采用与细胞生物分类同样的阶元或等级对非细胞性病毒进行分类：域（realm）、界（kingdom）、门（phylum）、纲（class）、目（order）、科（family）、属（genus）、种（species）。科和属是病毒分类最主要的分类单位。种是最基本的分类单位，它是表型特征高度相似，亲缘关系极其相近，与同属内其他种有着明显差异的一组病毒。病毒是非细胞生物，其分类困难，在没有特定目的时，科是最高的病毒分类等级。

病毒分类的依据包括病毒基因组特性、亲缘关系、形态与结构、复制策略、理化特性等诸多方面，病毒全基因组测定或部分测序是目前病毒分类的主流方向。

病毒的名称不采用拉丁文双名法，采用英文或者英语化的拉丁文，只用单名。凡是被ICTV正式认定的病毒名称，其名称用斜体书写，一般通用名则为正体。目、科、属分别用拉丁文后缀 "-virales"、"-viridae" 及 "-virus"。种及以上分类，全部斜体且首字母大写。病毒的亚种、株、变异株等主要用于诊断、疫苗研发等实际应用目的，对病毒分类并不重要。

病毒的种指一组具有共同祖先的病毒，其所拥有的特性不同于其他种的病毒，可以通过多种方法进行鉴别。

毒株是具有独特和稳定表型特征的病毒。

变异株是指与参考毒株序列不同的病毒。

分离株是受感染的宿主样品在适当细胞中培养后获得的病毒。

亚病毒是一类比病毒结构更为简单、具有感染性的非细胞生物体，不具有完整的病毒结构，仅有核酸或蛋白质，主要包含类病毒、卫星病毒和朊病毒。

properties and phylogenetic relationships so as to better understand their shared and individual characteristics. The International Committee on Taxonomy of Viruses (ICTV) is the recognized authority in virus classification and nomenclature.

ICTV uses a taxonomic scheme for acellular virus similar to the classification systems used for cellular organisms at the levels of realm, kingdom, phylum, class, order, family, genus, and species. Family and genus are the principal units of classification, while species is the most basic taxon and represents a group of viruses whose properties, such as phenotypes or phylogenetic relationship, are highly similar among themselves, but distinct from other species in the same genus. As viruses are non-cellular organisms that are often hard to be classified, family is the highest classification taxon if no specific objective is defined.

The criteria for virus taxonomy include the genome characteristics, phylogenetic relationships, replication strategies, structure of the virions, physicochemical properties and so on. Sequencing or partial sequencing of the viral genome provides powerful taxonomic information to guide virus classification.

The viruses are not named by the Latin binomial method. They are given only single names in English or anglicized Latin. The names of all viruses officially recognized by ICTV are written in italics, while the general common names are in block letters. The Latin suffixes "-virales", "-viridae" and "-virus" are used for the order, family and genus, respectively. For species and higher taxa, the names are italicized, and the first letters are capitalized. Lower levels, including subspecies, strains, and variants, which are established for practical purposes such as diagnostics and vaccine development, are not important from taxonomical perspective.

A species is a monophyletic group of viruses whose properties can be distinguished from those of other species by multiple criteria.

Strain refers to a virus with unique and stable phenotypic characteristics.

Variant refers to a virus isolate whose genome sequence is different from that of the reference virus strain.

Isolate is a virus that is derived from the sample of an infected host by cultivation in an appropriate cell line.

Subvirus refers to any non-cellular infective microorganism that is simpler than viruses, lacks a complete viral structure, and contains either nucleic acids or proteins, such as viroids, satellite viruses and prions.

类病毒是一类裸露的、单链、环状闭合、不编码蛋白质、依赖宿主聚合酶进行复制的RNA分子，目前仅在植物中发现，有些具有致病性。

朊病毒是指中枢神经系统内正常细胞蛋白质PrP^C因错误折叠形成的一类具有感染性的蛋白质分子PrP^{SC}，朊病毒"prion"一词由此得名，而PrP则是朊蛋白的缩写。PrP^{SC}有感染性，能诱导与其接触的PrP^C发生错误折叠，由此积累的PrP^{SC}发生寡聚化，并形成纤维样蛋白，从而导致传染性海绵状脑病，可发生于人、绵羊、山羊和牛，统称朊病毒病。

卫星病毒是一类不编码复制相关基因，必需依赖与其共同侵染寄主细胞的辅助病毒进行复制的核酸分子，其核酸可为DNA或RNA。

辅助病毒是指能够辅助缺陷病毒（如卫星病毒），使之增殖出完整病毒的病毒。即在共感染细胞内，能为缺陷病毒提供所需要的聚合酶或其他功能蛋白，使缺陷病毒能复制产生完整的子代病毒。

Viroids are circular, single-stranded and naked RNA molecules that do not code for proteins and depend on host polymerases for duplication. So far, viroids are only found in plants and some are pathogenic.

Prion, a word derived from the term proteinaceous infectious particle with suffix "-on" denoting substance, refers to a misfolded infectious form of prion proteins PrP^{SC} from normal cellular proteins PrP^C that are rich in the central nervous system. PrP is an abbreviation of the prion protein. PrP^{SC} is infectious and induces misfolding of PrP^C molecules when they encounter. Accumulated PrP^{SC} will become oligomeric and tend to assemble into fibrils, thus leading to transmissible spongiform encephalopathy that may occur in humans, goats, sheep and cattle, generally called as prion diseases.

Satellite viruses refer to the subviral pathogens that are entirely dependent upon the replication machinery of co-infected helper viruses for gene replication and expression. Their nucleic acids can be either DNA or RNA.

Helper viruses refer to those that help the defective viruses, such as satellite viruses, to form complete virus particles, that is to help compensate for the deficiency of the defective viruses by providing polymerases or other functional proteins in the co-infected cells for them to produce progeny viruses.

思考题　Questions

1. 简述病毒核酸及蛋白质的特点。

2. 如何确定病毒是否有囊膜？

3. 病毒粒子有哪些基本的特性？

4. 病毒粒子的化学组成是什么？什么是病毒样颗粒？

5. 病毒核酸有哪些类型？

6. 简述病毒的分类系统、分类等级、分类依据。

7. 简述病毒种、变异株、分离株、毒株的概念。

8. 简述病毒的命名原则和病毒名称的书写原则。

1. Briefly describe the characteristics of nucleic acids and proteins of viruses.

2. How to determine whether a virus has an envelop or not?

3. What are the basic properties of virions?

4. What are the main chemical components of a virion? What are VLPs?

5. What are the major types of viral nucleic acids?

6. Briefly describe the virus taxonomic system, the taxon order and the classification criteria.

7. Please describe the definitions of species, variants, isolates and strains.

8. Please describe the principles for virus nomenclature and writing.

第20章 病毒的复制
Chapter 20 / Virus replication

内容提要 病毒只有侵入活细胞内才能复制，包括吸附、穿入与脱壳、生物合成、组装与释放等步骤。

Introduction Virus replication occurs only after invasion into living cells and includes stepwise processes of adsorption, penetration and uncoating, biosynthesis, assembly and release.

第1节 一步生长曲线
Section 1 One-step growth curve

病毒增殖：病毒在活细胞内，以自身基因组为模板合成新的病毒基因组，借助宿主的蛋白质翻译系统合成病毒蛋白质，再组装成完整的子代病毒颗粒，这一过程称为病毒复制。

一步生长曲线：在所有细胞同时感染病毒的"一步"感染条件下，感染后不同时间收集细胞和上清用于测定病毒数量，以时间为横坐标，以病毒数量为纵坐标绘制的曲线，为一步生长曲线，包括隐蔽期、成熟期和释放期三个时期（**图20-1**）。

病毒的复制周期：从病毒吸附和入侵宿主细胞开始，经过基因组复制和病毒蛋白质合成，至释放出子代病毒的全过程，称为一个复制周期，包括吸附、穿入与脱壳、生物合成、组装和释放5个步骤（**图20-1**）。

Multiplication: replication of viruses is a continuous process that occurs in a living cell where their own genomes are used as templates to synthesize new viral genomes and the protein translation machinery of the host cells is employed to generate viral proteins. The genomes and proteins thus produced are reassembled into complete progeny virus particles.

One-step growth curve: under "one-step condition" in which all cells are infected simultaneously, the cells and their culture supernatants at different time points after infection are then collected for titration of the virus particles. When these titers in Y-axis are plotted as a function of time in X-axis, a one-step growth curve is obtained in which three periods can be recognized: eclipse period, maturation period, and release period (**Figure 20-1**).

Replication cycle: referring to the full process that starts from virus attachment to and entry into the host cell to genome replication, virus protein synthesis and release of progeny virus particles in the end. The cycle includes five consecutive steps of adsorption, penetration and uncoating, biosynthesis, assembly, and release (**Figure 20-1**).

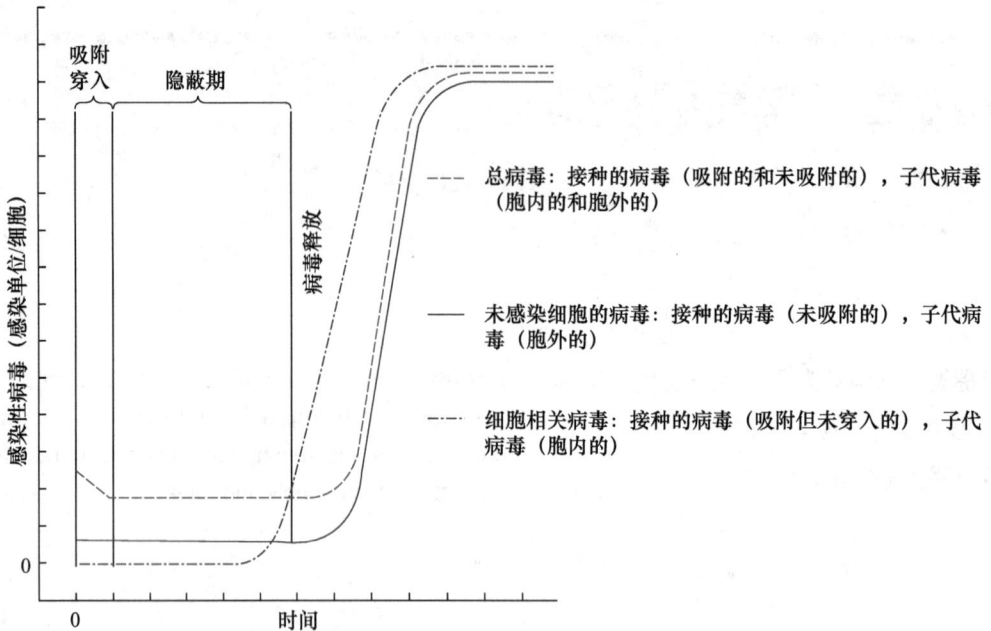

纵坐标：感染性病毒（感染单位/细胞）

横坐标：时间

图例：
- 总病毒：接种的病毒（吸附的和未吸附的），子代病毒（胞内的和胞外的）
- 未感染细胞的病毒：接种的病毒（未吸附的），子代病毒（胞外的）
- 细胞相关病毒：接种的病毒（吸附但未穿入的），子代病毒（胞内的）

图20-1　无囊膜病毒的一步生长曲线

病毒吸附和穿入细胞之后，即进入2～12 h的隐蔽期，在此期间无法检测到与细胞相关的传染性。病毒在随后的几个小时内逐渐成熟。细胞裂解时，非囊膜病毒的病毒粒子释放通常延迟和不完全。囊膜病毒粒子的释放与细胞膜出芽成熟同时发生。图像来源见英文。

Figure 20-1　One-step growth curve of a non-enveloped virus

Attachment and penetration are followed by an eclipse period of 2-12 h during which cell-associated infectivity cannot be detected. This is followed by a period of several hours during which virus maturation occurs. Virions of non-enveloped viruses are often released late and incompletely, when the cell lyses. The release of enveloped virions occurs concurrently with maturation by budding from the plasma membrane. Source: Maclachlan NJ, Dubovi EJ, Barthold SW, et al. 2016. Fenner's Veterinary Virology. 5th ed. Amsterdam: Elsevier Inc.

第2节　吸附、穿入与脱壳
Section 2　Attachment, penetration and uncoating

吸附：吸附就是病毒与细胞表面受体结合的过程。分两个阶段，非特异性静电吸附和特异性受体吸附。前者是指病毒与细胞接触之后进行静电结合，该过程可逆。后者是指病毒表面蛋白特定位点与宿主细胞膜相应的受体结合，是病毒感染的开始（**图20-2**）。

Attachment: the process of virus binding to receptors on cells that occurs in two stages: nonspecific electrostatic attachment and specific receptor-mediated attachment. The former refers to electrostatic binding between the virus and the cell which is often reversible, and the latter to the binding between the particular region of a protein on the viral surface and the specific receptor on the host cell membrane, a critical step in initiating viral infection (**Figure 20-2**).

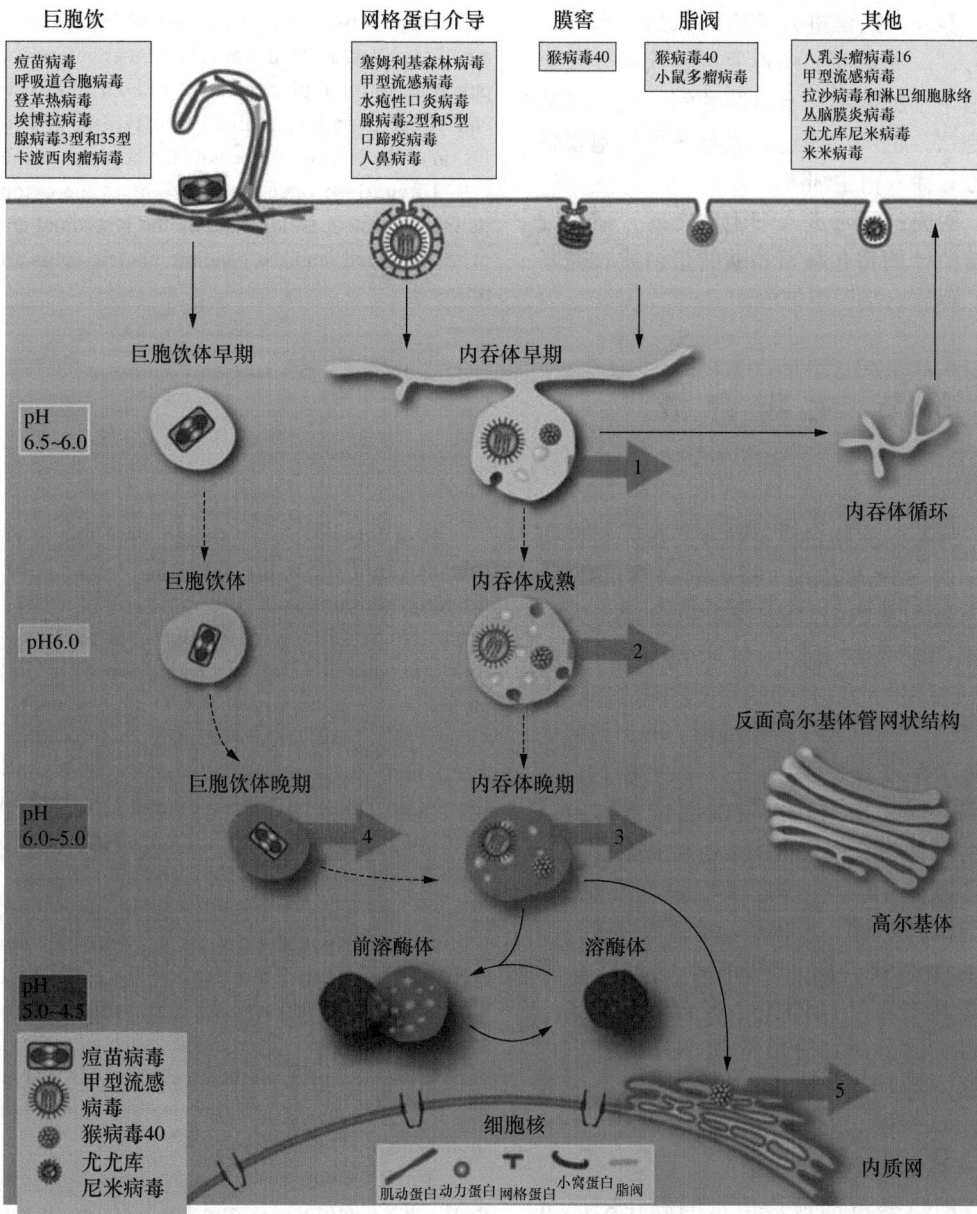

图20-2　病毒的内吞网络

病毒与细胞表面受体结合后，可通过多种途径内化：包括巨胞饮、网格蛋白介导的胞吞作用、膜窖，以及网格蛋白或膜窖蛋白独立的机制。初级内吞小泡和液泡将病毒颗粒分别运送至内吞网络不同位置并进行分类，然后进入细胞质中。其中，猴病毒40被运至内质网进行穿入和脱壳而甲型流感病毒和痘苗病毒分别在成熟和晚期内体和巨胞饮体酸性环境中发生膜融合。病毒逃逸的位置和时间通常由病毒颗粒穿入的pH阈值决定。图像来源见英文。

Figure 20-2　Viruses in the endocytic network

After binding to cell-surface receptors, viruses are internalized through a variety of endocytic processes including macropinocytosis, clathrin-mediated endocytosis, caveolae, and clathrin- or caveolin-independent mechanisms. The primary endocytic vesicles and vacuoles formed ferry incoming virus particles into the endocytic network, where they undergo sorting and eventually penetration into the cytosol from different locations within the vacuolar network. Of these, SV40 is transported to the endoplasmic reticulum for uncoating and penetration, whereas influenza A and vaccinia virus undergo acid-activated membrane fusion in maturing and late endosomes and macropinosomes, respectively. The location and timing of escape are often determined by the pH threshold for penetration of the virus particles. Source: Cossart P, Helenius A. 2014. Endocytosis of viruses and bacteria. Cold Spring Harb Perspect Biol, 6 (8): a016972.

穿入：病毒进入细胞内部的过程称为"穿入"。无囊膜的病毒通过胞饮的方式穿入宿主细胞内部。有囊膜的病毒既可以通过胞饮，也可以通过囊膜与宿主细胞膜融合方式穿入宿主细胞。

脱壳：病毒进入细胞质之后，宿主细胞溶酶体酶将包裹着核酸的蛋白质衣壳降解，释放出基因组核酸。

Penetration: the process by which a virus enters a cell is called as "penetration". Non-enveloped viruses penetrate through pinocytosis, while enveloped viruses can either penetrate through pinocytosis or enter via fusion of the virus envelope with the host cell membrane.

Uncoating: once the virus enters the cytoplasm, its capsid protein is degraded by the lysosomal enzymes of the host cell, and the genomic nucleic acids are then released.

第3节　生物合成
Section 3　Biosynthesis

病毒的生物合成：病毒的基因组在宿主细胞内释放之后，将启动病毒mRNA、蛋白质和基因组DNA或RNA合成。大多数DNA病毒的复制（包括mRNA转录和DNA复制）与真核细胞相似。基因组较大的双链DNA病毒（如痘病毒、非洲猪瘟病毒等）在细胞质中复制，可以利用其自身编码的DNA聚合酶以及DNA依赖的RNA聚合酶实现基因组DNA复制和mRNA合成。疱疹病毒、腺病毒等双链DNA病毒在细胞核内复制，其基因组复制依赖自身编码的DNA聚合酶，但mRNA合成需要宿主细胞DNA依赖的RNA聚合酶Ⅱ。细小病毒、圆环病毒等单链DNA病毒在核内复制，需要依赖宿主DNA聚合酶合成双链DNA，然后利用宿主细胞DNA依赖的RNA聚合酶Ⅱ合成mRNA。

RNA病毒的遗传信息编码在RNA分子中，而且RNA病毒又有单链或双链、正义（链）或反义（链）、单片段或分节段之分，基因组复制和mRNA转录的差别较大。单链正链RNA病毒的基因组RNA可以直接作为mRNA用于蛋白质翻译。单链负链及双链RNA病毒必须先转录合成mRNA，而且由于真核细胞没有RNA依赖的RNA聚合酶，单链负链及双链RNA病毒必须自身编码单链RNA或双链RNA依赖的RNA聚合酶。

Biosynthesis: once the genome of the virus is released into the cytoplasm, the synthesis of viral mRNAs, proteins and genomic RNA or DNA starts. Most of the DNA viruses perform replication of their genomic DNA and transcription of the mRNAs similar to the eukaryotic cells. Double-stranded DNA viruses that have large genomes, such as poxviruses and asfaviruses, replicate in the cytoplasm and utilize their own DNA polymerases and DNA-dependent RNA polymerases for synthesis of genomic DNA and mRNA, respectively. Other double-stranded DNA viruses, such as herpesviruses and adenoviruses, replicate their genomes in the nucleus by their own DNA polymerases, but depend on cellular DNA-dependent RNA polymerase Ⅱ to synthesize mRNA. The single-stranded DNA viruses, such as parvoviruses, circoviruses, *etc.*, rely on both cellular DNA polymerase and DNA-dependent RNA polymerase Ⅱ to synthesize their genomic DNA and mRNA.

The RNA viruses contain their genetic information in the RNA molecules and have different types of genomes, single-stranded or double-stranded, positive-sense or negative-sense, monopartite or segmented, thus having different strategies for production of mRNA. For the positive-sense, single-stranded RNA viruses, the genome RNA could be directly employed for translation. In negative-sense, single-stranded RNA viruses or double-stranded RNA viruses, viral RNA must first be transcribed into mRNA. Furthermore, these viruses must carry their own RNA-dependent RNA polymerases because the eukaryotic cells do not contain such type of enzymes.

单顺反子病毒mRNA的翻译与真核细胞一样，但也有部分病毒存在多顺反子mRNA，可通过多种方式翻译成各种病毒蛋白。

Monocistronic viral mRNAs are translated into proteins in the same fashion as cellular mRNAs. However, there are polycistronic mRNAs in some viruses that are translated in a number of ways into different functional proteins.

第4节 组装与释放
Section 4　Assembly and release

组装：mRNA的合成和蛋白质的翻译几乎是同步进行的。蛋白质的翻译是先有非结构蛋白后有结构蛋白。当蛋白质翻译完毕之后，病毒核酸与衣壳和非结构蛋白等一起装配形成病毒子。绝大多数DNA病毒均在细胞核内组装，RNA病毒和痘病毒在细胞质内组装。

释放：病毒粒子在细胞内装配完成之后，转移到细胞外的过程，称为释放。绝大多数无囊膜病毒释放时破坏宿主细胞膜，使宿主细胞迅速死亡；绝大多数有囊膜病毒成熟后以出芽方式成熟，释出时利用宿主细胞核膜、内质网膜或细胞膜形成病毒的囊膜，宿主细胞逐渐走向死亡。

Assembly: synthesis of mRNAs and translation of proteins are almost simultaneous. Protein translation starts first with nonstructural proteins and then with structural proteins. The newly translated capsids and other proteins as well as viral nucleic acids are assembled together to form virions. Most DNA viruses are assembled in the nucleus. RNA viruses and poxviruses are assembled in the cytoplasm.

Release: after virus particles are assembled in host cells, they will be transferred to the outside of the cell. This process is called release. Most non-enveloped viruses are released by destroying the host cell membrane, leading to rapid cell death. Most mature enveloped viruses are released via budding through which the viruses obtain host cell nuclear membrane, endoplasmic reticulum membrane or cell membrane as part of the viral envelope, and the cells gradually die.

思考题　Questions

1. 简述病毒的一步生长曲线。

2. 简述囊膜病毒的释放方式。

1. Please describe one-step viral growth curve.

2. Please describe how the enveloped viruses are released.

第21章 病毒的变异和演化
Chapter 21 / Virus variation and evolution

内容提要 病毒的变异是病毒的遗传物质发生改变，包括突变、基因重组和病毒基因产物间的相互作用三个方面，诱变多指人工方法有目的地进行基因突变。

Introduction The variation of a virus is a change in the genetic material. Variations of the virus include mutation, mutagenesis, recombination and interactions among viral products. Mutagenesis usually refers to artificial induction of mutations for some specific purposes.

第1节 突 变
Section 1 Mutation

突变是指基因组中核苷酸序列的变化，可以是一个核苷酸的点突变，也可为多个核苷酸的缺失或插入。此类突变可引起编码蛋白及其可观察的功能变化，即表型变化。

缺损型干扰颗粒（DIP）是一种缺失突变的病毒，自身不能复制，只有在亲本野生株作为辅助病毒存在时才能复制，但又干扰亲本病毒的复制，导致后者数量减少（**图21-1**）。

由于DIP对病毒复制酶的亲和性更高，且DIP颗粒基因组小，复制快。当亲本病毒和DIP共同存在时，DIP的数量会越来越多。

病毒准种：包括冠状病毒在内的RNA病毒，由于其RNA聚合酶在复制时的低保真性，加之竞争和选择压力，以某种病毒基因组RNA为母序列进行复制时，可形成与之密切相关但又不完全相同的基因组序列突变的多个变异株，共同组成这种病毒的异质性群体，称为病毒准种（**图21-2**）。

Mutation refers to the alterations of nucleotides in the genome. It can be a point mutation of a single nucleotide or deletion/insertion of multiple nucleotides. Such mutations could lead to changes of the coding proteins and their functionality, referred to as observable phenotypic changes.

Defective interfering particle (DIP) is a mutated virus unable to replicate by itself. It can replicate only when the wild parent virus coexists as a helper virus. However, DIP interferes with replication of its parental virus, leading to reduced number of progeny virions **(Figure 21-1)**.

When DIP and its parental virus and DIP coexist, the number of DIP will increase. This is because DIP has higher affinity with viral replicase, and its genome is small for quick replication.

Virus quasispecies: in RNA viruses including the coronaviruses, quasispecies are defined as a heterogeneous group of closely related variants of a certain virus species with various mutations distinct from its parental genome as a result of relatively low fidelity of its RNA polymerase in genome replication, competition and selective pressure **(Figure 21-2)**.

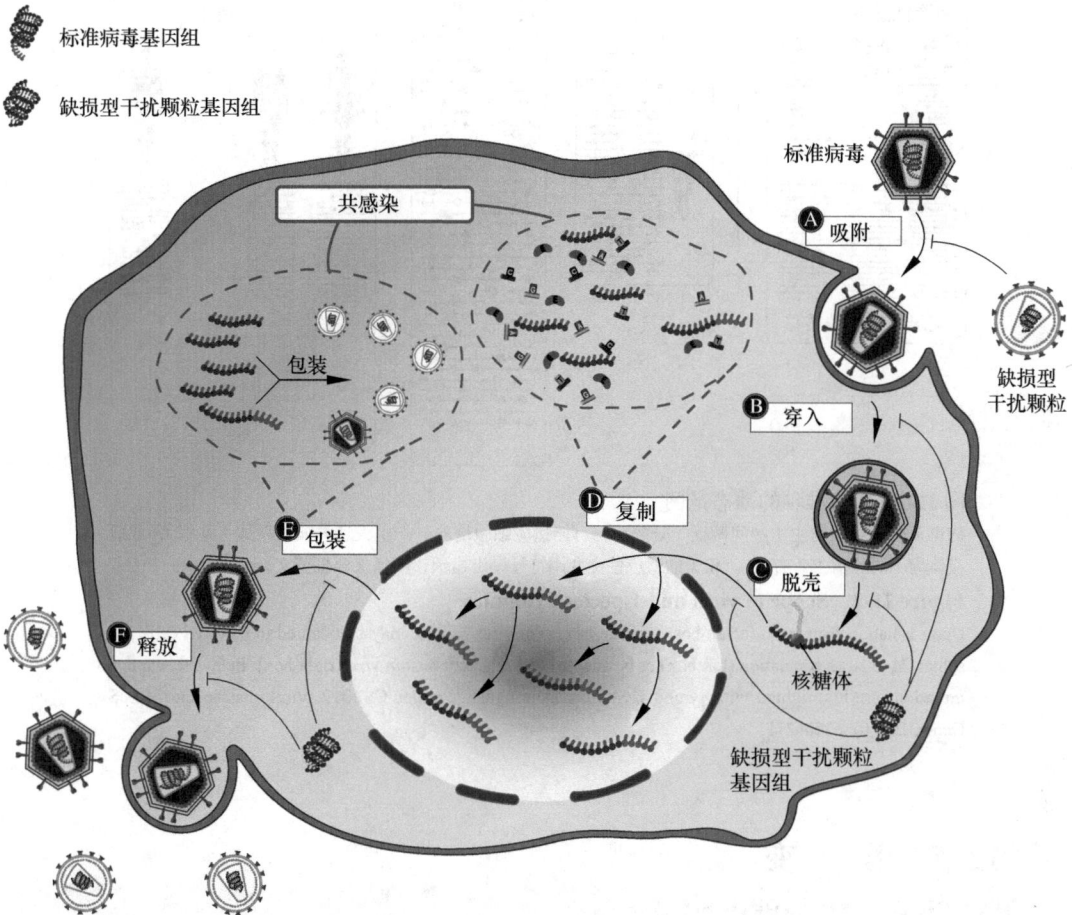

图21-1 缺损型干扰颗粒基因组抑制标准病毒（亲本病毒）的原理

与标准的亲本病毒相比，缺损型干扰颗粒基因组的关键部分缺失。平头箭头代表了缺损型干扰颗粒的抑制作用。通过与免疫系统互作，标准病毒的生命周期包括（A）吸附，（B）穿入，（C）脱壳，（D）复制，（E）包装，（F）释放等步骤受到抑制。由于缺损型干扰颗粒基因组相对较短，能够在单位时间内复制更多拷贝，数量众多的缺损型干扰颗粒基因组将竞争性结合病毒组装需要的原料，因此缺损型干扰颗粒干扰标准病毒（即亲本病毒）的复制和组装。图像来源见英文。

Figure 21-1 DIP genomes suppress the cycle of standard viruses

DIP genomes have deletion in critical parts compared to standard viruses. The flat arrows represent the inhibitory effect of DIPs. By interaction with immune system, the life cycle of standard viruses including (A) adsorption, (B) penetration, (C) uncoating, (D) replication, (E) packaging, (F) release is under suppression. With relatively shorter genes, DIP genomes are able to replicate more copies in per unit time and outnumbered DIP genomes will compete for packaging materials, and thus DIP genomes interfere with replication and packaging of standard viruses. Source: van de Sandt CE, Kreijtz JH, Rimmelzwaan GF. et al. 2019. The antiviral and antitumor effects of defective interfering particles/genomes and their mechanisms. Front Microbiol, 10: 1852.

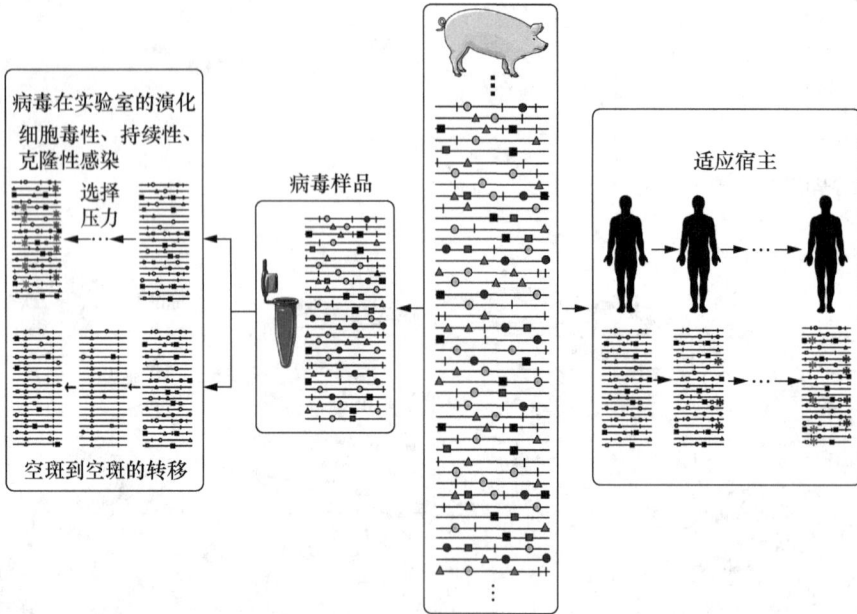

图21-2 病毒准种的动态范围

病毒从受感染的宿主（中间两框）分离后，可能适应细胞培养，产生大量异质性群体（左框），或适应体内不同的宿主（右框）。相关的适应性突变用符号突出显示。图像来源见英文。

Figure 21-2 Scope of viral quasispecies dynamics

Upon isolation from an infected host (middle boxes), a virus sample may be adapted to cultured cells and subjected to large population (left box) or be adapted to a different host *in vivo* (right box). Relevant adaptive mutations are highlighted with symbols. Source: Domingo E, Perales C. 2019. Viral quasispecies. PLoS Genet, 15 (10): e1008271.

第2节 诱 变
Section 2 Mutagenesis

定点诱变：病毒除了在自然界或宿主体内发生变异之外，也可以人为地用物理或化学方法使病毒基因组发生改变。比如，用紫外线、X射线或亚硝基胍等化学试剂处理，进行随机诱变；也可以采用PCR或基因工程方法，将病毒基因组任何既定部位的核苷酸替换，或使之缺失，或插入另一段核苷酸，称为定点诱变。

Site-directed mutagenesis: in addition to spontaneous mutations in nature or within the host, the viral genomes can also be randomly changed by physical or chemical methods, such as ultraviolet rays, X-rays or nitrosoguanidine and other chemicals. The nucleotides in any given location of the viral genome can be replaced, deleted or inserted with some other nucleotides by PCR and genetic engineering, hence the term site-directed mutagenesis.

第3节 基因重组
Section 3 Recombination

基因重组是指两种不同的病毒或同一病毒的两个不同毒株同时感染同一细胞时，在复制过程中发生核苷酸的交换，包括分子内重组、重配和激活的现象（**图21-3**）。

Recombination refers to the nucleotide exchange processes including intramolecular recombination, reassortment and reactivation between two different viruses or two different strains of the same virus when they infect the same cell and replicate simultaneously (**Figure 21-3**).

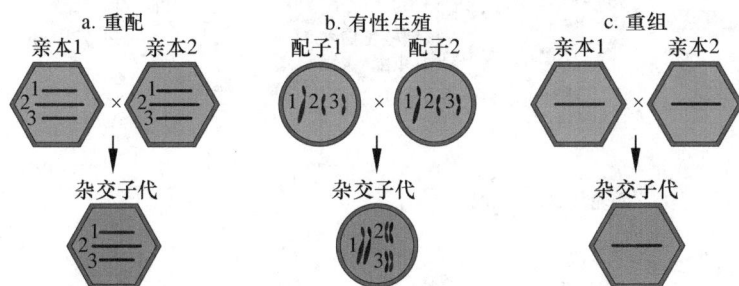

图21-3 重配、有性生殖和重组

a. 分节段RNA病毒重配。两个病毒粒子，每个含三个完整的基因组片段；重配后，形成包含来自双亲部分片段的杂交后代；b. 有性生殖。两个亲本配子细胞，每个细胞含三个染色体片段组成的单倍体基因组。在双亲交配后，产生包含来自双亲每条染色体副本的杂交二倍体后代；c. 非节段单链RNA病毒重组。两个病毒粒子之间重组后，产生了来自亲本基因组成的嵌合基因组。图像来源见英文。

Figure 21-3 Reassortment, sexual reproduction and recombination

a. Reassortment in segmented RNA viruses. Two virus particles are shown, each with a full complement of three viral genome segments. Following reassortment, hybrid progeny can be formed that contain segments derived from both parents; b. Sexual reproduction. Two parent gamete cells are shown, each with a haploid genome of three chromosomal segments. Following sex between the two parents, a hybrid diploid progeny is produced that contains one copy of each chromosome from each parent; c. Recombination in non-segmented, single-stranded RNA viruses. Following recombination between two virus particles, chimeric genomes are produced that have regions derived from each parent. Source: McDonald SM, Nelson MI, Turner PE, et al. 2016. Reassortment in segmented RNA viruses: mechanisms and outcomes. Nat Rev Microbiol, 14 (7): 448-460.

分子内重组是指两个有密切相关但生物学性状不同的病毒感染同一个宿主细胞时发生核苷酸片段互换，产生兼有两亲本特性的子代的现象。同科或不同科的病毒之间均可以发生，比如东部马脑炎病毒和辛德比斯样病毒发生分子内重组之后，产生一种新的病毒——西部马脑炎病毒。

重配是指亲缘关系相近的分节段RNA病毒感染同一细胞时，可交换其基因组片段，产生稳定的或不稳定的重配毒株。重组针对的是核酸不分节段的病毒，而重配针对的是核酸分节段的病毒，如砂粒病毒、双RNA病毒、似布尼亚病毒、正黏病毒和呼肠孤病毒（**图21-4**）。

Intramolecular recombination refers to the exchange of nucleic acids when two closely related but biologically different viruses infect the same host cell and generate a progeny virus with both parental characteristics. It can occur between viruses of the same or different families. For example, intramolecular recombination between Eastern equine encephalitis virus and Sindbis-like virus produces a new virus—Western equine encephalitis virus.

Reassortment refers to generation of recombinant viruses, whether stable or unstable, as a result of exchange of RNA segments when two segmented RNA viruses of close phylogenetic relatedness infect the same cell. Recombination occurs in the non-segmented RNA viruses, while reassortment occurs in the segmented RNA viruses, such as arenavirus, birnavirus, peribunyavirus, orthomyxovirus and reovirus (**Figure 21-4**).

a. 假单胞菌噬菌体φ6基因组
3个基因组片段（双链RNA）

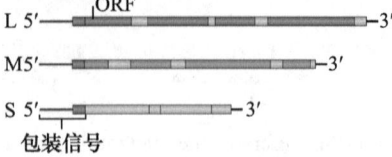

b. 假单胞菌噬菌体重配

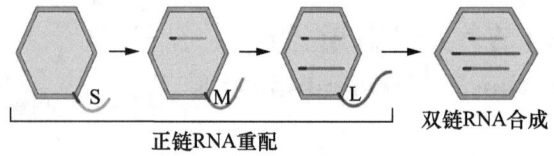

c. 甲型流感病毒基因组
8个片段（负链RNA）

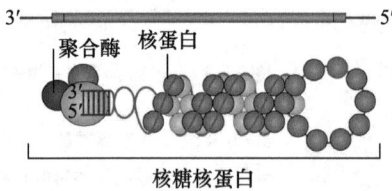

d. 甲型流感病毒重配

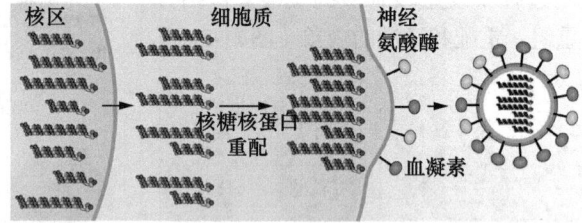

e. 轮状病毒A基因组
11个片段（双链RNA）

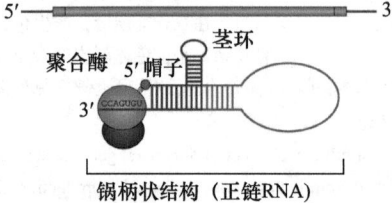

f. 轮状病毒A重配

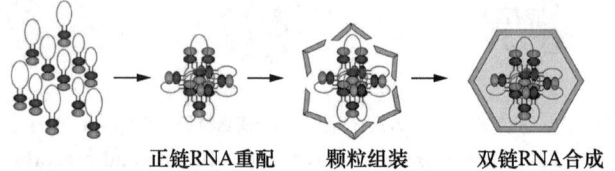

图21-4 假单胞菌噬菌体φ6、甲型流感病毒和轮状病毒的重配

a. 假单胞菌噬菌体φ6基因组由三个双链RNA片段组成：小（S）、中（M）和大（L）；b. φ6基因组片段重配和包装模型。φ6正链RNA先按顺序依次组装。在包装所有三个正链RNA片段后，前衣壳蛋白核心扩大，触发三个正链RNA通过病毒聚合酶转化为双链RNA基因组片段；c. 甲型流感病毒基因组由8个负链RNA片段组成。图中为一个线性负链RNA分子示例（顶部）和核糖核蛋白（RNP；底部），负链RNA由异源三聚体聚合酶复合物和核衣壳蛋白缠绕；d. 甲型流感病毒基因组片段重配和包装模型。甲型流感病毒8个RNP在细胞核中合成，并输出至细胞质中相互配对；e. 轮状病毒A的基因组由11个双链RNA片段组成，图中为一个线状正链RNA片段示例（顶部）及折叠成盘状的结构（底部）；f. 轮状病毒基因组片段重配和包装模型。11个正链RNA，每个均结合聚合酶-加帽酶复合物，相互配对形成一个超级复合物后组装成病毒粒子。图像来源见英文。

Figure 21-4 Pseudomonas phage φ6, influenza A virus and rotavirus assortment

a. The Pseudomonas phage φ6 genome consists of three double-stranded RNA segments: small (S), medium (M) and large (L); b. A model of φ6 genome segment reassortment and packaging. φ6 (+) RNAs are packaged sequentially. Following packaging of all three (+) RNA segments, the procapsid core expands, which triggers the conversion of the three (+) RNAs into dsRNA genome segments by viral polymerases; c. The influenza A virus genome comprises eight negative-sense RNA segments. A representative segment is shown as a linear (−) RNA molecule (top) and as a ribonucleoprotein (RNP; bottom), in which the (−) RNA is bound by a heterotrimeric polymerase complex and nucleocapsid protein; d. A model of genome segment reassortment and packaging in influenza A viruses. Eight influenza A virus RNPs are synthesized in the nucleus and individually exported into the cytosol, where they pair up with each other; e. The genome of rotavirus A is composed of 11 dsRNA segments, one of which is shown as a (+) RNA precursor in linear form (top) and folded into a putative panhandle shape (bottom); f. A model of genome segment reassortment and packaging in rotaviruses. Eleven (+) RNAs, each bound to a polymerase-capping enzyme complex, pair with each other to form a supercomplex that assembles into virion. Source: McDonald SM, Nelson MI, Turner PE, et al. 2016. Reassortment in segmented RNA viruses: mechanisms and outcomes. Nat Rev Microbiol, 14 (7): 448-460.

增殖性激活是指用同一株产生不同程度致死性突变的若干病毒粒子同时感染某一细胞时，致死突变的病毒将重新获得感染性，称为增殖性激活。

Multiplicity reactivation: when several virions from the same strain with different lethal mutations are used to infect the same cell at the same time, the virions with lethal mutations will regain their infectivity, hence the term multiplicity reactivation.

第4节　病毒组分之间的相互作用
Section 4　Interactions among the virus components

补偿作用是指同种或异种病毒感染的同一细胞中，病毒蛋白质之间由于相互作用，拯救了一种或两种病毒或增加了病毒的产量。例如，腺病毒和腺联病毒共感染同一细胞时，腺病毒可为腺联病毒提供后者不能合成的基因产物，从而导致腺联病毒产量增加。

表型混合是指两种病毒或同一病毒两种毒株共感染同一细胞时，形成的子代病毒粒子混有两种或两株病毒的衣壳或囊膜蛋白。两种基因组不分节段的病毒可以发生4种不同的表型混合。

表型互换是指两种病毒共感染同一细胞时，一种病毒的衣壳或囊膜包裹了另一种病毒的基因组。

基因型混合是指两种病毒的核酸，偶尔混合装在同一病毒衣壳内或两种病毒的核衣壳偶尔包在一个囊膜内，但它们的核酸都未发生重组。

Complementation refers to rescuing of one virus or two, or generation of a high-yielding progeny virus as a result of the interactions among viral proteins of the same or different viruses in the same infected cell. For example, when adenovirus and adeno-associated virus co-infect the same cell, adenovirus can provide the latter with the gene products that can't be synthesized by itself, leading to increased yield of the adeno-associated virus.

Phenotypic mixing occurs when two viruses or two strains of the same virus co-infect the same cell. The progeny virus may carry the capsid or envelope proteins of both viruses or strains. Four different phenotypic mixings can occur for two non-segmented viruses.

Phenotypic exchanging refers to the accidental assembly of the genome of one virus into the capsid or envelope of the other virus when both viruses co-infect the same cell.

Genotype mixing refers to the nucleic acids of two viruses, occasionally mixed within the same viral capsid or with the nucleocapsid of one virus occasionally enveloped within a membrane. However, the nucleic acids have not undergone recombination.

第5节　遗传变异与病毒演化
Section 5　Genetic variation and viral evolution

病毒的上述突变可以引起其表型变异，包括主要由病毒表面蛋白变化引起的毒力变异、抗原性变异和抗体逃逸变异，以及空斑变异。

毒力变异是指病毒基因组发生的变异影响其对宿主致病性的变异毒株。

抗原漂移指的是由基因组发生突变导致抗原的小幅度变异，属于量变，不产生

The viral mutations described above can cause phenotypic variations, including virulence mutations primarily caused by the changes of viral surface proteins, antigenic mutations, antibody escape mutations, as well as clear-plague mutations.

Virulence-related mutants refer to variants with mutations in the viral genome that affect pathogenicity to the host.

Antigenic drift refers to small variations of the

新的亚型。例如，甲型流感病毒发生抗原漂移，可引发局部区域流行（**图 21-5**）。

抗原转换通常指编码抗原的基因组重排，引起的抗原变异幅度较大，属于质变，产生新的亚型。例如，甲型流感病毒发生抗原转换，可引发流感大流行（**图 21-5**）。

antigenic structure caused by mutations in the genome. The changes are quantitative, and no new subtypes are produced. With influenza A viruses, such changes could cause endemic outbreaks (**Figure 21-5**).

Antigenic shift refers to significant variations of the antigenic structure, mostly due to genetic reassortment. The changes are qualitative and new subtypes are generated. For instance, antigenic shift of influenza A viruses often causes a pandemic (**Figure 21-5**).

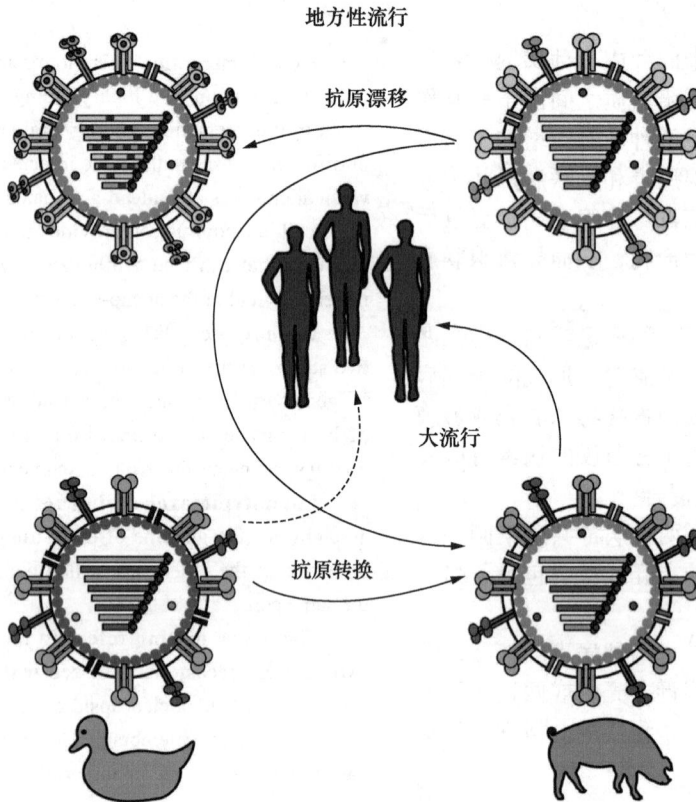

图 21-5 抗原漂移和抗原转换

流感病毒在HA蛋白球状头部可变区逐渐发生突变，导致流感病毒逃避中和抗体并导致季节性暴发，该现象称为抗原漂移。若在人群中引入一种新的亚型则称为抗原转换，当病毒在人与人之间有效传播时，可能会在未免疫人群中引起大暴发，因为没有针对该新亚型的抗体。过去的大流行是由两种或两种以上流感毒株（如禽流感和人流感）之间的基因片段发生交换（重配）引起。然而，最近对雪貂的研究表明，禽流感病毒，如H5N1可直接从动物传给人，只需要少量适应性突变，如图虚线所示。图像来源见英文。

Figure 21-5　Antigenic drift and antigenic shift

The gradual accumulation of mutations, mainly in the highly variable globular head region of HA, causes the influenza virus to escape recognition by virus neutralizing antibodies and allows it to cause seasonal epidemic outbreaks. This phenomenon is called antigenic drift. The introduction of a novel subtype into the human population is called antigenic shift and may cause a pandemic outbreak in the naïve human population when the virus is efficiently transmitted from human to human, since antibodies directed against the novel subtype are absent. Past pandemic outbreaks were caused by exchange (reassortment) of gene segments between two or more influenza strains, *e.g.*, avian influenza viruses and human influenza viruses. However, recent studies on ferrets have suggested that avian influenza viruses, like H5N1, can be transmitted directly from animal reservoirs to humans with only a few adaptive mutations, as indicated by the dashed line in the diagram. Source: van de Sandt CE, Kreijtz JH, Rimmelzwaan GF. 2012. Evasion of influenza A viruses from innate and adaptive immune responses. Viruses, 4 (9): 1438-1476.

抗体逃逸变异是指流行病毒株抗原基因变异发生在抗体结合区，使疫苗免疫产生的抗体不能有效结合和中和，抗原漂移和抗原转换均可导致免疫逃逸。

空斑变异指能产生细胞病变的病毒基因变异引起其在感染细胞半固体琼脂上的空斑大小发生变化，空斑的大小可反映病毒的释放和能力。

病毒演化：病毒（特别是RNA病毒）具有突变频率高、世代时间短、感染个体内的病毒群体大等特性，使其较短时间内能在被感染的宿主群体内通过突变实现演化，以适应宿主环境，由此造成同一种病毒群体的遗传多样性。演化会显著增加病毒的血清型，逃避宿主的免疫清除，改变病毒的感染能力，甚至引发病毒的跨物种传播和人畜共患病暴发。

病毒谱系动力学是联合应用流行病学与进化生物学方法，根据病毒的基因组序列研究其在特定宿主群体中的演化。这种方法可以解答一些重要的流行病学问题，如"病毒什么时候出现？""哪个毒株是导致流行的始祖？"，以及"病毒跨国界传播的时间？"。

无论DNA病毒还是RNA病毒，其变异反映的是对宿主和环境的一种适应能力。例如，新型冠状病毒、非洲猪瘟病毒和甲型流感病毒都对宿主和环境产生了极高的适应性，并在全球暴发。一般来说，具备以下特征的病毒（包括因变异获得的特征）可能引起大流行：①快速复制的能力，达到较高的滴度；②在特定组织内具有较强的增殖力；③长期排毒；④在环境中生存能力强；⑤能逃逸宿主的防御；⑥能垂直传播。

Antibody escape mutants denote epidemic virus strains with mutations in the antibody-binding sites of the antigens that can help the strain escape from binding and neutralization by antibodies generated from vaccination. Both antigenic drift and antigenic shift lead to escape from antibody-binding.

Clear-plaque mutants mean mutations that affect the changes in plaque size of the virus-infected cells on the semi-solid agar. Plaque size is the reflection of virus release and spread.

Viral evolution: viruses, especially RNA viruses, possess characteristics such as high mutation rates, short generation times, and large viral populations within infected individuals. These characteristics enable them to undergo rapid evolution through mutations within the infected host population, allowing them to adapt to the host environment. As a result, genetic diversity arises within the population of the same virus. Evolution significantly increases the serotypes of the virus, allowing it to evade immune clearance, alter its infectivity, and even cause cross-species transmission and outbreaks of zoonotic diseases.

Viral phylodynamics represents a collaboration between epidemiology and evolutionary biology to investigate evolution of a virus in a host population based on the viral genomic sequences. Such approach has made it possible to address a range of important epidemiological questions such as: "when did a virus emerge?" "what is the progenitor strain of a circulating epidemic?" and "what is the timing of the spread of a virus across countries?".

Regardless of either DNA viruses or RNA viruses, their variations reflect a kind of adaptability to the host and external environment. For example, SARS-CoV-2, African swine fever virus and influenza A virus are highly adaptable to the host and external environment, which allows them to spread globally causing pandemic. Generally speaking, a virus may cause a pandemic as long as it has the following characteristics, including those arising from genetic mutations: ①be able to replicate rapidly to reach high titers; ②be able to proliferate efficiently in specific tissues; ③be able to secrete or discharge virus persistently; ④be able to survive in the environment; ⑤be able to evade from the host defense; ⑥be able to vertical transmission.

思考题　Questions

1. 简述抗原漂移和抗原转换的本质。

2. 简述准种的概念及其意义。

3. 病毒演化的生物学意义是什么？

1. Please describe the nature of antigenic drift and antigenic shift.

2. Please describe the concept and significance of quasispecies.

3. What are the biological significances of viral evolution?

第22章 病毒与细胞的相互作用

Chapter 22 Interaction between virus and cell

内容提要 病毒是严格的细胞内寄生微生物。当细胞与病毒接触后，病毒将进入细胞内部进行生长和繁殖，受感染的细胞会产生形态变化甚至死亡。这种病毒感染导致的细胞损伤，包括细胞膜和细胞骨架发生的变化，统称为细胞病变效应（CPE）。

Introduction Viruses are strict intracellular parasitic microorganisms. The virus will enter the cell for growth and reproduction when they get contact with each other. The infected cell will demonstrate morphological changes or even die. The cell damage caused by virus infection, including the changes of cell membrane and cytoskeleton, is collectively referred to as cytopathic effect (CPE).

第1节 病毒的细胞培养
Section 1　Cell culture of viruses

病毒的细胞培养：病毒是严格的细胞内寄生微生物，培养病毒必须使用细胞，根据病毒的不同可以选用动物接种、鸡胚接种或细胞培养法。其中病毒的细胞培养是最常用的方法，用于实验室病毒分离鉴定或疫苗生产。

细胞培养的特点：细胞培养具有以下特点：细胞群体中的每个细胞生理特性基本一致，对病毒易感性相等；没有实验动物的个体差异，不涉及动物保护问题；可严格执行无菌操作；细胞培养本身能显示病毒的生长特征；用空斑技术可进行病毒的纯化。

细胞的类型：根据细胞来源、染色体特征可传代性，用于病毒培养的细胞可分为三种类型：原代细胞、二倍体细胞株和传代细胞系。原代细胞是指动物组织经胰蛋白酶消化处理，使细胞分散，获得单个

Cell culture of the virus: since viruses are strictly intracellular parasitic microorganisms, they must be cultured in cells. According to different viruses, animal inoculation, chicken embryo inoculation or cell culture can be selected. Cell culture is the most commonly used method for virus isolation and identification in the laboratory or for vaccine production.

Characteristics of cell culture: cell culture has the following features: each cell in the population has the same physiological characteristics and the same susceptibility to the virus; there is neither individual differences as compared to animal inoculation, nor ethical concerns involving animals; aseptic operation can be strictly carried out; cell culture can show the growth characteristics of the virus; virus can be purified by the plaque technique.

Types of cells: the cells used for culture can be divided into three types: primary cells, diploid cells and cell lines according to the source, chromosome characteristics and availability for repeated passaging. Primary cells are temporarily culturable cells obtained from animal tissues by trypsin digestion and dispersing

细胞，并可短暂培养。原代细胞中存在各种类型的未分化细胞，对大多数动物病毒都是敏感的，但因其不能持续培养传代，限制了其在疫苗生产及实验室诊断中的使用价值。二倍体细胞株是将长成单层的原代细胞继续培养传代后，大部分细胞退化，少数细胞能继续传下来，且染色体数与原代细胞一样，保持其二倍染色体数目的细胞（**图22-1**）。传代细胞系指能在体外可持续增殖传代的细胞，大多数由

treatment. There are various types of undifferentiated cells in the primary cell population that are susceptible to most animal viruses. However, primary cells cannot be passaged continuously, thus limiting their use in vaccine production and laboratory diagnosis. Majority of the primary cells become degenerated during successive passages and eventually die out, while some are still culturable into monolayers and can be passaged. Such cells are called diploid cells that have the same numbers of chromosomes as the primary cells (**Figure 22-1**). The cell lines are those that can proliferate and passage continuously *in vitro*, most of which are made from

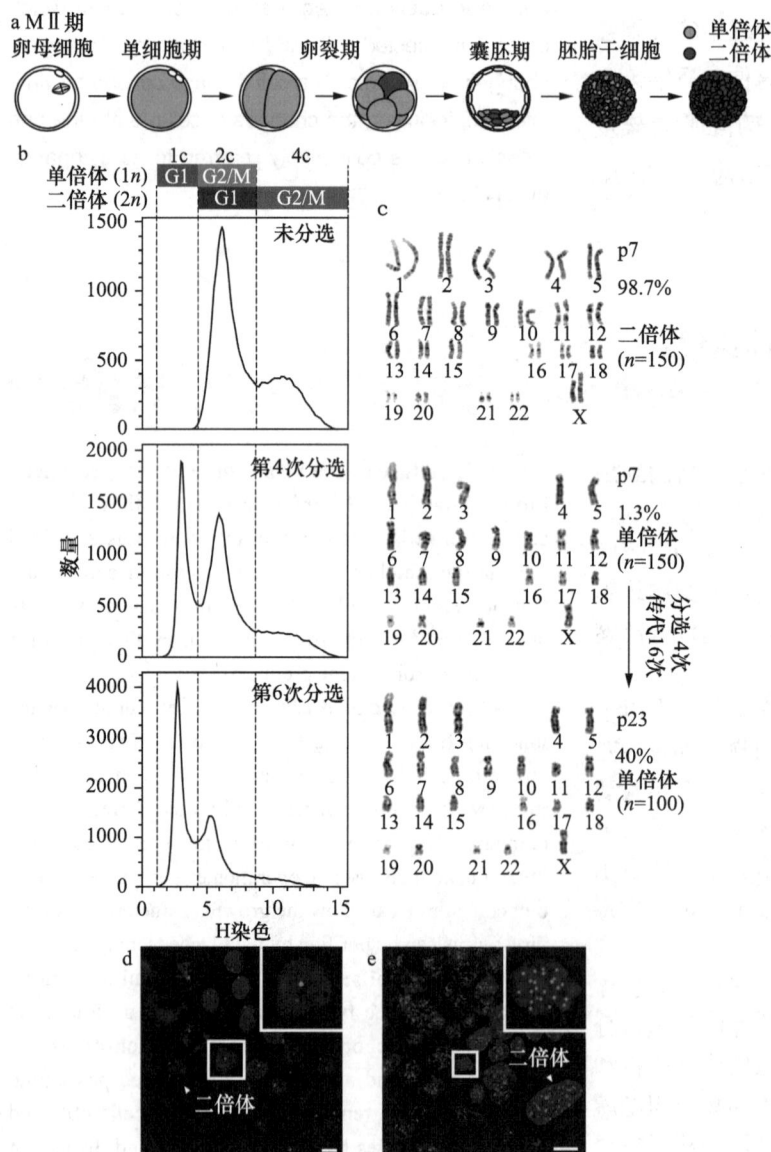

图22-1 从单倍体到二倍体细胞

a. 生殖胚胎干细胞单倍体分裂示意图；b. 反复富集1c细胞的单倍体pES10细胞DNA丰度，自上而下为未分选的二倍体细胞，部分纯化（第4次分选）和大部分纯化（第6次分选）的单倍体细胞；c. 1c细胞富集前后的pES10细胞核型。p表示传代。d、e. 单倍体富集pES10细胞DNA原位杂交（d）和着丝粒染色（e）。标尺：10 μm。图像来源见英文。

Figure 22-1 Derivation from haploid to diploid

a. Schematic of putative haploidy in parthenogenetic ES (pES) cells; b. DNA content profiles of haploid pES10, established by repeated enrichment of 1c cells. Top to bottom, unsorted diploid cells, partially purified (fourth sort) and mostly purified (sixth sort) haploid cells; c. pES10 karyotypes before and after 1c-cell enrichment. p, passage. d, e. DNA FISH (d) and centromere staining (e) in haploid-enriched pES10 cells. Scale bars: 10 μm. Source: Sagi I, Chia G, Golan-Lev T, et al. 2016. Derivation and differentiation of haploid human embryonic stem cells. Nature, 532 (7597): 107-111.

癌细胞或二倍体细胞突变而成（如HeLa、Vero、BHK-21、PK-15细胞系等）。

细胞培养的方法有静置培养、旋转培养、悬浮培养、微载体培养等。

静置培养： 将消化分散的细胞悬液装瓶后放于5% CO_2 温箱静置培养，细胞沉降并贴附在细胞培养皿面上生长分裂，最后长成单层细胞，实验室原代细胞培养和传代细胞培养常用静置培养方法。

旋转培养： 将细胞转入玻瓶或塑料瓶内，不断缓慢旋转瓶体，细胞贴附于瓶四周生长成单层细胞，此方法的细胞量多，病毒产量高，常用于大规模生产疫苗。

悬浮培养： 此方法适用于不需贴壁的细胞，如淋巴瘤细胞。将分散的细胞加入培养基稀释后分瓶即可，培养过程中每天摇动瓶体数次使细胞分散，可促进细胞生长。

微载体培养： 将对细胞无害的微载体颗粒加入生物反应器的培养液中作为载体，使细胞在微载体表面附着生长，同时设置反应器的持续搅动速度使微载体始终保持悬浮状态。微载体是指能适用于贴壁细胞生长的微珠，一般由天然葡聚糖或者其他人工合成的聚合物制成。该方法单位体积培养液的细胞产率高，把悬浮细胞和贴壁细胞培养融合在一起，兼有两者的优点，适用于产业化细胞培养和疫苗生产。

cancer cells or by mutation of the diploid cells, such as HeLa, Vero, BHK-21, PK-15 cell lines, *etc.*

The methods of cell cultivation include static culture, rotational culture, suspension culture, microcarrier culture, *etc.*

Static culture: the dispersed cell suspension obtained from digestion is dispensed into the culture flask and then placed in a 5% CO_2 incubator for static culture. The cells settle and attach to the surface for growth and division, and finally grow into a cell monolayer. Static culture method is commonly used in the laboratory for primary cells and passage cell lines.

Rotational culture: the cells are dispensed into the glass or plastic bottle which is then rotated continuously but slowly to allowed the cells to attach to the inner surface of the bottle and grow into monolayer cells. This method can increase the virus yield due to increased cell population and is often used in large-scale production of vaccines.

Suspension culture: a method suitable for non-adherent cells, such as lymphoma cells. The cells in suspension are briefly dispersed and diluted with the culture medium which are then aliquoted into suitable bottles for incubation. The bottles should be shaken several times a day to disperse the cells to promote their growth.

Microcarrier culture: the microcarrier particles innocuous to the cells are added into a bioreactor containing the culture medium as the carrier for cell attachment and growth. The bioreactor should be set for constant agitation at a certain speed to keep the cell-containing particles suspended. Microcarriers are the microspheres that are suitable for attachment and growth of adherent cells and made of natural dextran or various synthetic polymers. This method combines the advantages both suspension culture and adherent cell culture for high cell yield per unit volume of medium and is, therefore, suitable for industrial cell culture and vaccine manufacturing.

第2节　病毒与细胞的相互作用
Section 2　Virus-cell interactions

病毒在细胞水平上主要产生杀细胞效应、细胞融合效应、包涵体形成及诱导宿主细胞凋亡、转化等效应，从而对宿主细胞造成损伤。

杀细胞效应 是指病毒感染引起的宿主细胞病变效应（CPE），包括细胞形态、结构和功能变化（**图22-2**）。很多病毒感染体外培养细胞后都能使细胞产生病变效

Viruses can cause damages to the host cells by inducing cytotoxic effects, cell fusion effects, formation of inclusion bodies, induction of host cell apoptosis, transformation, and others.

Cytotoxic effects refer to the cytopathic effects (CPEs) imposed by the viruses on the host cells where there are morphological, structural and functional damages (**Figure 22-2**). Many viruses can produce CPE in the cells infected *in vitro* that can be visualized under

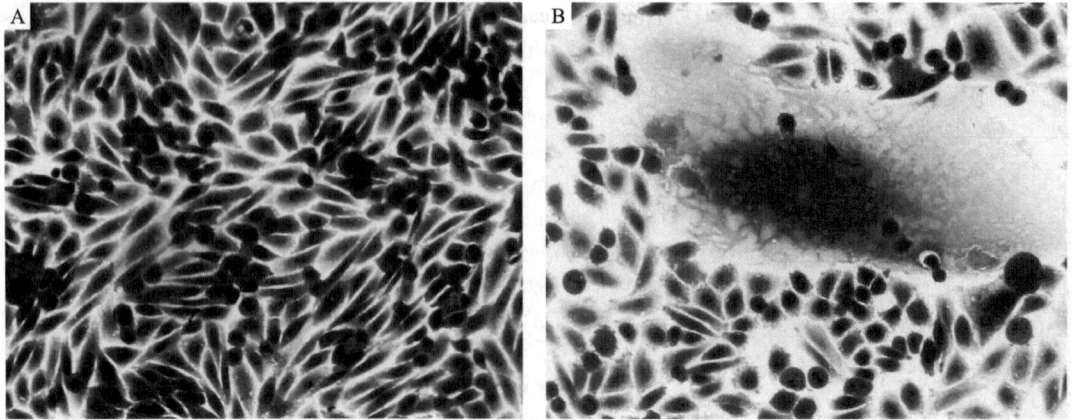

图22-2 人类疱疹病毒诱导的合胞体形成

将表达CD46（CHO-CD46）的中国仓鼠卵巢（CHO）细胞不接种（A）或接种人类疱疹病毒（B）10 h后将细胞固定并用HE染色。图像来源见英文。

Figure 22-2 Human herpesvirus induced syncytium formation

Chinese hamster ovary (CHO) cells expressing CD46 (CHO-CD46) were mock infected (A) or inoculated with human herpesvirus (B). After 10 h, the cells were fixed and stained with HE. Source: Mori Y, Seya T, Huang HL, et al. 2002. Human herpesvirus 6 variant A but not variant B induces fusion from without in a variety of human cells through a human herpesvirus 6 entry receptor, CD46. J Virol, 76 (13): 6750-6761.

彩图

应，在光学显微镜下能观察到细胞圆缩或肿大（如痘病毒、呼肠孤病毒等感染）、细胞聚合（如腺病毒感染）、细胞膜融合形成多核巨细胞（即合胞体，如副黏病毒和疱疹病毒的感染）、包涵体（如巨细胞病毒）等。

杀细胞效应的途径：病毒将宿主细胞破坏到一定程度时就会使机体产生严重的病理变化。产生杀细胞效应的途径有多种，但病毒复制直接对细胞造成杀伤是重要途径，其他途径包括以下几点。

（1）抑制宿主细胞的生物大分子合成：很多病毒能抑制宿主细胞DNA合成、mRNA转录或加工，或将宿主细胞的蛋白质合成机制据为己有，此类代谢紊乱最终导致细胞死亡。

（2）细胞膜功能障碍：某些病毒感染可增加宿主细胞膜的通透性，从而导致细胞内外的离子平衡失调，也影响细胞内外分子的正常交换。

（3）影响细胞溶酶体和其他细胞器的功能：细胞代谢紊乱往往导致膜系统通透性增高，若溶酶体膜裂解，则其内各种酶

light microscope, *e.g.,* cell shrinkage or swelling (such as poxvirus or reovirus infections), cell aggregation (such as adenovirus infection), cell membrane fusion to form multinucleated giant cells called syncytium (such as paramyxovirus or herpesvirus infections), inclusion bodies (such as cytomegalovirus infections).

Mechanisms of the cytotoxic effects: when viruses destroy host cells to a certain extent, serious pathological changes could occur in the body. There are quite a number of mechanisms behind the cytotoxic effects. However, direct killing of the cells by replicating viruses is often an important mean. The other means include the following.

(1) Inhibition of biomacromolecules synthesis in host cells: many viruses can inhibit host cell DNA synthesis, mRNA transcription or processing, or hijack the host cell protein synthesis machinery for synthesis of their own proteins. Such metabolic disorders eventually lead to cell death.

(2) Dysfunction of the cell membrane: infections by some viruses could increase the permeability of host cell membrane, leading to imbalance of ions across the cell membrane and disorders in exchange of intracellular and extracellular molecules.

(3) Dysfunction of lysosomes and other organelles: cell metabolism dysfunction often leads to increased permeability of the membrane system. When lysosomal membrane is damaged, various enzymes therein will be released into the cytoplasm and cause cell autolysis. In

扩散到细胞质中引起细胞自溶。此外，还有些病毒能影响内质网、高尔基体的功能，抑制细胞内各种物质的正常运转。

（4）影响细胞骨架：细胞骨架由微丝、中间丝、肌动蛋白丝、微管蛋白等组成，参与细胞形态维持、细胞器胞内迁移等功能，部分病毒感染可以抑制细胞骨架的形成，在细胞裂解前的细胞形态变化中起主要作用。

病毒杀细胞效应的应用：杀细胞病毒感染易感细胞后，通过观察 CPE 可以对病毒的毒力进行简单的量化计算。50% 组织细胞感染量（$TCID_{50}$）指能够使半数细胞在一定条件下发生细胞病变的病毒量，用于判定病毒的毒力和毒价。空斑测定方法：将适当浓度的病毒悬液加入单层细胞培养中，当病毒吸附细胞后，再覆盖一层融化的琼脂后进行培养，病毒在细胞内复制后产生局限性病灶，病灶逐渐扩大，形成肉眼可见的空斑（**图 22-3**）。空斑是由一个感染性病毒粒子复制形成的，病毒悬

addition, some viruses can affect the functions of the endoplasmic reticulum and Golgi complex, and inhibit normal metabolism of various substances in the cell.

(4) Derangement of the cytoskeleton: the cell cytoskeleton consists of microfilaments, intermediate filaments, actin filaments, tubulin proteins, and other components, which are involved in maintaining cell morphology, intracellular organelle transportation and other functions. Some viral infections can inhibit the formation of the cell cytoskeleton and can play a major role in the morphological changes of cells before cell lysis.

Application of cytotoxic effect of the virus: after a cytotoxic virus infects susceptible cells, the virulence of the cytotoxic virus can be simply quantified by observing CPE. 50% tissue culture infective dose ($TCID_{50}$) refers to the amount of the virus that can cause cytopathic changes in half of the cultured cells under certain conditions. It is used to determine viral virulence or titer. The plaque assay is performed as follows: a suitable concentration of virus suspension is added into a cell monolayer. After the virus adsorbs to the cells, the infected cell monoplayer is covered with a layer of melted agar and then incubated. Replication of the virus in the cells will produce localized CPEs, which gradually expand with continuing viral replication to form plaques that are visible by naked eyes (**Figure 22-3**).

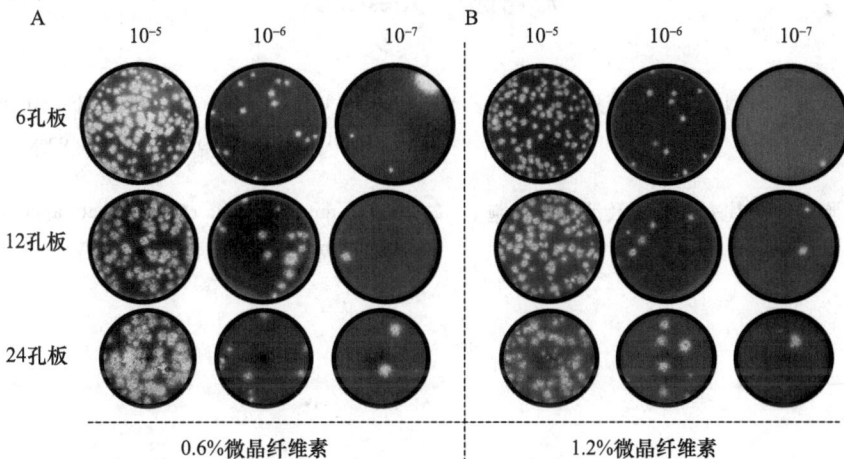

图 22-3　利用微晶纤维素（MCC）替代琼脂进行的新冠病毒空斑测定
利用倍比稀释的新冠病毒感染 Vero E6 细胞，并在无血清 DMEM 中覆盖 0.6%（A）或 1.2%（B）微晶纤维素。72 h 后，去除微晶纤维素覆盖层，固定单层并用结晶紫染色，观察空斑。图像来源见英文。

Figure 22-3　Microcrystalline cellulose (MCC) is a suitable alternative as an overlay medium for SARS-CoV-2 plaque assays
Vero E6 cells were infected with serially diluted SARS-CoV-2 stock virus and overlaid with 0.6% (A) or 1.2% (B) MCC in serum-free DMEM. After 72 h, the MCC overlay was removed and monolayer were fixed and stained with crystal violet to visualize plaques.
Source: Jureka AS, Silvas JA, Basler CF. 2020. Propagation, inactivation, and safety testing of SARS-CoV-2. Viruses, 12 (6): 622.

液中的病毒粒子数可以根据空斑数进行定量，表示为每毫升含有的空斑形成单位（PFU），类似于细菌的菌落形成单位。PFU 与 $TCID_{50}$ 之间的换算关系：1 PFU=0.7 $TCID_{50}$。空斑测定也可用于体外病毒中和抗体效价的检测。

Each plaque is formed by replication of a single infectious virus particle. Therefore, the amount of virus particles in a suspension could be quantified from the number of plaques and expressed as plaque-forming unit (PFU) per milliliter, just like colony-forming units in bacterial quantification. The conversion between PFU and $TCID_{50}$ is 1 PFU = 0.7 $TCID_{50}$. Plaque assay can also be used to determine the viral neutralization titers *in vitro*.

第3节 病毒引致的非杀细胞变化
Section 3　Virus-induced noncytocidal changes

非细胞杀伤病毒通常不会导致宿主细胞死亡，相反往往会产生持续性感染，感染的宿主细胞持续产生和释放病毒粒子，但细胞总体的代谢几乎不受影响。他们主要是通过一些非杀伤细胞机制间接致病的，包括免疫病理损伤、影响细胞正常合成和分泌功能、宿主细胞转化等。

Noncytotoxic viruses do not usually kill host cells. On the contrary, they often cause persistent infection. Infected host cells continuously produce and release virions, but the overall metabolism of the cells is minimally affected. They mainly exert indirect pathogenic effects on the host through some noncytotoxic mechanisms, such as immunopathological damages, disturbances of the cellular synthesis and secretion, transformation of the host cells, *etc*.

思考题　Questions

1. 细胞培养有哪些类型？用于培养病毒各有何适用范围？

1. What are the types of cell culture? What are the scopes of application of these culture types for virus propagation?

2. 什么是细胞病变？其表现形式及涉及的细胞结构如何？

2. What is the cytopathic effect? What happens in the cell structure during CPE?

第23章 病毒的致病机理
Chapter 23 / Pathogenic mechanisms of viruses

内容提要 病毒通过呼吸道、胃肠道、皮肤伤口等途径入侵宿主，并在宿主体内进一步扩散；感染期间宿主可以排毒。不同病毒感染对机体的影响各不相同，可以是急性感染、慢性感染、持续性感染、潜伏感染。

Introduction Viruses invade hosts through respiratory tract, intestinal tract or skin, and further spread in the host. The hosts shed viruses during infection, and the infection may be acute, chronic, persistent or latent depending on the type of viruses.

第1节 病毒的入侵、扩散和排放
Section 1 Virus invasion, spreading and shedding

病毒的入侵： 病毒可以通过多种途径入侵机体，包括呼吸道途径、胃肠道途径、皮肤途径、生殖道途径、眼结膜等。冠状病毒、流感病毒、腺病毒、鼻病毒等是通过呼吸道入侵机体，病毒可与呼吸道黏膜上皮细胞的特异性受体结合，逃避纤毛和巨噬细胞的清除作用，从而感染宿主；轮状病毒、冠状病毒、细小病毒、星状病毒等可通过胃肠道途径入侵机体，此类病毒是引起动物腹泻的主要病原；节肢动物的叮咬和动物的咬伤是病毒通过皮肤入侵的有效途径，例如，马传染性贫血病毒、兔出血热病毒等均通过昆虫叮咬进行传播，狂犬病毒通过动物咬伤的方式进行传播。此外，疱疹病毒可以通过生殖道、腺病毒可以通过眼结膜感染动物。

病毒的扩散和靶器官感染： 病毒入侵机体后，在机体局部增殖，也可以进一步扩散和入侵其他组织和器官，此过程称为

Virus invasion: viruses can invade the body through many ways, including the respiratory tract, intestinal tract, skin, reproductive tract and conjunctiva. Coronavirus, influenza virus, adenovirus, rhinovirus invade the body through the respiratory tract. Viruses can bind to specific receptors of the respiratory epithelial cells and escape ciliary and macrophage scavenging effects. Rotavirus, coronavirus, parvovirus and astrovirus cause infection through the intestinal tract. These viruses are the main pathogens that cause diarrhea in animals. Bites from arthropods and animal bites are effective pathways for viral invasion through the skin. For example, viruses such as equine infectious anemia virus and rabbit hemorrhagic fever virus are transmitted through insect bites. Rabies virus is transmitted through animal bites. Additionally, herpes viruses can be transmitted through the reproductive tract, and adenoviruses can infect animals through ocular conjunctiva.

The spread of virus and infection of target organs: after viruses invade into the body, they can replicate locally in the body, or further spread and invade other tissues and organs. This process is called as the

病毒在体内的扩散。

在上表皮表面局部扩散：部分病毒在上皮细胞进行复制，并可扩散感染相邻细胞，产生局部性感染，病毒往往从这些部位直接排入环境。

侵入皮下组织并扩散入淋巴组织：病毒突破上皮屏障入侵皮下组织，并扩散至淋巴组织。这一过程与多种因素有关，包括感染的吞噬细胞，尤其是树突状细胞和巨噬细胞的迁移。皮下组织有丰富的树突状细胞，如移行的树突状细胞可被蓝舌病毒感染，随后从上皮表面迁移至邻近的淋巴结。

通过血流扩散：局部感染的病毒可以直接通过毛细血管进入血液或者通过感染内皮细胞后反复向血液内释放病毒而造成血流扩散。病毒血症是其特点，病毒感染后初次进入血液，称为初始病毒血症，通常无明显的临床表现。当病毒在其靶器官复制，持续产生大量病毒，出现第二次病毒血症，身体其他部位受到感染，表现出相关临床症状。例如，细小病毒、肠病毒、披膜病毒、黄病毒等可随血液在体内循环，也可以进入或依附于白细胞、血小板或红细胞。被白细胞携带的病毒不易被抗体和其他血浆成分清除，并可随细胞的迁徙而到达相关组织或器官。

神经扩散：某些病毒在局部感染后，可通过感染部位的神经末梢入侵神经细胞进行扩散，这种扩散方式称为神经扩散。例如，狂犬病毒能够以鼻中的嗅神经末梢作为入侵部位，病毒进入嗅神经上皮细胞，引起局部感染，子代病毒经过嗅神经直接到达大脑嗅球。

病毒的排放：病毒从宿主体内排出的过程称为病毒的排放，有利于病毒传播给易感动物。呼吸道分泌物是大多数呼吸道感染病毒的传播媒介，流感病毒和冠状病毒排入黏液，通过咳嗽和打喷嚏的方式排出。狂犬病毒从感染动物的唾液腺排入口腔，动物通过舔舐伤口或咬伤被感染。轮

spread of virus in the body.

Local spread on epithelial surfaces: some viruses replicate in epithelial cells and can spread to adjacent cells, causing localized infections. The virus often sheds directly into the environment from these sites.

Subepithelial invasion and lymphatic spread: viruses can break through the epithelial barrier, invade the subcutaneous tissues and spread to lymphoid tissues. This process may be attributed to a number of factors, including migration of infected phagocytic cells, especially dendritic cells and macrophages. There are abundant dendritic cells in the subcutaneous tissue. The migrating dendritic cells can be infected by virus, such as bluetongue virus, and then migrate from the epithelial surface to adjacent lymph nodes.

Hematogenous dissemination: viruses causing localized infections can directly enter the bloodstream through capillaries or repeatedly release viruses into the blood by infecting endothelial cells, leading to hematogenous dissemination. Viral viremia is characteristic of this process. After initial infection, the virus enters the bloodstream, known as primary viremia, which typically has no obvious clinical manifestations. As the virus replicates in its target organs and continues to produce a large number of viruses, a second viremia occurs, resulting in infection in other parts of the body and the manifestation of related clinical symptoms. For example, picornaviruses, enteroviruses, paramyxoviruses, and flaviviruses can circulate in the body through the bloodstream. They can also enter or attach to white blood cells, platelets, or red blood cells. Viruses carried by white blood cells are less susceptible to clearance by antibodies and other plasma components. They can travel with migrating cells to reach relevant tissues or organs.

Spread via nerves: some viruses can invade nerve cells through the nerve endings at the infection sites, which is called as neural spread. Rabies virus can invade the olfactory nerve endings in the nose. The virus enters into the olfactory nerve epithelial cells and causes local infection. The progeny viruses directly reach the olfactory bulb of the brain through the olfactory nerve.

Virus shedding: the process of virus excretion from the host is called as virus shedding, which is conducive to viral transmission to susceptible animals. Respiratory secretions are the vectors of most respiratory viruses. Influenza virus and coronavirus are excreted into the mucus, which are then excreted by coughing and sneezing. Rabies virus is excreted from the salivary glands of infected animals into the mouth. Animals are infected by wound licking or

状病毒通过粪便排出体外而造成传播。口蹄疫病毒从破裂的水疱排出体外，通过接触感染动物。犬传染性肝炎病毒在犬肾管状上皮细胞中复制，通过尿液排出体外。

biting. Rotavirus is excreted from the body through feces to spread infections. Foot and mouth disease virus is excreted from ruptured blisters and infects other animals through contact. Canine infectious hepatitis virus replicates in canine renal tubular epithelial cells and is excreted in urine.

第2节 病毒的持续性感染
Section 2　Persistent virus infection

持续性感染：不论是否发病，感染性病毒始终存在，这种感染称为持续性感染。例如，牛病毒性腹泻病毒，可垂直传播，幸存的犊牛长期带毒，由于免疫耐受，犊牛对病毒不产生有效的免疫应答，抗体阴性，各种排泄物带毒；有时也可以发展为临床疾病。持续性感染可以再次激活，引起疾病复发，并能引致免疫病理疾病，还与肿瘤的形成有关。病毒的持续性感染可以分为四种类型（**图23-1**）。

Persistent infection means the infectious viruses are present regardless of whether symptoms are present. For example, bovine viral diarrhea virus can be vertically transmitted, and surviving calves can carry the virus long-term. Due to immune tolerance, the calves do not mount an effective immune response against the virus, resulting in negative antibody status and viral shedding in various excretions. In some cases, this can progress to clinical disease. Persistent infection can be reactivated, leading to disease relapse and can also be associated with immune-mediated diseases and tumor formation. Persistent viral infection can be categorized into four types (**Figure 23-1**).

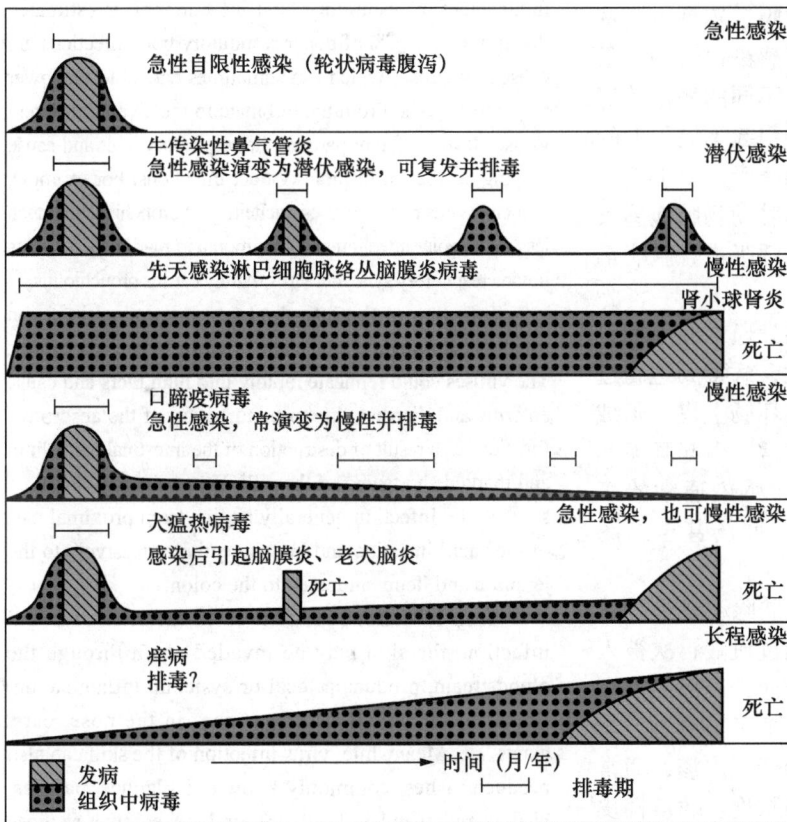

图23-1　急性自限性感染和各种持续性感染中病毒的排毒和疾病临床症状

时间尺度为概念性，假定疾病持续时间相似。图像来源见英文。

Figure 23-1　The shedding of virus and the occurrence of clinical signs in acute self-limited infections and various kinds of persistent infection

The time scale is notional and the duration of various events approximate. Source: Maclachlan NJ, Dubovi EJ, Barthold SW, et al. 2016. Fenner's Veterinary Virology. 5th ed. Amsterdam: Elsevier Inc.

潜伏感染：又称隐性感染，除非激活，体内一般检测不到感染性病毒粒子。例如，牛疱疹病毒，若不引起奶牛发病，在潜伏感染的奶牛体内不能分离到病毒。

慢性感染：病毒长期存在，逐渐增多，渐进性致死。例如，朊病毒在感染神经组织若干年后才能检出，直到动物死亡时，脑内病毒滴度才比较高。

迟发性临床症状的急性感染：病毒复制与疾病进程无关。例如，猫泛白细胞减少症病毒，在妊娠阶段就被感染，直到青年猫阶段才表现出小脑综合征，在神经损伤出现时并不能分离到病毒。

Latent infection: it is also called silent infection. Unless activated, infectious virions cannot be found in the host. For example, bovine herpes virus cannot be isolated in latently infected cows without clinical symptoms.

Chronic infection: the virus persists over a long period, accumulates over time and gradually progress to death. For example, it takes several years for Prion viruses to be detected in infected neural tissue, and it is only at the time of animal death that the viral load in the brain becomes relatively high.

Acute infection with delayed clinical manifestations: viral replication is independent of disease progression. For example, feline panleukopenia virus infects the fetus during cat pregnancy. However, infected kittens do not show cerebellar syndromes until they are adolescent. The viruses cannot be isolated when neurological signs occur.

第3节　病毒感染对宿主组织和器官的损伤
Section 3　Damage to host tissues and organs by virus infection

呼吸道感染：动物呼吸道与外界环境直接接触是临床呼吸道感染的重要诱因。据估计，90%以上的上呼吸道感染由病毒感染引起，上呼吸道病毒感染也延伸至下呼吸道。从鼻甲到肺泡，不同的病毒偏好呼吸道的不同部位，并会引起不同程度的呼吸道感染。例如，流感病毒主要引起气管炎、支气管炎，引起哺乳动物肺泡感染和肺炎的程度较低；而牛呼吸道合胞病毒主要引起细支气管炎。

胃肠道感染：肠道感染通常经由饲料或饮水直接摄入病毒引发，病毒到达肠道上皮后迅速大量复制，破坏肠上皮，肠绒毛缩短，肠道吸收功能异常，引起肠炎和腹泻。幼龄动物较易感。感染通常始于十二指肠，并可向空肠、回肠甚至结肠蔓延。

皮肤感染：动物体皮肤除了是最初感染的部位以外，还可能通过血液再次被入侵，产生局部或全身性炎症，局部的炎症主要发生在鼻子、耳朵、爪子等地方。与此同时病毒感染皮肤也会产生皮疹，通常被称为斑、丘疹、水疱和脓疱。病毒性皮

Infection of the respiratory tract: direct exposure of the animal's airway to the environment is an important inducement of respiratory tract infections. It is estimated that more than 90% of upper respiratory tract infections are caused by viruses which can sometimes extend to the lower respiratory tracts. From the turbinate to the alveoli, different viruses favor different parts of the respiratory tract and cause varying degrees of respiratory tract infections. For example, influenza virus mainly causes tracheitis and bronchitis, but much less for alveolar infection and pneumonia in mammals, whereas bovine respiratory syncytial virus mainly causes bronchiolitis.

Infection of the intestinal tract: intestinal infection usually occurs upon direct intake of viruses via feed or water. The viruses could replicate rapidly into high titers and cause enteritis and diarrhea due to perturbation of the absorptive functions as a result of destruction of the intestinal epithelium and marked shortening of the villi. Young animals are more susceptible. Infection generally begins from proximal part of the small intestine and may spread progressively to the jejunum and ileum and even to the colon.

Skin infection: in addition to the sites of initial infection, the skin may be invaded again through the bloodstream, producing local or systemic inflammation. Local inflammation mainly occurs in the nose, ears, claws, *etc.* Meanwhile, virus infection of the skin can also produce rashes, commonly known as plaques, papules, blisters and pustules. There are viral rashes, such as those

疹，如在各种动物中由痘病毒引起的皮疹和口蹄疫病毒引起的水疱。

中枢神经系统感染：病毒可以通过神经末梢传播到大脑，也可以从血液传播到大脑，后者首先必须穿越血脑屏障，这种屏障由具有紧密连接的毛细血管内皮细胞、血管基底层、具有紧密连接的脉络丛和室管膜上皮组成。例如，披膜病毒、黄病毒科的一些病毒可引起脑炎、脑脊髓炎等神经系统感染，以神经细胞坏死、噬神经现象和血管周围积聚炎性细胞为特性。

胎儿感染：妊娠动物感染病毒产生的影响与孕龄有关，包括流产、死胎、生长发育受阻、胎儿感染等。一般来说，妊娠早期感染最具有破坏性，导致流产、死胎或畸胎。例如，感染牛病毒性腹泻病毒和蓝舌病毒会使先天感染的反刍动物出现致畸性脑缺陷。在怀孕后期，病毒感染会受到胎儿免疫应答的影响，有效的免疫应答会清除胎儿体内的病毒，不会对胎儿的发育产生影响。

其他组织器官感染：几乎所有的器官都可以通过血流扩散被病毒感染，但是大多数病毒感染都有明确的靶器官和组织嗜性。各种器官和组织感染病毒的临床症状与其引起的生理机能异常有关。

caused by poxviruses in various species of animals and blisters caused by foot and mouth disease virus.

Infection of the central nervous system: viruses can be transmitted to the brain through nerve endings. They can also spread from blood to the brain by crossing the blood-brain barrier which consists of capillary endothelial cells with tight junctions, the vascular basal layer, choroid plexus and ependymal epithelium with tight junctions. Neurological infections, including encephalitis, encephalomyelitis, caused by some togaviruses and flaviviruses are characterized by neuronal necrosis, neuronophagia and accumulation of inflammatory cells around blood vessels.

Fetal infection: depending on the gestational age, viral infection can have varying effects on pregnancy outcomes including spontaneous abortion, intrauterine fetal demise, intrauterine growth restriction, and fetal infection. Generally, infection in early pregnancy is more detrimental, often shown as abortion, fetal death or fetal malformation. For example, infections of cattle at early pregnancy with bovine viral diarrhea virus or bluetongue virus can cause teratogenic brain defects in newborn ruminants. In late pregnancy, virus infections will be affected by maternal and fetal immune responses that can effectively clear the virus in the fetal body without affecting its development.

Infection of other organs: almost all organs can be infected by viruses through the bloodstream, but most viral infections have specific target organs and tissue tropism. The clinical symptoms of various organs and tissues caused by viruses are related to the resulting physiological dysfunctions.

第4节 病毒感染对免疫系统的损伤
Section 4 Damage to the immune system by virus infection

病毒可感染并激活免疫细胞，感染细胞的类型因病毒种类而有差异（**图23-2**）。感染T淋巴细胞的病毒主要有猫白血病病毒、猫免疫缺陷病毒等，猫白血病病毒可感染CD4⁺和CD8⁺T细胞；感染B淋巴细胞的病毒主要有马疱疹病毒2型、传染性法氏囊病毒等，传染性法氏囊病毒在腔上囊内的前淋巴细胞选择性地复制，破坏囊内B细胞，从而造成免疫抑制；感染单核细胞的病毒主要有犬瘟热病毒、甲型流感

Viruses can infect and activate immune cells, and the types of cells infected vary with the types of viruses (**Figure 23-2**). The main viruses that infect T lymphocytes are feline leukemia virus and feline immunodeficiency virus. Feline leukemia virus can infect CD4$^+$ and CD8$^+$ T cells. The main viruses infecting B lymphocytes are equine herpesvirus type 2 and infectious bursal disease virus (IBDV). IBDV selectively replicates in the pre-lymphocytes of the bursal cava and destroys B cells in the bursa of Fabricius, resulting in immunosuppression. The main viruses that infect monocytes are canine distemper virus, influenza A virus, parainfluenza virus, goat arthritis

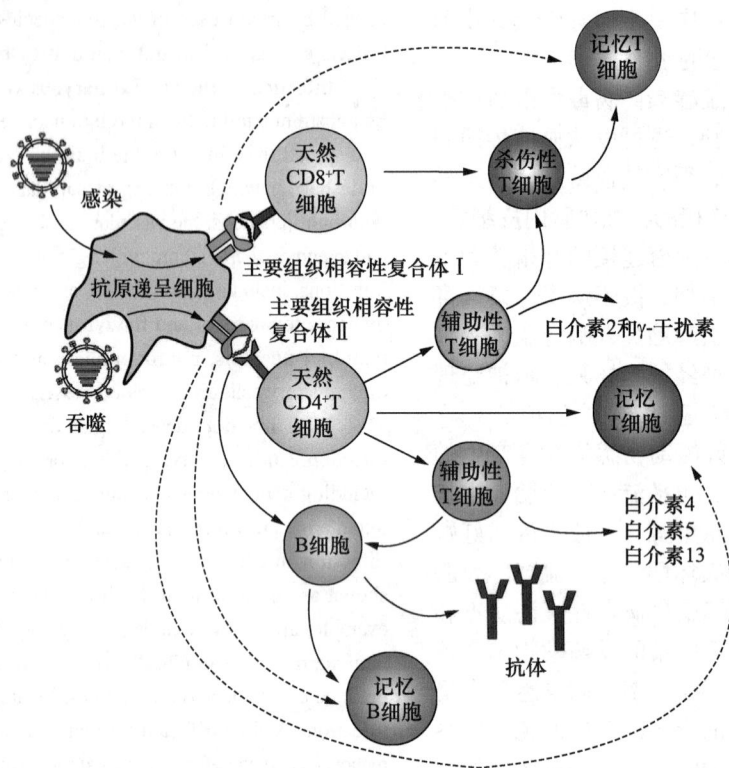

图23-2　体液免疫和细胞免疫

首次感染甲型流感病毒产生的免疫反应用实心箭头表示；第二次感染甲型流感病毒时产生的免疫反应用点状箭头表示，此时病毒激活特异性记忆细胞群的速度更快。图像来源见英文。

Figure 23-2　Induction of humoral and cellular immunity

Induction of immune responses after a primary influenza A virus infection is indicated by solid arrows. The more rapid activation of virus-specific memory cell populations upon secondary encounter with an influenza A virus are indicated by dotted arrows. Source: van de Sandt CE, Kreijtz JH, Rimmelzwaan GF. 2012. Evasion of influenza A viruses from innate and adaptive immune responses. Viruses, 4 (9): 1438-1476.

病毒、副流感病毒、山羊关节炎脑炎病毒、巨细胞病毒等；感染树突状细胞的病毒主要有委内瑞拉马脑脊髓炎病毒和人免疫缺陷病毒等；感染淋巴网状组织基质细胞的病毒有波纳病毒、巨细胞病毒等。

　　大部分病毒感染对宿主造成的损害并不是由于病毒的致细胞病变效应直接引起，而是由于病毒抗原刺激宿主的免疫应答反应，对机体造成间接伤害，是宿主为清除病毒而付出的代价。免疫应答在感染性疾病的致病机理中起到正反两方面的作用，对机体既有保护作用，也有损伤作用。例如，禽白血病病毒、马传染性贫血病毒、猫免疫缺陷病毒等诱导的免疫应答的后果是致命的。

encephalitis virus, cytomegalovirus, *etc.* Venezuelan equine encephalomyelitis virus and human immunodeficiency virus are the main viruses that infect dendritic cells. Viruses that infect the stromal cells of the lymphoid reticular tissue are bornavirus and cytomegalovirus, *etc.*

　　The majority of damage caused by viral infections to the host is not directly due to the cytopathic effects of the virus on cells. Instead, it is primarily caused by the viral antigens stimulating the host's immune response, resulting in indirect harm to the body. This harm is the cost paid by the host in its efforts to eliminate the virus. Immune responses play both positive and negative roles in the pathogenesis of viral infections, which correspond respectively to both protective and damaging effects on the body. For example, the consequences of immune responses induced by avian leukosis virus, equine infectious anemia virus, and feline immunodeficiency virus can be fatal.

思考题　Questions

1. 猪繁殖与呼吸障碍综合征病毒会引起猪繁殖与呼吸障碍综合征，该病毒的排毒方式可能有哪几种？

2. 急性感染与持续性感染对病毒的增殖和传播有何意义？

3. 举例说明病毒感染对中枢神经系统的影响。

1. Porcine respiratory and reproductive syndrone virus (PRRSV) causes porcine reproductive and respiratory syndrome. What are the possible routes of PRRSV shedding?

2. What is the significance of acute infection and persistent infection for virus proliferation and transmission?

3. Give examples of the effects of viral infection on the central nervous system.

第24章 病毒的检测
Chapter 24 / Virus detection

内容提要 动物病毒检测包括分离与鉴定，涉及病料的采集与准备、分离培养、形态学观察、理化特性测定、血清学和分子生物学鉴定等基本过程。

Introduction Virus detection is to isolate and identify the virus, and includes basic processes such as collection and preparation of samples from diseased animals, isolation and cultivation, morphological observation, determination of physical and chemical characteristics, and serological and molecular identification.

第1节 病毒的分离和鉴定
Section 1 Virus isolation and identification

病毒的分离与鉴定包括病料采集和准备、病毒的分离培养、病毒的形态学观察、病毒的理化特性测定、血清学及分子生物学鉴定等基本过程。病毒的形态学观察主要采用电子显微镜。

病毒的采集和准备

病料的采集需要遵循"尽早采样、双抗要放、部位适宜、冷藏速递"的16字原则。

尽早采样是指应在发病初期采集样品，此时宿主的免疫反应还没有显现，样品中容易检测到病毒。

双抗要放是指应在采样缓冲液中加入青霉素和链霉素，目的是防止细菌污染。

部位适宜是指应在感染或排毒部位采样，因为不同的病毒感染的靶器官和排毒方式是不一样的。

Virus isolation and identification include basic processes, such as samples collection and preparation, isolation and cultivation, morphological observation, determination of physiological and biochemical characteristics, serological identification, molecular identification and so on. The morphological observation of the virus is mainly done by electron microscopes.

Sample collection and preparation

The collection of samples needs to follow the principles of "early sampling, addition of double antibiotics, appropriate sampling location and cold-chain express".

Early sampling refers to the collection of samples at the early stage of the disease when the host immune response is not effectively mounted and viruses in the samples could be readily detected.

Addition of double antibiotics refers to the addition of penicillin and streptomycin into the sampling buffer to prevent bacterial contamination.

Appropriate sampling location refers to taking samples from the sites of infection and detoxification which may vary with the type of viruses.

冷藏速递是指应用装有冰块或干冰的专用容器寄送样品。

应在疾病的急性期从活动物采集标本。根据具体的疾病过程，排泄物或分泌物、体腔或液体（淋巴或血液）拭子，以及活检收集的组织都是适合病毒分离的样本。在实验室，组织标本用添加有抗生素的平衡盐溶液处理成10%或20%（m/V）的匀浆。严重污染的标本可过滤以去除其他微生物。病毒分离采用细胞培养，避免污染，必要时在细胞培养基中添加一定浓度的广谱抗生素。用于病毒分离的标本应放置在样品运输培养基中（如含抗生素的平衡盐溶液），并置于密封容器中以确保安全。样品外包装应该标识清楚，并冷藏（4℃）或冷冻（-20℃）存放。

病毒的分离培养

病毒分离培养的方法主要有细胞培养、SPF鸡胚或实验动物。

细胞培养：细胞培养是用于病毒分离培养最常用的方法，将组织匀浆适当稀释后离心，上清液过滤除菌后置于单层细胞上，在35～37℃吸附1 h或更长时间，将培养液去除，并添加新鲜培养基，培养观察7～10 d。大多数致细胞病变的病毒通常在24～72 h之间表现出细胞病变效应（**图24-1**）；不致细胞病变的病毒可在培养若干天后，用特异性抗体进行免疫荧光检测。

9～11日龄SPF鸡胚：鸡胚可用于分离多种哺乳动物病毒和禽类病毒。根据目标病毒的不同，待检测样本可接种于羊膜腔、尿囊腔、卵黄囊或绒毛尿囊膜。

动物接种：对易感实验动物的接种仍是鉴定某些病毒的有用程序，特别是那些非常挑剔和难以在其他系统中增殖的病毒。

Cold-chain express refers to fast sample delivery in special containers containing ice or dry ice.

Specimens should be collected from live animals at the acute phase of the disease. Depending on the specific disease, excretions, secretions, swabs from body orifices, liquids (lymph or blood) and tissues collected by biopsy all are suitable specimens for viral isolation. In the laboratory, tissue specimens are processed into a 10% or 20% (m/V) homogenate in a balanced salt solution with antibiotics. Heavily contaminated specimens may be filtered to remove other microorganisms. Virus isolation should be conducted in cell culture and contamination should be avoided. Appropriate concentrations of broad-spectrum antibiotics can be added, where necessary, into the cell medium. Specimens for viral isolation should be placed into virus transport media (e.g., balanced salt solution containing antibiotics) in sealed containers for safety during handling. The containers should be clearly identified by appropriate labeling and kept on ice (4 ℃) or frozen (-20 ℃).

Virus isolation and culture

Virus isolation and culture methods mainly include cell culture, SPF chicken embryos, and experimental animals.

Cell culture: it is the most commonly used method for virus isolation and culture. To isolate the virus, tissue homogenates should be dilution appropriatly and centrifugation. After filtration and sterilization, the supernatant samples should be placed onto monolayer cells to allow absorption for 1 h or longer at 35-37 ℃. The inocula are then removed and fresh medium added. The cells are observed for 7-10 d and the viral cytopathic effect (CPE) of the cells is usually evident between 24 h and 72 h for most cytopathic viruses (**Figure 24-1**). For the viruses that is not cytopathic, immunofluorescence can be performed with specific antibodies after days of incubation.

9-11 days old SPF chicken embryos: a lot of mammalian and avian viruses can be isolated using embryonated chicken eggs (ECEs). According to the virus of interest, the diagnostic specimen is inoculated into the amniotic cavity, the allantoic cavity or the yolk sac, or onto the chorioallantoic membrane.

Animal inoculation: the inoculation of susceptible laboratory animals remains a useful procedure for identification of some viral pathogens, particularly those that are highly fastidious and difficult to propagate in other systems.

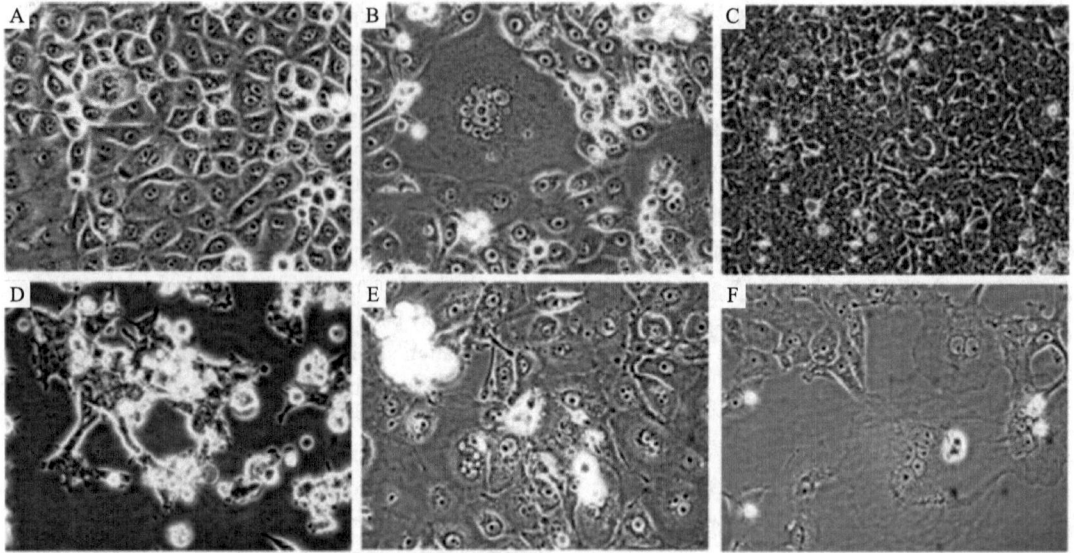

图24-1 接种冠状病毒不同毒株产生的细胞病变效应

A. 未感染病毒的LLC-MK2细胞；B. 感染冠状病毒HCoV-NL63/RPTEC/2004 pp A的LLC-MK2细胞，出现合胞体、圆缩和无细胞区域；C. 未感染病毒的HEK-293细胞；D. 感染冠状病毒HCoV-NL63/RPTEC/2004 pp A的HEK-293细胞，出现细胞病变效应；E. 感染冠状病毒HCoV-NL63/Amsterdam-1的LLC-MK2细胞，出现细胞分离、无细胞区、空泡化和小合胞体；F. 感染冠状病毒HCoV-NL63/Amsterdam-1的Vero细胞，出现空泡、漂浮的死细胞、无细胞胞区和小合胞体。图像来源见英文。

彩图

Figure 24-1   Cytopathic effects formed by HCoV-NL63/RPTEC/2004 pp A and by HCoV-NL63/ Amsterdam-1

A. Non-infected LLC-MK2 cells; B. LLC-MK2 cells infected with HCoV-NL63/RPTEC/2004 pp A, showing a syncytium, rounding of some cells, and areas of clearing; C. Non-infected HEK-293 cells; D. Advanced cytopathic effects in HEK-293 cells with HCoV-NL63/RPTEC/2004 pp A; E. LLC-MK2 cells infected with HCoV-NL63/Amsterdam-1. Detached cells, areas of clearing, vacuolation, and a small syncytium are visible; F. Vero cells infected with HCoV-NL63/Amsterdam-1. Vacuolation, a few floating dead cells, large areas of clearing, and a small syncytium are visible. Source: Lednicky JA, Waltzek TB, McGeehan E, et al. 2013. Isolation and genetic characterization of human coronavirus NL63 in primary human renal proximal tubular epithelial cells obtained from a commercial supplier, and confirmation of its replication in two different types of human primary kidney cells. Virol J, 10: 213.

病毒的理化特性测定

理化特性对病毒，尤其是未知病毒的分离和鉴定尤为重要。理化鉴定包括病毒核酸型鉴定、溶脂性试验、耐酸性试验等。

病毒核酸型鉴定确定病毒为DNA病毒还是RNA病毒。

溶脂性试验确定病毒是否为囊膜病毒。

耐酸性试验确定病毒对酸等消毒剂的敏感性。

Determination of virus physiological and biochemical characteristics

Physical and chemical properties are important for isolation and identification of viruses, especially unknown viruses. The physiological and biochemical identification includes determination of the viral nucleic acid type, sensitivity testing to organic solvents, and acid resistance test.

Identification of the viral nucleic acid type is to determine whether the virus is DNA virus or RNA virus.

Sensitivity testing to organic solvents is to determine whether the virus is an enveloped virus.

Acid resistance test is to determine sensitivity of the virus to acid and other disinfectants.

病毒的血清学及分子生物学鉴定

病毒分离后，可用已知的抗病毒血清或单克隆抗体，对分离毒株进行血清学鉴定，以确定病毒的种类、血清型及亚型。常用的血清学试验有血清中和试验、血凝抑制试验等。此外，可采用一些血清学技术如免疫沉淀技术和免疫印迹技术分析病毒的结构蛋白成分。

分子生物学技术已广泛应用于病毒的鉴定，包括病毒基因的PCR扩增及其序列分析、核酸杂交技术、病毒全基因组序列测定分析等，从而可获得分离毒株的基因组信息，依据基因组序列绘制遗传进化树，分析比较分离毒株的遗传变异情况，确定分离毒株的基因型。

Serological and molecular identification of viruses

After the virus isolation, the antiviral serum or monoclonal antibodies can be used to perform serological identification of the isolated strains to determine the type, serotype and subtype of the virus. Commonly used serological tests include serum neutralization test and hemagglutination inhibition test. In addition, some serological techniques such as immunoprecipitation and immunoblotting techniques can be used to analyze the structural protein components of the virus.

Molecular biological technology has been widely used in virus identification, including PCR amplification of virus genes and sequence analysis, nucleic acid hybridization, virus genome sequence determination and analysis. In this way, the genome information of the isolated strains can be obtained, and the genetic evolution tree can be drawn. Genotypes of the isolated strains can be determined based on comparison of the genetic variations.

第2节 病毒滴度的测定
Section 2 Determination of virus titration

常用病毒滴度的测定方法有空斑测定、终点稀释法和血凝试验。

空斑测定：测定样本中病毒的滴度可以采用空斑试验，其原理是用倍比稀释的病毒感染已形成致密单层状态的宿主细胞，经过一定时间培养使每个感染细胞周围的细胞逐渐感染崩溃，形成肉眼可见的空斑。

终点稀释法：用于测定几乎所有种类的病毒滴度，包括某些不能形成空斑的病毒（用特异抗体进行免疫荧光显示），并可用以确定病毒对动物的毒力或毒价。将病毒做系列稀释，选择4～6个稀释度，接种一定数量的细胞、鸡胚或动物，每个稀释度做3～6个重复。使用细胞培养，可通过CPE来判定组织培养半数感染量（$TCID_{50}$）。在鸡胚或动物中，以死亡或发病来确定。以感染发病

Commonly used methods for determining the virus titers include the plaque assay, end-point dilution method, and hemagglutination test.

Plaque assay: one of the methods determining the titers of the virus in samples is the plaque assay. The principle is to use the doubly diluted virus to infect host cells in a dense monolayer. After a certain period of cultivation, the cells around the infected cells gradually become infected and ultimately rupture, forming visible plaques.

End-point dilution method is used to determine the titers of almost all types of viruses, including some that do not form plaques (specific antibody-based immunofluorescence). It is also used to determine virulence to animals in the form of virulence titers. The virus is diluted in series. 4-6 dilutions are chosen to inoculate a certain number of cells, chicken embryos or animals with 3-6 replicates per dilution. With cell culture, CPE can be used to determine the tissue culture infectious dose 50 ($TCID_{50}$). Which were determined by the number of death or morbidity in chicken embryos or animals. When the onset of infection is used as an indicator, the median infective dose (ID_{50})

作为指标时，可计算半数感染量（ID_{50}）；以体温反应作为指标时，可计算半数反应量（RD_{50}）；用鸡胚测定时，可计算鸡胚半数致死量（ELD_{50}）或鸡胚半数感染量（EID_{50}）。

血凝试验：许多动物病毒，如正黏病毒科、副黏病毒科、腺病毒科的成员，能够凝集某些种类动物（如鸡、小鼠、豚鼠、人）的红细胞。这些病毒含有可与红细胞结合的蛋白质，如禽流感病毒的囊膜上有一种称为血凝素的糖蛋白，可与红细胞表面的 N-乙酰神经氨酸糖蛋白结合，引起红细胞凝集。利用这种特性，可做血凝试验来检测这些病毒的存在以及病毒的含量。一般将病毒液在血凝反应板上做倍比稀释，加入红细胞，未凝集的红细胞呈圆点或纽扣状沉于孔底，而凝集的红细胞呈弥漫状覆盖整孔。

can be calculated. When the body temperature response is used as an indicator, median reaction dose (RD_{50}) can be calculated. When chicken embryos are used, median embryo lethal dose (ELD_{50}) or median embryo infective dose (EID_{50}) can be calculated.

Hemagglutination (HA) test: many animal viruses, such as members of *Orthomyxoviridae*, *Paramyxoviridae* and *Adenoviridae*, can agglutinate the red blood cells of certain species of animals (such as chicken, mice, guinea-pig, and human). These viruses contain proteins that can bind to red blood cells. For example, there is a glycoprotein called as hemagglutinin on the envelope of avian influenza virus, which can bind to the *N*-acetylneuraminic glycoprotein on the surface of red blood cells and cause hemagglutination. Hemagglutination test is based on this property and can be used to detect the presence and the virus content of these viruses. Generally, the virus solution is double diluted on a hemagglutination reaction plate, and then the red blood cells are added. Unagglutinated RBCs are round, dot or button-like sunk at the bottom of the pore, while the agglutinating RBCs showed a diffuse pattern covering the whole hole.

第3节　病毒颗粒的检测
Section 3　Detection of virus particles

病毒颗粒的观察与检测可以借助电子显微镜。

电子显微镜术：电子显微镜技术（EM）可用于识别标本中或在细胞培养中分离出的任何病毒的形态和大小。不同的病毒在电子显微镜下呈现的形态不同，如狂犬病病毒为子弹状，痘病毒为椭圆形或砖头状。

免疫电镜术：免疫电子显微镜技术（IEM）通过使特异性抗体与病毒反应，增强电镜对组织、细胞或粪便样本中病毒的检测。

The observation of virus particles can be performed by electron microscope.

Electron microscopy: electron microscopy (EM) can be used to identify the morphology and size of any virus isolated in specimens or in cell culture. Different viruses show different morphologies under the electron microscope. For example, rabies virus is bullet-like, while poxvirus is oval or brick-like.

Immune electron microscopy: immune electron microscopy (IEM) enhances detection of viruses in tissues, cells, or fecal specimens by EM through reaction of specific antibodies with corresponding viruses.

第4节　病毒感染的血清学检测
Section 4　Serologic detection of viruses

病毒血清学检测可针对病毒或病毒蛋白，主要包括病毒中和试验、血凝抑制试

Virus serological testing may target viruses or viral proteins, including virus neutralization test (VN),

验、补体结合试验和病毒蛋白检测。

病毒中和试验： 大多数病毒在细胞培养中产生可见的细胞病变效应（CPE）。CPE常用于测定血清中是否存在保护性抗体或病毒中和抗体。

血凝抑制试验： 血凝抑制试验广泛应用于那些能凝集一种或多种红细胞的病毒，如节肢动物携带的病毒、流感病毒和副流感病毒，可用于检测病毒或抗体（**图24-2**）。

hemagglutination inhibition (HI) test, complement fixation test (CFT) and detection of viral proteins.

Virus neutralization test: most viruses produce visible CPE in cell cultures. CPE is used as an indicator to determine the presence of protective or virus-neutralizing antibodies in the serum.

Hemagglutination inhibition tests are widely used in those viruses that can agglutinate one or more types of red blood cells, such as the arthropod-borne viruses, influenza virus, and parainfluenza virus and can be used to detect viruses or antibodies (**Figure 24-2**).

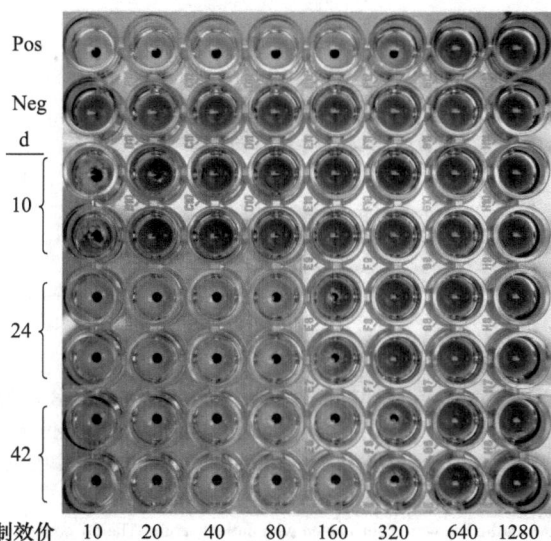

彩图

图24-2　分别在第10、24和42天，通过血凝抑制试验检测病人2009年甲型H1N1流感大流行时的抗体效价
Pos: 阳性对照孔，Neg: 阴性对照孔。图像来源见英文。

Figure 24-2　Antibody titer against influenza A pandemic (H1N1) 2009 by hemagglutination inhibition (HI) test on days 10, 24, and 42 of life of the patient
Pos: Positive, Neg: Negative. Source: Dulyachai W, Makkoch J, Rianthavorn P, et al. 2010. Perinatal pandemic (H1N1) 2009 infection, Thailand. Emerg Infect Dis, 16 (2): 343-344.

补体结合试验： 补体是血清中存在的特定蛋白质，其有两个特点：第一是可以和抗原抗体复合物结合，第二是可以凝集绵羊红细胞，从而导致溶血。如果抗原抗体复合物和补体同时存在，三者会形成复合物，补体的结合位点被封闭，不再结合绵羊血细胞，因此不产生溶血现象，即补体结合试验阳性（**图24-3**）。

免疫荧光或荧光抗体染色 是一种抗原检测试验，包括直接免疫荧光和间接免疫荧光，主要用于冷冻组织切片、细胞涂片或培养细胞中特定抗原成分的检测。

酶联免疫吸附试验（ELISA） 是一

Complement fixation test (CFT): complement is a specific protein in the serum. They have two characteristics. First, they can bind to the antigen-antibody complex. Second, they can agglutinate sheep's red blood cells and cause hemolysis. When the antigen-antibody complex and complement coexist, these three will form a complex. Then, the binding site of complement will be blocked and unavailable to bind to sheep blood cells. Therefore, hemolysis will not occur, indicating positive reaction of CFT (**Figure 24-3**).

Immunofluorescence or fluorescent antibody staining: an antigen detection test, including direct immunofluorescence and indirect immunofluorescence, mainly used for specific detection of antigens in frozen tissue sections, cell smears or cultured cells.

Enzyme-linked immunosorbent assay (ELISA) is a highly specific and sensitive immunoassay (**Figure 24-4**), in which the specificity of the reaction

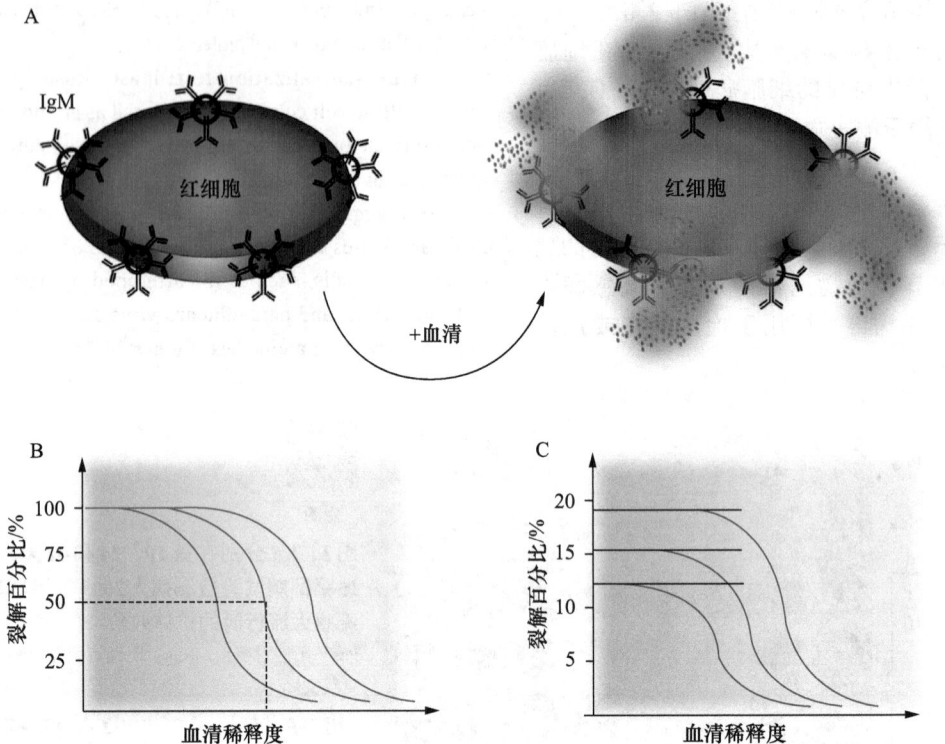

图24-3 检测补体经典途径（CP）功能的溶血试验

A. 溶血试验：包被IgM抗体的绵羊红细胞与患者血清共同孵育后，C1复合物结合并启动补体经典途径形成转化酶，激活C3，组装C5b-9复合物，红细胞裂解；B. 半数补体溶血量CH_{50}测定：即确定溶解50%指定和固定数量红细胞所需的血清量。C. 单管补体经典途径测定：补体的活性与溶解的细胞数量成正比，因此可在过量红细胞中进行测定。图像来源见英文。

Figure 24-3 Haemolytic assay for complement classical pathway function

A. Haemolytic assay. Sheep erythrocytes coated with IgM antibodies were coincubated with patient serum. The C1 complex binds and initiates formation of the CP convertase, activate C3, assemble the C5b-9 complex, and subsequent erythrocyte lysis; B. CH_{50} assay. Titration of the amount of serum needed to lyse 50% of a specified limited and fixed quantity of cells in the CH_{50} assay; C. Single CP assay. Since the activity of complement is proportional to the quantity of cells that are lysed, this assay is performed in an excess of erythrocytes. Source: Ekdahl KN, Persson B, Mohlin C, et al. 2018. Interpretation of serological complement biomarkers in disease. Front Immunol, 9: 2237.

种特异性好、灵敏度高的免疫分析方法（**图24-4**），可通过使用纯化的抗原或抗体增强反应的特异性。常采用间接法和阻断法检测抗体水平。

免疫层析指抗原样品在过滤基质以横向流动形式迁移，与已标记的抗体（如胶体金标记）结合后继续迁移，直至被另一种固定在基质中的未标记的特异抗体结合（双抗夹心）后因胶体金纳米颗粒聚集而呈现红色条带；多余的标记抗体-抗原复合物继续迁移至抗金标抗体线，亦出现红

can be enhanced by using purified antigen or antibody. Indirect ELISA and blocking ELISA are commonly used to determine the level of antibody responses.

Immunochromatography refers to the migration of antigens in samples through a filter matrix via lateral flow. When the antigen migrates to an area where there is a labeled specific antibody (*e.g.*, colloidal gold label), antigen-antibody complexes form. These complexes continue to migrate to an area where another unlabeled specific antibody is immobilized in the support matrix, forming labeled antibody-antigen-antibody complexes (double antibody sandwich) shown as a red line due to aggregation of the colloidal gold nanoparticles. Parts of the labeled antibody-

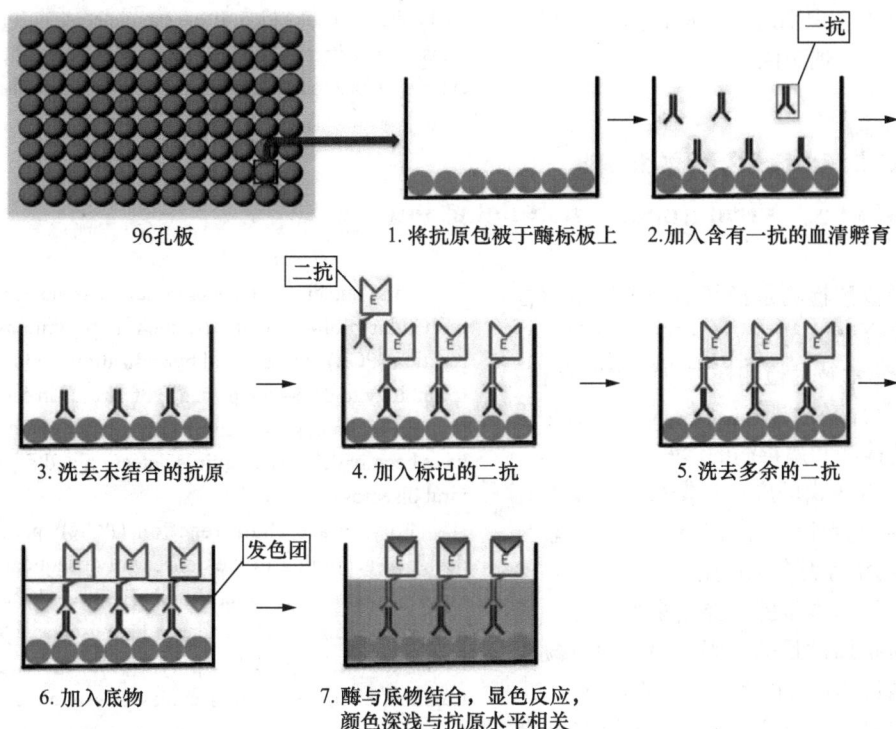

96孔板　　　　1. 将抗原包被于酶标板上　　2. 加入含有一抗的血清孵育

3. 洗去未结合的抗原　　4. 加入标记的二抗　　5. 洗去多余的二抗

6. 加入底物　　7. 酶与底物结合，显色反应，
颜色深浅与抗原水平相关

图 24-4　酶联免疫吸附试验（ELISA）技术——以检测给定样本中的抗原为例
将抗原（液相）加入酶标板孔中进行包被，然后加入一抗与抗原特异性结合；添加酶标二抗与一抗反应；最后通过酶与底物产生的颜色反应定量或定性检测抗原。图像来源见英文。

Figure 24-4　Enzyme-linked immunosorbent assay (ELISA) technique used to detect an antigen in a given sample
The antigen (in liquid phase) is added to the wells, where it adheres to the walls. Primary antibody binds specifically to the antigen. An enzyme-linked secondary antibody is added that reacts with a chromogen, producing a color change to quantitatively or qualitatively detect the antigen. Source: Gan SD, Patel KR. 2013. Enzyme immunoassay and enzyme-linked immunosorbent assay. J Invest Dermatol, 133 (9): e12.

色条带，为对照线。此即胶体金试纸条检测原理。

　　免疫扩散通常被用作为一种诊断工具，用于检测各种动物疾病（如蓝舌病、马传染性贫血、牛白血病、山羊关节炎、传染性法氏囊病等）中的特定病毒。该检测的基础是某些可溶性病毒抗原具有在半固体琼脂中扩散的能力并与特异性抗血清形成沉淀线。

　　免疫印迹分析可以利用特异抗体检测病毒蛋白。当病毒样品经适当处理后通过凝胶电泳使病毒蛋白分离，再转膜（如硝酸纤维素膜），先后用特异性抗体和酶标二抗进行孵育和显色，若样品中有病毒，

antigen complexes move further to the line where a specific antibody to the labeled antibody is immobilized, forming another red line as a positive control. This is the principle of the immunogold lateral test strip.

　　The immunodiffusion is routinely used as a diagnostic tool to monitor the specific viral pathogens in various animal diseases (*e.g.*, bluetongue, equine infectious anemia, bovine leukosis, caprine arthritis, infectious bursal disease, *etc.*). The basis of the test is that certain soluble viral antigens can diffuse in a semi-solid agar and form a precipitation line with specific antisera.

　　The Western blot can detect viral proteins that react with specific antibodies. When the virus sample is properly treated and subjected to gel electrophoresis to separate the viral proteins. The proteins on the gel are then blotted onto the nitrocellulose strip, followed by probing with primary and enzyme-labeled

特异性抗体会识别对应的病毒蛋白，在其适当的位置呈现颜色。

antibodies, and coloration. If the sample contains a particular virus, one of its proteins will be recognized by the specific antibody with a colorized band in an appropriate position.

第5节 病毒核酸的检测
Section 5   Viral nucleic acid detection

病毒核酸检测是指利用分子生物学方法，如聚合酶链式反应（PCR）、核酸杂交和DNA-芯片技术检测样本中病毒的存在。目前，聚合酶链式反应等技术已广泛应用于病毒的检测和病毒病的诊断。

聚合酶链式反应：根据GenBank中病毒保守序列设计引物，以待检病毒核酸为模板，在DNA聚合酶作用下，经变性、退火、延伸等基本步骤，经过多次循环，最后获得所扩增的目的基因片段。通过凝胶电泳检测目的片段的大小，甚至测序。

核酸杂交：分为DNA杂交和RNA杂交。DNA杂交用于检测病毒DNA，DNA样本经限制性内切酶消化，凝胶电泳，变性后转移到膜（如硝酸纤维素膜）上，然后用放射性同位素标记的病毒核酸序列探针检测结合到固相膜上的DNA。目前放射性同位素标记的核酸探针很少使用，而大多用非放射性标记探针（如生物素标记）进行检测。斑点杂交是一种改良的方法，可用于样本中病毒核酸的快速检测。RNA杂交用于病毒RNA的检测，基本过程与DNA杂交相似。核酸杂交技术可用于细胞、组织中病毒基因组或转录产物的定位检测，称为原位杂交。

Viral nucleic acid detection refers to the use of some molecular biology methods, such as polymerase chain reaction (PCR), nucleic acid hybridization, and DNA chip technology to detect the presence of viruses in samples. At present, technologies such as polymerase chain reaction have been widely used in virus detection and diagnosis of viral diseases.

Polymerase chain reaction (PCR): primers are designed according to the conservative sequence of the virus of interest deposited in GenBank, and the nucleic acids of the samples to be detected are used as a template. Under the action of DNA polymerase, the target gene fragment is amplified after cycles of sequential steps of denaturation, annealing and extension. Finally, the size of the target fragment is detected by gel electrophoresis or even by sequencing.

Nucleic acid hybridization: nucleic acid hybridization is divided into DNA hybridization and RNA hybridization. DNA hybridization or called Southern blotting is used to detect viral DNA. The DNA sample is digested with restriction enzymes and subjected to gel electrophoresis. Then, it is denatured and transferred onto the nitrocellulose membrane, and immobilized DNA is detected using a radioisotope-labeled DNA probe specific to the sequence of the virus of interest. Currently, non-radiolabeled probes (such as biotin-labeled) are preferred to radioisotope-labeled nucleic acid probes used for detection. A simplified method, called as dot blotting hybridization, can be used for rapid detection of viral nucleic acids in samples. RNA hybridization, or Northern blotting, is used for detection of viral RNA. The basic process is similar to that of DNA hybridization. Nucleic acid hybridization technology can be used to detect the viral genomes or transcription products in cells and tissues, which is called as *in situ* hybridization.

思考题　Questions

1. 如何采集病毒检验样本？病料采集的原则以及注意事项有哪些？

1. How to collect test samples for detection of viruses? What are the principles and precautions for collection of clinical materials?

2. 简要说明血凝试验和血凝抑制试验的原理。

2. Briefly explain the principles of hemagglutination test and hemagglutination inhibition test.

3. 病毒血清学检测主要包括哪些方法？请对各方法进行简要描述。

3. What are the main methods of virus serological testing? Please give a brief description of each method.

4. 酶联免疫吸附试验和免疫层析各有什么特点？

4. What are the characteristics of enzyme-linked immunosorbent assay and immunochromatographic?

第25章 双链DNA病毒
Chapter 25 / Double-stranded DNA viruses

内容提要 基因组为双链DNA的病毒包含痘病毒科、非洲猪瘟病毒科、腺病毒科、虹彩病毒科、疱疹病毒科、乳头瘤病毒科、多瘤病毒科、杆状病毒科等。其中，痘病毒科、非洲猪瘟病毒科、腺病毒科是兽医病毒学研究的重要对象。

Introduction Viruses with double-stranded DNA genomes include the families *Poxviridae*, *Asfarviridae*, *Adenoviridae*, *Iridoviridae*, *Herpesviridae*, *Papillomaviridae*, *Polyomaviridae*, *Baculoviridae, etc.* Among them, *Poxviridae*, *Asfarviridae* and *Adenoviridae* are important subjects of veterinary virology.

第1节 痘病毒科
Section 1 *Poxviridae*

痘病毒科是病毒粒子最大的一类DNA病毒，结构复杂。痘病毒是引起人和动物局部或全身化脓性皮肤损害的重要病原之一。英国医生爱德华·琴纳用牛痘病毒预防人类天花，从而在全世界消灭了天花，是病毒免疫的里程碑事件。不同属间的痘病毒有一定的抗原交叉反应，但禽痘病毒与哺乳动物痘病毒之间不能交叉感染和交叉免疫。

形态结构：痘病毒的病毒粒子呈砖形或大卵圆形，有囊膜，相对分子质量为 $8.5 \times 10^7 \sim 2.4 \times 10^8$，大小约为 360 nm × 270 nm × 250 nm。其中，副痘病毒属的病毒粒子比较特殊为纺锤状，大小为 260 nm × 160 nm。大多数痘病毒的病毒粒子外层上有类似管状物的凸起结构（**图25-1**），而副痘病毒粒子的外观如线团。核衣壳复合对称，病毒颗粒的外层结构之内是哑铃样

Poxviridae virion is the largest DNA virus with complex structure. Poxviruses are one group of the important pathogens causing local or systemic purulent skin lesions in humans and animals. British doctor Edward Jenner used cowpox virus to prevent human smallpox, leading to its eradication in the world, which is a milestone in virus immunization. Poxviruses in different genera have some antigenic cross-reactions, but there are no cross-infections or cross-protection between avian poxviruses and mammalian poxviruses.

Morphological structure: poxvirus virions are usually brick-shaped or ovoid, and enveloped. The relative molecular mass is 8.5×10^7-2.4×10^8. The size is about 360 nm × 270 nm × 250 nm. *Parapoxvirus* virions are fusiform, with a size of 260 nm × 160 nm. Most poxviruses have projecting tubular structures on the surface of their virions (**Figure 25-1**), while *Parapoxvirus* virions look like a ball of yarn. The nucleocapsid of the poxviruses is symmetric. The outer structure of the virus particles is composed of a dumbbell-shaped core and two lateral bodies. The core

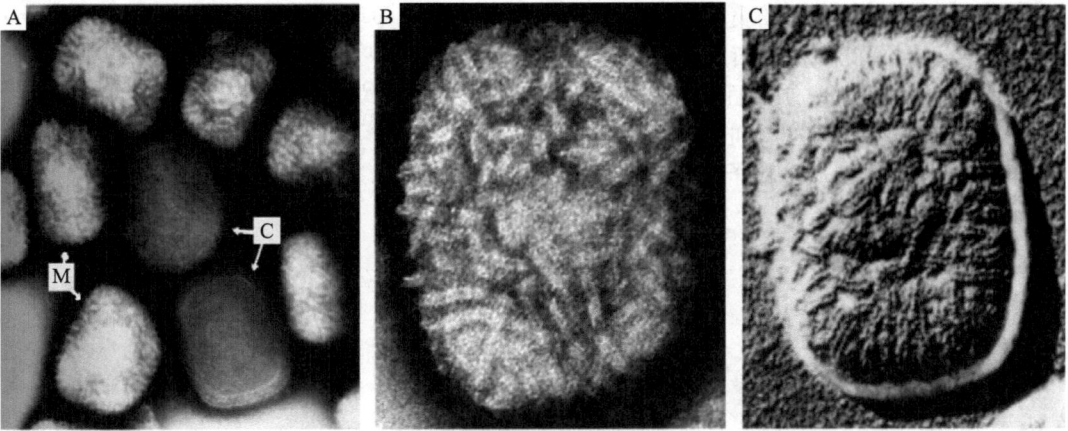

图25-1　痘苗病毒的表面特征

A. 痘苗病毒电镜照片，负染，可见"桑葚"（M）和"胶囊"（C）状；B. 痘苗病毒粒子的电镜照片，负染；C. 痘苗病毒粒子的冷冻蚀刻照片。图像来源见英文。

Figure 25-1　Vaccinia virus surface features

A. Electron micrograph of purified virus, whole mount, negative stain. "Mulberry" (M) and "capsule" (C) forms are shown; B. Electron micrograph of a purified virion, whole mount, negative stain; C. Freeze etch electron micrograph of a purified virion. Source: A: Westwood JC, Harris WJ, Zwartouw HT, et al. 1964. Studies on the structure of vaccinia virus. J. Gen. Microbiol, 34: 67-78. B: Wilton S, Mohandas AR, Dales S. 1995. Organization of the vaccinia envelope and relationship to the structure of intracellular mature virions. Virology, 214: 503-511. C: Nermut MV. 1973. Freeze-drying and freeze-etching of viruses//Benedetti EL, Favard P. Freeze-Etching Techniques and Applications. Paris: Societe Francaise de Microscopie Electronique, 135-150.

的芯髓和两个侧体，芯髓内含病毒DNA。

基因组：痘病毒基因组由单分子的线状双链DNA组成，不同种属的基因组大小在130～375 kb之间，包含328个开放阅读框（ORF），编码约200多种蛋白质，其中100多种是结构蛋白。基因组序列分析表明，位于基因组中间约100 kb的核苷酸序列在所有痘病毒中高度保守，两端存在多个高变的末端反向重复序列（ITR），大小0.1～13 kb，而在多数痘病毒成员的ITR中又包含许多复制基因。某些基因与宿主范围有关，被称作宿主范围基因。目前已经证实，在不同的痘病毒中至少有16个宿主范围基因。

感染及复制：痘病毒的成员表现出不同的宿主范围，即使在同一属中亲缘关系较近的种间，其宿主范围具有很大的差异。正痘病毒属的天花病毒仅感染人，而同属的牛痘病毒和猴痘病毒却能感染许多种哺乳动物，具有相对广泛的宿主谱。痘

contains viral DNA.

Genome: genomes of poxviruses are composed of a single molecule of linear double-stranded DNA with 130-375 kilobase (kb) in length which contains up to 328 open reading frames (ORFs) length depending on the virus species. The genomes encode more than 200 proteins in which 100 are structural proteins. Sequence analysis has shown that the 100 kb nucleotide sequence in the middle of the genome is highly conserved in all poxviruses, and there are several highly variable inverted terminal repeats (ITRs) at both ends, being 0.1-13 kb in length. The ITRs in most poxviruses contain many replication-related genes. Some of these genes are related to host ranges and are called host range genes (Hrg). At least 16 Hrg genes have been identified so far in different poxviruses.

Infection and replication: members of the poxviruses display different host ranges, even among closely related species within the same genus. Variola virus (VARV) of the genus *Orthopoxvirus* infects only people, while cowpox virus (CPXV) and monkeypox virus (MPXV) of the same genus can infect many kinds of mammals and display a wide host range. During their infection,

病毒在感染的过程中，病毒囊膜与细胞膜融合并将病毒核酸释放到细胞中，在胞质中完成病毒的全部复制过程（**图 25-2**）。痘病毒复制完成后，通过胞吐方式出芽，而非细胞裂解释放出病毒粒子。大多数痘

the viral envelope fuses with the cell membrane and releases viral nucleic acids into cells, and the virus completes the whole replication cycle in the cytoplasm (**Figure 25-2**). After the poxviruses finish replication, they exit the cell via exocytosis rather than cell lysis. Characteristic circular or ovoid inclusion bodies are

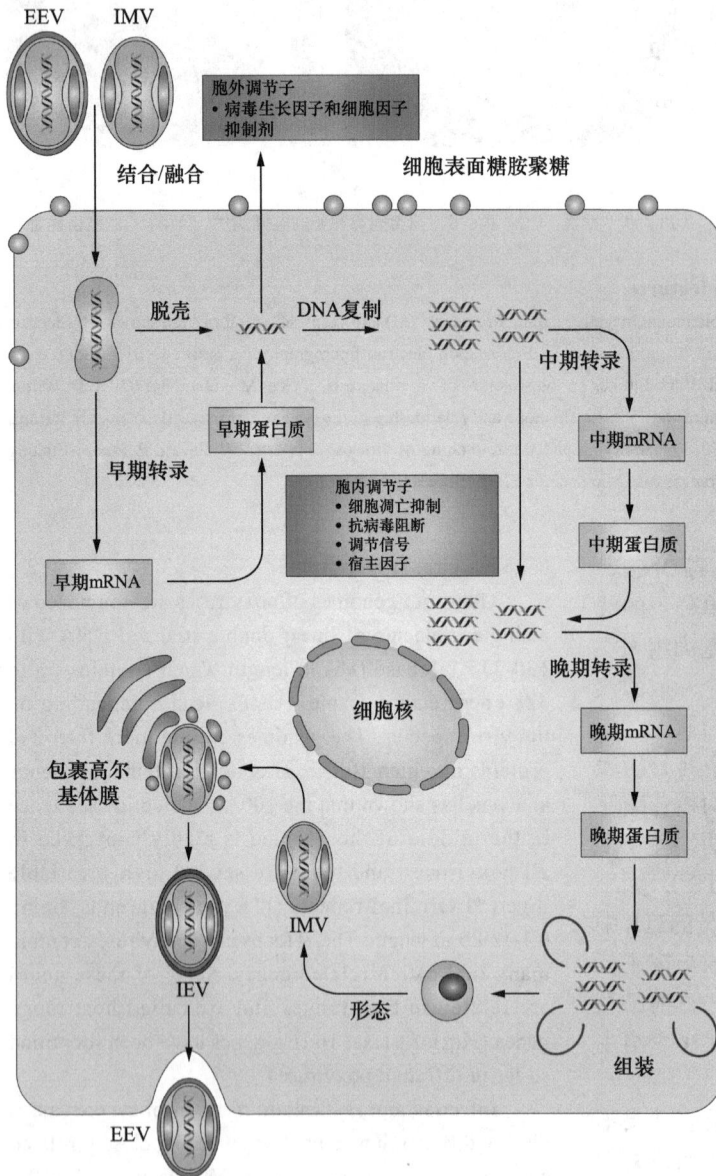

图 25-2　痘病毒的复制周期

痘病毒在感染细胞的细胞质中通过复杂但保守的形态途径进行复制。痘病毒有两种不同的传染性病毒颗粒：细胞内成熟病毒（IMV）和细胞外囊膜病毒（EEV）。IMV和EEV病毒粒子的区别在于表面糖蛋白和包裹囊膜的数量不同。完整的病毒复制以病毒mRNA和蛋白质合成为特征，包括早期、中期和晚期三个阶段，随后病毒颗粒发生形态变化。最初的IMV通过微管运输，并被高尔基体衍生膜包裹，之后称为细胞内囊膜病毒（IEV）。IEV又与细胞表面膜融合形成细胞相关囊膜病毒（CEV），该病毒通过肌动蛋白尾部聚合通过胞吐释放形成游离EEV。EEV也可通过IMV直接出芽并绕过IEV步骤释放到胞外。图像来源见英文。

Figure 25-2　Poxvirus replication cycle

All poxviruses replicate in the cytoplasm of infected cells by a complex, but largely conserved, morphogenic pathway. Two distinct infectious virus particles, the intracellular mature virus (IMV) and the extracellular enveloped virus (EEV), can initiate infection. The IMV and EEV virions differ in their surface glycoproteins and in the number of wrapping membranes. Fully permissive viral replication is characterized by three waves of viral mRNA and protein synthesis (known as early, intermediate and late), which are followed by morphogenesis of infectious particles. The initial IMV is transported via microtubules and is wrapped with Golgi-derived membrane, after which it is referred to as an intracellular enveloped virus (IEV). The IEV fuses to the cell surface membrane to form cell-associated enveloped virus (CEV), which can either be extruded away from the cell by actin-tail polymerization or is released to form free EEV. EEV might also form by direct budding of IMV, therefore bypassing the IEV form. Source: McFadden G. 2005. Poxvirus tropism. Nat Rev Microbiol, 3 (3): 201-213.

病毒感染的特征，是在感染细胞内形成圆形或卵圆形包涵体。

理化性质：痘病毒对热抵抗力不强，55 ℃作用20 min或37 ℃作用24 h丧失感染力。对冷及干燥有一定的抵抗力，冷冻干燥可保存3年。在干燥痂皮中的病毒至少能存活几个月，在正常条件下的土壤中可生存几周。在pH 3的环境下，病毒可逐渐失去感染能力。直射日光或紫外线可导致病毒灭活。对氯化剂和氧化剂敏感，0.5%福尔马林、3%石炭酸、0.01%碘溶液、3%苯酚、3%硫酸及盐酸都可在数分钟内使病毒失去感染力。正痘病毒属和禽痘病毒属的成员可耐受乙醚，其他属则对乙醚敏感。

传播：痘病毒在动物间的传播途径包括皮肤的伤口感染，由污染环境的直接或间接传染。例如，口疮病毒通过呼吸道感染，绵羊痘病毒、猪痘病毒、鸡痘病毒及黏液瘤病毒可通过昆虫叮咬感染。

培养：痘病毒可在发育的鸡胚绒毛尿囊膜上生长，多数可在膜上形成痘斑，如山羊痘病毒、禽痘病毒、牛痘病毒等。痘斑的形态、颜色、大小及形成的时间因痘病毒的种类而异。黏液瘤病毒引起的痘斑比较小，独立而分散分布，呈灰白色。而牛痘病毒、痘苗病毒常侵害血管形成出血坏死性的痘斑。可根据痘斑特征，对病毒做出初步鉴定。各种痘病毒均可在同种动物的单层细胞（肾、胚胎组织、睾丸等）生长良好，引起CPE或肉眼可见的空斑。

formed in the cytoplasm of most poxviruses infected cells.

The physical and chemical properties: poxviruses have low resistance to heat. 55 ℃ for 20 min or 37 ℃ for 24 h will lead to loss of their infectivity. They have certain resistance to coldness and dryness and can be stored for 3 years in a freeze-drying environment. Viruses can survival for at least a few months in dry scabs and for several weeks in normal soil conditions. In an environment with pH 3, they may gradually lose their infectivity. Direct sunlight or ultraviolet can inactivate the viruses. They are sensitive to chlorination agents and oxidants, and 0.5% formalin, 3% carbolic acid, 0.01% iodine solution, 3% phenol, and 3% sulfuric acid or hydrochloric acid can lose their infectivity in a few minutes. Members of the genera *Orthopoxvirus* and *Avipoxvirus* are tolerant to ether, while others are sensitive.

Transmission: poxviruses are transmitted among animals by infecting wounds on the skin or by direct or indirect contamination of the environment. For example, aphthous ulcer virus is infected through respiratory tract, and sheeppox virus, swinepox virus, fowlpox virus and myxoma virus can be infected through insect biting.

Cultivation: poxviruses, such as goat pox virus, fowl pox virus, and cowpox virus, grow on the chorioallantoic membrane of the developing chicken embryo, and most of them can form blotches on the membrane. The shape, color, size, and time of formation of the blotches vary with the species of poxviruses. The blotches caused by myxoma virus are relatively small, independent and scattered, and grayish-white, while vaccinia virus and cowpox virus often invade blood vessels to form hemorrhagic necrotizing spots. The virus can be preliminarily identified according to the features of the blotches or spots. All poxviruses can grow well in monolayer cells (kidney, embryonic tissue, testis, *etc.*) of the same animal species, causing CPE or visible plaques.

第2节　非洲猪瘟病毒科
Section 2　*Asfarviridae*

非洲猪瘟病毒（ASFV）的复制特性与其他核质大DNA病毒家族的病毒类似。2005年，国际病毒分类委员会将ASFV归类于非洲猪瘟病毒科，非洲猪瘟病毒属，

The replication characteristics of African swine fever virus (ASFV) are similar to those of other nucleocytoplasmic large DNA virus families. In 2005, African swine fever virus was classified into *Asfarviridae* and named as *Asfivirus* (only one member of *Asfarviridae*) by International Committee of

ASFV是唯一成员。

我国于2018年8月在辽宁省沈阳市发生第一例非洲猪瘟，随后传遍全国，严重影响我国养猪业经济效益。非洲猪瘟病毒自1921年在肯尼亚首次发现以来，一直存在于撒哈拉以南的非洲国家，1957年先后传播到欧洲和拉丁美洲，死亡率高达100%；2007年出现于俄罗斯。世界动物卫生组织（WOAH）将非洲猪瘟列为法定报告动物疫病，我国也将其列为一类动物疫病，是重要的检疫对象。在没有疫苗或治疗方法的情况下，扑杀病猪是控制疫情暴发的最有效方法。

形态结构： 非洲猪瘟病毒是一种大型二十面体DNA病毒，病毒粒子直径为175～215 nm。病毒粒子具有独特的多层结构。外层是含脂质的囊膜，为ASFV出芽时获得于细胞的质膜。囊膜下是典型的二十面体对称的蛋白质衣壳。ASFV的衣壳由2760个伪六聚体壳粒和12个五聚体壳粒构成。p72是关键的保护性抗原之一，在病毒衣壳上形成同源三聚体，每个单体采用双果冻卷结构，从而构成伪六聚体壳粒（也称为p72壳粒）。内膜主要来源于内质网膜。清华大学饶子和院士团队使用优化的图像重建策略，解析了精细度高达4.1埃的ASFV衣壳结构，该结构由17280个蛋白质构成，包括一个主要衣壳蛋白p72和四个次要衣壳蛋白（M1249L、p17、p49和H240R），共同组成五重对称和三重对称结构（**图25-3**）。

基因组： ASFV是一种双链的核质大DNA病毒，基因组由一个线性的双链DNA分子组成，全长170～193 kb，主要由中央保守区（C区）、左可变区、右可变区组成，基因组编码151～167个ORF，这些ORF存在于双链DNA链上，不同ORF之间间隔序列小于200 bp，部分间隔区还存在短的串联重复。ASFV的ORF编码约68种结构蛋白和约100种非结构蛋白。根据p72结构蛋白基因3'端序列可将ASFV

Taxonomy of Viruses (ICTV).

In August 2018, the first case of African swine fever occurred in Shenyang, Liaoning province, and subsequently spread throughout the country, causing serious economic losses to the pig industry in China. ASFV has been present in sub-Saharan Africa since it was first discovered in Kenya in 1921 and spread to Europe and Latin America successively in 1957, with a mortality rate of 100%. It appeared in Russia in 2007. ASF is listed by World Organisation for Animal Health (WOAH) as a statutorily reportable animal disease, and by China as a class A animal disease. ASF is an important disease subject to quarantine. In the absence of vaccines or treatments, slaughtering sick pigs is the most effective way to control outbreaks of ASF.

Morphology: ASFV is a large icosahedral DNA virus with a size of 175-215 nm in diameter. ASFV virions are characterized by their unique multi-layered structure. The outer layer is a lipid envelope derived from the membrane of the host cell during the budding process. Underneath the membrane, there is a complex icosahedral protein capsid, which consists of 2760 pseudo-hexamers and 12 pentamers. The protein p72 is one of the key protective antigens. On the viral capsid, p72 forms a homotrimer, with each monomer in a double jelly-roll structure that forms pseudo-hexameric capsomeres. The inner membrane mainly comes from the endoplasmic reticulum. Using an optimized image reconstruction strategy, Zihe Rao's group from Tsinghua University solved the ASFV capsid structure up to 4.1 angstroms, which is built from 17280 proteins, including one major (p72) and four minor (M1249L, p17, p49, and H240R) capsid proteins organized into pentasymmetrons and trisymmetrons (**Figure 25-3**).

Genome: ASFV is a large nucleoplasmic DNA virus that contains a linear double-stranded DNA genome of 170-190 kb in size. It mainly consists of the central conserved region (C region) in the middle and the left and right variable regions (VL and VR region) at both ends. The genome contains 151-167 ORFs that are closely spaced (less than 200 bp, some with short tandem repeats) and read from both DNA strands. Of the total ORFs, approximately 68 code for structural proteins and about 100 for non-structural proteins. According to the 3' end coding sequences of the p72 structural protein, ASFV can be divided into 24 genotypes. The virulence of different genotypes or

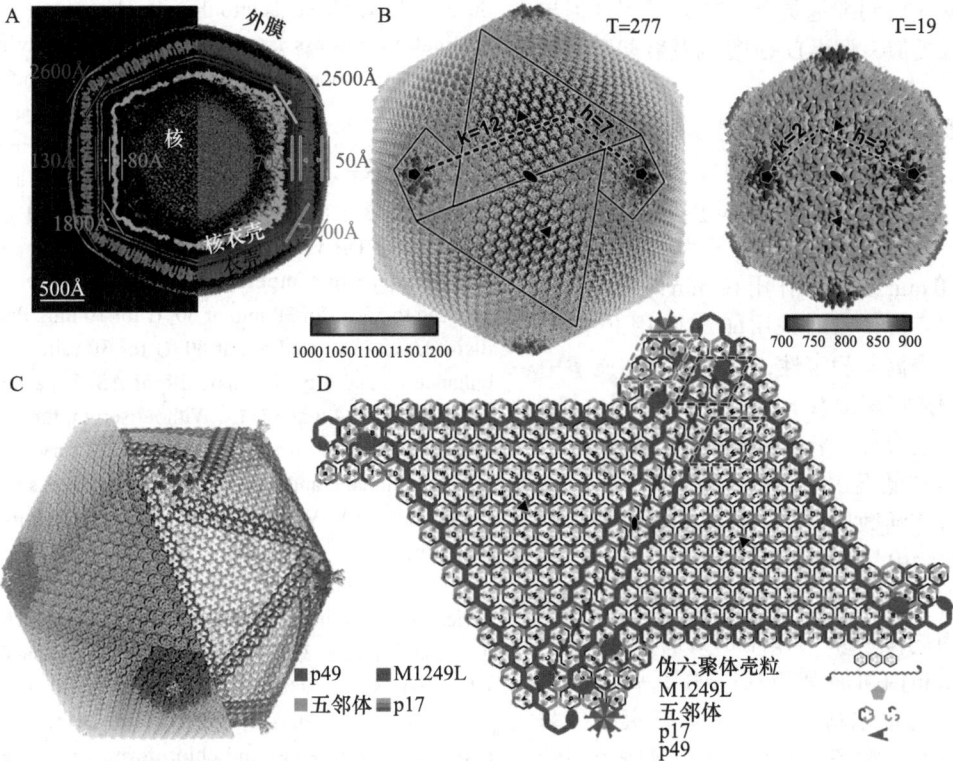

图25-3 ASFV病毒粒子的精细结构

A. ASFV病毒粒子二十面体结构的中央切片（左）和横截面（右）；B. ASFV衣壳和核衣壳的放射图；C. ASFV衣壳的冷冻电镜重建图。左侧为三重对称和五重对称结构，右侧为次要衣壳蛋白，包括去除外壳后的五聚体蛋白；D. 从衣壳内部视野看到的次要衣壳蛋白和壳粒结构。图像来源见英文。

Figure 25-3　Architecture of the ASFV virion

A. The central slice (left) and cross section (right) of the icosahedral ASFV virion structure; B. Radially colored representations of the ASFV capsid and core shell; C. Cryo-EM reconstruction of the ASFV capsid. The left half shows the trisymmetron and pentasymmetry organization. The right half shows the density of the minor capsid proteins, including the penton proteins after removing the outer capsid shell; D. Diagrammatic organization of the minor capsid proteins and capsomeres as viewed from inside the capsid. Source: Wang N, Zhao D, Wang J, et al. 2019. Architecture of African swine fever virus and implications for viral assembly. Science, 366 (6465): 640-644.

分为24个基因型，不同基因型或同一基因型的不同分离株毒力差异较大，亚洲主要流行基因Ⅱ型。ASFV基因组大，且复杂、易变，这导致对ASFV的研究充满了挑战。

感染及复制： ASFV主要经呼吸道和消化道途径侵入猪体，在鼻咽部或扁桃体增殖，然后迅速通过淋巴和血液遍及全身，单核巨噬细胞是其主要复制场所。ASFV主要通过胞饮和胞吞两种方式进入宿主细胞，网格蛋白介导的胞吞作用是主要入侵方式。病毒进入宿主细胞后，首先是脱外衣壳层，然后内囊膜与内体融合，

different isolates of the same genotype varies greatly, and the genotype Ⅱ is the prevalent one in Asia. The large, complex and changeable ASFV genome poses a great challenge for the study of ASFV.

Infection and replication: ASFV mainly infects pigs via respiratory and digestive tracts. The virus proliferates in the nasopharynx or tonsils and then rapidly spreads throughout the body through lymph and blood. The primary replication site of ASFV after infection is mononuclear macrophages. ASFV enters host cells mainly through macropinocytosis and endocytosis. Clathrin-mediated endocytosis is the main way of its invasion. Once inside the cell, the virus sheds the outer capsid layer and its inner envelope fuses with the endosome, releasing

裸核酸释放到细胞质中。DNA复制过程与痘病毒类似，病毒DNA复制及装配发生于核周。

理化性质： 非洲猪瘟病毒粒子在Percoll细胞分离液和CsCl中的浮密度分别为1.095 g/cm^{3}和1.190～1.240 g/cm^3，可耐受pH 4～13环境，但对温度较为敏感，55 ℃作用30 min或60 ℃作用10 min即可将其灭活，但实际消毒时采用60 ℃作用30 min。血清能提高其稳定性，存在于25%血清中的非洲猪瘟病毒粒子，37 ℃半衰期为24 h；但当无血清时，半衰期缩短至8 h；血液中的病毒在低温阴暗条件下可存活6年，室温下可存活数周。非洲猪瘟病毒粒子还可在分泌物中长期存活。

病毒对不同灭活剂的抗性：0.05%的β-丙内酯在4 ℃作用24 h或在37 ℃作用60 min内可将其灭活；2%氢氧化钠需24 h、1%福尔马林需6 d才将其灭活。非洲猪瘟病毒粒子对乙醚、氯仿等脂溶剂敏感。带囊膜的病毒粒子能够明显抵抗蛋白酶的作用，但易被胰酶灭活。氢氧化钠、次氯酸钠溶液、过硫酸氢钾、戊二醛、甲醛、高锰酸钾等是猪场常用消毒剂。

传播： 感染猪、野猪和软蜱是非洲猪瘟病毒的自然宿主和重要传染源。非洲猪瘟病毒可以通过受感染的动物、猪肉制品或受污染的物品（如衣服、车辆、靴子）和易感动物之间的直接或间接接触传播。健康动物可能通过接触感染动物的血液、分泌物、粪便、排泄物等直接感染。

作为目前为止唯一已知的DNA虫媒病毒，它的重要传染源软蜱受到了广泛的关注。非洲猪瘟病毒首先是西班牙实验室从游走钝缘蜱中分离到的，前期研究表明，病毒主要寄生在蜱的中肠部位和血细胞。在蜱吸血后，中肠上皮组织中的消化细胞采食了附着有病毒的红细胞而获得感染。病毒在消化细胞吞噬体内进行复制，2～3周后再转移到唾液腺和生殖组织中复制，再次吸血时将病毒传播给易感猪，或者蜱

the naked nucleic acids into the cytoplasm. ASFV DNA replication process is similar to that of poxviruses. Viral DNA replication and assembly occur around the nucleus.

Physical and chemical properties: the buoyant density of ASF virions in Percoll cell isolation solution and CsCl is 1.095 g/cm^3 and 1.190-1.240 g/cm^3, respectively. The virus is stable in pH 4-13 environment, but sensitive to temperature, and inactivated when exposed to 55 ℃ for 30 min or 60 ℃ for 10 min. Practical disinfection is carried out at 60 ℃ for 30 min. Sera can enhance its stability. The half-life of ASFV particles in 25% serum is 24 h at 37 ℃. Without serum, the half-life is decreased to 8 h. Viruses in the blood can survive for up to 6 years in cold and dark conditions and weeks at room temperature. ASFV can also survive in secretions for a long time.

The resistance of the virus to different inactivators varies: it can be inactivated by 0.05% β-propiolactone at 4 ℃ for 24 h or at 37 ℃ for 60 min. It can be inactivated when exposed to 2% sodium hydroxide for 24 h and 1% formalin for 6 d. ASFV virions are sensitive to lipid solvents, such as ether and chloroform. The enveloped ASF virions can resist the action of protease but are easily inactivated by trypsin. Sodium hydroxide, sodium hypochlorite solution, potassium bisulfite, glutaraldehyde, formaldehyde and potassium permanganate are commonly used disinfectants in pig farms.

Transmission: infected pigs, wild boars and soft ticks (*Argasidae*) are the natural hosts and important sources of transmission of ASFV. The virus can be transmitted through direct or indirect contact between infected animals or between pork products or fomites (such as clothing, vehicles, boots) and susceptible animals. Healthy animals may be infected directly through contact with the blood, secretions and feces of the infected animals.

As the only known DNA arbovirus so far, the important vector, soft tick, has also attracted wide attention. ASFV was first isolated from *Ornithodoros* ticks in a Spanish laboratory. Previous studies have shown that the virus mainly parasitizes in the midgut and blood cells of the ticks. After the tick feeds the infected blood, digestive cells in the midgut epithelium tissues ingest the virus-attached red blood cells and become infected. The virus replicates in the phagosomes of digestive cells, and then moves to salivary glands and reproductive tissues for replication 2-3 weeks later. The virus is

自身经卵、交配等方式而传播。钝缘蜱虽在贮藏病毒方面发挥着重要的作用，但是对于非洲猪瘟病毒的长距离传播不起主要作用。

培养：非洲猪瘟病毒体内感染的主要细胞类型为单核巨噬细胞，包括组织巨噬细胞和网状内皮细胞，也能在鸡胚卵黄囊和骨髓细胞中增殖。在体外，病毒也可在培养的巨噬细胞和内皮细胞中复制。一些分离株已适应于组织培养细胞系，适应猪肾细胞的毒株也能在牛肾细胞、鸡肾细胞、BHK-21、MS细胞、CV细胞和Vero细胞上生长，并产生包涵体样的细胞病变，但也有部分毒株不产生细胞病变。

transmitted to susceptible pigs during blood feeding, or vertically through egg transmission and mating. Although *Ornithodoros* ticks play an important role as virus reservoirs, they do not play a major role in the long-distance transmission of ASFV.

Cultivation: the main cell type infected by ASFV *in vivo* is mononuclear macrophages, including tissue macrophages and reticuloendothelial cells, and the virus can also proliferate in yolk sac of chicken embryo and bone marrow cells. The virus can also replicate *in vitro* in cultured macrophages and endothelial cells. Some isolates have been adapted to tissue culture cell lines. The strains adapted to porcine kidney cells can also grow in bovine kidney cells, chicken kidney cells, BHK-21, MS cells, CV cells and Vero cells, and produce inclusion body-like cytopathy, but some do not produce cytopathy.

第3节 腺病毒科
Section 3 *Adenoviridae*

腺病毒最早是在1949年从牛结节性皮炎病料接种的鸡胚中分离得到。基于腺病毒倾向于感染上皮细胞，且于腺组织中首次分离的事实，将其命名为腺病毒。腺病毒科成员主要引起家畜和家禽上呼吸道感染。人类、哺乳动物及禽类的许多腺病毒具有高度的宿主特异性，通常存在于上呼吸道，有时在肠道。腺病毒科多数成员产生亚临床感染，偶尔引致上呼吸道疾病，但犬传染性肝炎病毒及鸡的减蛋综合征病毒具有重要致病意义。

形态结构：腺病毒为双链线状DNA病毒，无囊膜，中等大小，直径为90～100 nm，由252个呈二十面体对称的壳粒组成，有240个六邻体（非顶点壳粒）、12个五邻体（顶点壳粒）及12根纤突。12根纤突以五邻体蛋白为基底由衣壳表面伸出，纤突顶端形成头节区。五邻体和纤突的头节区可与细胞表面的受体结合，在病毒感染细胞过程中起着非常重要的作用。

基因组：腺病毒为线状双链DNA病毒，大小为26～45 kb，多肽Ⅶ和X（mu）

Adenovirus was first isolated in 1949 by inoculating samples from bovine nodular dermatitis into chicken embryos. Since adenovirus tends to infect epithelial cells and was first isolated from glandular tissues, it is named as adenovirus. Members of *Adenoviridae* mainly cause the infections of the upper respiratory tract in domestic animals and poultry. Human, mammalian and avian adenoviruses are highly host-specific, usually residing in the upper respiratory tract and sometimes in the intestinal tract. Most members of the family produce subclinical infections that occasionally lead to diseases in the upper respiratory tract. However, infectious canine hepatitis virus (ICHV) and Tembusu virus (or egg drop syndrome virus, EDSV) are of important pathogenic significance.

Morphological structure: adenovirus, 90-100 nm in diameter, is double-stranded linear DNA viruses with no envelope. It is composed of 252 capsomeres with icosahedral symmetry. There are 240 hexons (nonvertex capsomeres), 12 pentons (vertex capsomeres), and 12 fibers. The 12 fibers protrude from the capsid surface in which penton proteins act as the base, and the top of the cilia forms the cephalic ganglion regions. The cephalic ganglion regions of the pentons and fibers can bind to receptors on the cell surface and play a very important role in the process of viral infection.

Genome: adenovirus is linear double-stranded DNA

蛋白紧密地环绕在其周围，起到类组蛋白样的作用。另一种多肽 V 将这种 DNA-蛋白质复合物连接起来，并通过多肽 IV 与病毒衣壳连接在一起。DNA 分子两端都具有 40～200 bp 的末端反向重复序列（ITR），是复制的起始位点，其重复的次数和长度随病毒型和毒株的不同而异，并与传代次数有关。基因组左端 ITR 的 3′ 侧有一段长约 300 bp 的包装信号，介导腺病毒基因组包装入病毒衣壳。

感染及复制： 以犬腺病毒为例，分 1 型（CAV-1）和 2 型（CAV-2）两种，抗原高度交叉。1 型引起犬传染性肝炎和眼部病变，2 型引起幼犬传染性支气管炎。病毒经鼻咽或口腔黏膜途径进入体内，1 型自然感染康复期或接种弱毒疫苗后 8～12 d，因产生抗原抗体复合物而导致角膜水肿，因此又称犬蓝眼病。对犬进行弱毒苗接种是预防犬传染性肝炎最有效的方法。犬接种了 2 型腺病毒弱毒疫苗之后不久，不断将弱毒排出体外，其他犬与之接触而获得免疫，即群体免疫。腺病毒在细胞核内复制，其复制在很大程度上受宿主免疫应答的调控。腺病毒可使易感的宿主细胞圆缩，继而细胞染色体浓缩并边缘化，最终出现细胞核内包涵体。病毒颗粒通过细胞崩解释放。有趣的是，2 型犬腺病毒由于其可忽略的免疫原性、神经元优先转导特性、沿轴突运输的分布特性，以及在哺乳动物大脑中持续表达等优点，目前在神经生物学领域已被作为载体工具进行基础和应用研究。CAV-2 可选择性结合位于神经元突触的柯萨奇病毒-腺病毒受体（CAR），从而被神经元细胞内化。神经元内化和轴突运输由 CAR 介导，可加强 CAV-2 的生物分布。因此，神经元的优先转导和高水平的逆行运输特性可使 CAV-2 作为载体，成为研究神经退行性疾病的病理生理学和绘制体内复杂神经元网络的理想工具。

virus with a size of 26-45 kb. It is tightly surrounded by the polypeptide Ⅶ and X (mu) proteins which play a histone-like role. This DNA-protein complex is linked by another polypeptide V, which binds to the viral capsid via a polypeptide molecule. There are 40-200 bp inverted terminal repeats (ITR) at both ends of DNA molecules, which are the starting sites of replication. The number and length of repeats varies depending on the virus type and strain, and is related to the number of passages. There is a 300 bp long packaging signal on the 3′ side of the ITR at the left of the genome, which mediates the packaging of the adenovirus genome into the viral capsid.

Infection and replication: taking canine adenovirus as the example, the virus has two serotypes: serotype 1 (CAV-1) and serotype 2 (CAV-2). Their antigens highly cross-react with the antisera produced from either serotype. CAV-1 causes infectious canine hepatitis (ICH) and ophthalmopathy in canines and CAV-2, infectious bronchitis of puppies. The virus enters the body through the mucous membrane of nasopharynx and oral cavity. At the convalescent period of the natural infection of CAV-1 or at 8-12 d after inoculation of attenuated vaccines, corneal edema may appear due to the production of antigen-antibody complexes, hence the term canine blue eye disease. Vaccination with attenuated vaccines is the most effective way to prevent ICH in dogs. Shortly after the vaccination of CAV-2 attenuated vaccines, the canines will shed the vaccine virus particles continuously, with which the other canines get contacted and become immunized, leading to the development of herd immunity. Adenovirus replicates in the nucleus and its replication is largely regulated by the host immune responses. Adenoviruses can cause rounding of the infected cells, and condense and marginalize the chromosomes, resulting in formation of intranuclear inclusion bodies. Virus particles are released by rupture of cells. Interestingly, CAV-2 has been used as a powerful vector tool for fundamental and applied neurobiological research due to its negligible immunogenicity, preferential transduction of neurons, widespread distribution via axonal transport, and duration of expression in the mammalian brain. CAV-2 vectors are internalized in neurons by the selective use of coxsackievirus and adenovirus receptor (CAR), which is located at the presynapse in neurons. CAR-mediated neuronal internalization and axonal transport can potentiate vector biodistribution. The preferential transduction of neurons and the high level of retrograde transport make CAV-2 vectors ideal tools to study the pathophysiology of many neurodegenerative disorders and to map complex neuronal networks *in vivo*.

理化性质：腺病毒对酸的抵抗力较强，故能通过胃而继续保持活性。许多腺病毒就是从人和动物的粪便中获得的，这为口服腺病毒载体的研制提供了方便。由于无囊膜，腺病毒对有机溶剂不敏感，在丙酮中不稳定。腺病毒可在冷冻状态下保存，于4 ℃可存活70 d，在−20 ℃时可长期存活，56 ℃加热可灭活。适宜pH为5~6，pH在2以下或10以上时均不稳定。

腺病毒能凝集红细胞，具有宿主特异性。动物来源的腺病毒对不同动物红细胞的凝集表现不完全一致。牛腺病毒1、2、3型均能凝集大鼠红细胞，但不能凝集人、豚鼠和鸡的红细胞，其中2型和3型还可凝集小鼠的红细胞；猪腺病毒1型可凝集大鼠、小鼠、人和豚鼠的红细胞，不凝集鸡的红细胞；猪腺病毒2型和3型对上述几种动物的红细胞均不凝集，而猪腺病毒4型可凝集大鼠的红细胞；鼠类腺病毒对上述几种动物的红细胞都不凝集；减蛋综合征病毒只凝集鸡的红细胞，对其他动物的红细胞不表现凝集作用；犬传染性肝炎病毒可凝集人和豚鼠的红细胞，对鸡的红细胞的凝集情况不确定，不凝集大鼠和小鼠的红细胞。

传播：腺病毒感染人和动物以及禽类的眼、上呼吸道和消化道，患病动物的分泌物、排泄物内含有病毒粒子。病毒通过直接接触或间接接触传播，健康动物接触同属患病动物后，经过鼻腔、口腔黏膜等途径进入体内。有研究初步表明，禽腺病毒4型存在垂直传播风险。临床3~7日龄发病雏鸡可能通过垂直传播感染。

培养：腺病毒大多数具有较严格的宿主范围，一般不出现病毒感染不同属动物的情况。体外培养时，宿主来源的细胞最为敏感，其中上皮细胞比纤维细胞更为敏感。所以在体外培养时，应使用宿主来源的细胞，传代后可适应其他动物来源的细胞。

Physical and chemical properties: adenoviruses are resistant to acidic conditions, so they can remain active after passing the stomach. Many adenoviruses are collected from human and animal feces, which provides convenience for the development of oral adenoviral carriers. Due to the absence of envelopes, adenoviruses are insensitive to organic solvents and unstable in the acetone. Adenoviruses can be stored in the frozen state, survive at 4 ℃ for 70 d or at −20 ℃ for a long time, and can be inactivated at 56 ℃ . The suitable pH is 5-6, and the virus is not stable when the pH is below 2 or above 10.

Adenovirus can agglutinate erythrocytes with host specificity. Complete agglutination performance of animal-derived adenovirus against erythrocytes in different animals. Bovine adenovirus types 1, 2 and 3 can agglutinate rat erythrocytes, but not those of human, guinea pig and chicken. Adenovirus types 2 and 3 can also agglutinate mouse erythrocytes. Porcine adenovirus serotypes 1 can agglutinate the erythrocytes of rat, mouse and guinea pig, but not those from chicken while porcine adenovirus type 2 and 3 do not agglutinate the erythrocytes of the above animals. Porcine adenovirus type 4 can agglutinate the erythrocytes of rats. Adenoviruses of rat do not agglutinate the red blood cells from all the animal species mentioned above. Egg drop syndrome virus only agglutinates chicken erythrocytes but not those from other species. IHC virus can agglutinate erythrocytes of human and guinea pig, but not sure of those from chicken, and definitely not those from rat and mouse.

Transmission: adenoviruses can infect the eyes, upper respiratory tract or digestive systems of humans and animals. The infected animal can shed the virions via the secreta or excreta. The virus is transmitted through direct or indirect contacts. When other healthy animals in the same genus get contacted with the virus excreted from the infected animals, they can become infected via the mucous membrane of the nasal and oral cavities. It is reported that avian adenovirus type 4 has a high risk of vertical transmission. Clinical observations show that chicklings that are 3-7 days old can be infected by vertical transmission.

Cultivation: most adenoviruses have a strict host range and generally do not infect the animals from different genera. Host-derived cells are most ideal for *in vitro* culture, and epithelial cells are more sensitive compared to fibroblasts. Therefore, cells of host origin should be used for *in vitro* culture, and viruses can be adapted to the cells of other species after passage.

思考题 Questions

1. 痘病毒的增殖过程通常在哪里完成？

1. Where do poxviruses usually replicate?

2. 如何区别绵羊痘病毒和山羊痘病毒？

2. How to differentiate between sheep pox virus and goat pox virus?

3. 禽腺病毒属所有的亚种血清型都具有凝集红细胞的功能吗？

3. Do all subspecies of avian adenovirus serotypes have the function of agglutinating red blood cells?

4. 为什么选择腺病毒5型作为载体用于重组DNA疫苗及基因治疗？

4. Why is adenovirus type 5 selected as the vector for recombinant DNA vaccine and gene therapy?

5. 非洲猪瘟病毒具有什么样的结构？

5. What is the structure of African swine fever virus?

6. 当发生疑似非洲猪瘟疫情时，应采取的防控策略是什么？

6. What prevention and control strategies should be adopted in case of suspected outbreak of African swine fever?

第26章 疱疹病毒目
Chapter 26 / *Herpesvirales*

内容提要 疱疹病毒目包括3个科：疱疹病毒科、异样疱疹病毒科和贝类疱疹病毒科。疱疹病毒科包括鸟类、哺乳动物和爬行动物的疱疹病毒。

Introduction *Herpesvirales* contains 3 families: *Herpesviridae, Alloherpesviridae* and *Malacoherpesviridae*. *Herpesviridae* includes herpesviruses of birds, mammals and reptiles.

第1节 疱疹病毒科的主要特征
Section 1 Main features of *Herpesviridae*

疱疹病毒科是可引起人和动物疾病的一大类DNA病毒，统称为疱疹病毒。疱疹病毒几乎在所有被调查的物种中都被发现，在自然界分布广泛。除绵羊外，可引起各种家畜的重大疾病。

疱疹病毒形态相似，具有双链、线状DNA芯髓，二十面体衣壳由162个壳粒组成，包括150个六邻体和12个五邻体，外面围绕着被膜，被膜位于疱疹病毒的囊膜和核衣壳之间，由脂蛋白组成。核衣壳被脂质囊膜包围，有些病毒的囊膜表面有纤突（**图26-1**）。疱疹病毒的基因组很大，为125～290 kb，可以编码许多不同的蛋白质。

疱疹病毒的复制和装配在细胞核中进行，形成巨大的A型包涵体，然后经核膜以出芽方式获得囊膜，核内包涵体是其感染的特征（**图26-2**）。疱疹病毒可导致终身潜伏感染，间歇或连续排毒，在一定条件下可复发。

Herpesviridae is a large family of DNA viruses that cause diseases in animals and humans that are generally called herpesviruses. Herpesviruses have been found in virtually every species that have been investigated, widely existing in nature and causing significant diseases in domestic animal species except sheep.

Herpesviruses are similar in morphology, with a double-stranded DNA core and an icosahedral nucleocapsid consisting of 162 capsomeres, including 150 hexons and 12 pentons. The nucleocapsid is surrounded by a tegument which is a cluster of proteins that is encircled by lipoprotein envelope. There are spikes on the surface of the envelope in some herpesviruses (**Figure 26-1**). The genome of herpesviruses is large, up to 125-290 kb, and encodes many different proteins.

The replication and assembly of herpesviruses occur in the nucleus, forming huge type A inclusion bodies, and the envelope is obtained from the nuclear membrane by budding. Intranuclear inclusion bodies are a characteristic feature of the infection (**Figure 26-2**). Infection with herpesviruses results in lifelong latent infection with intermittent or continuous shedding of virus and recrudescence under certain conditions.

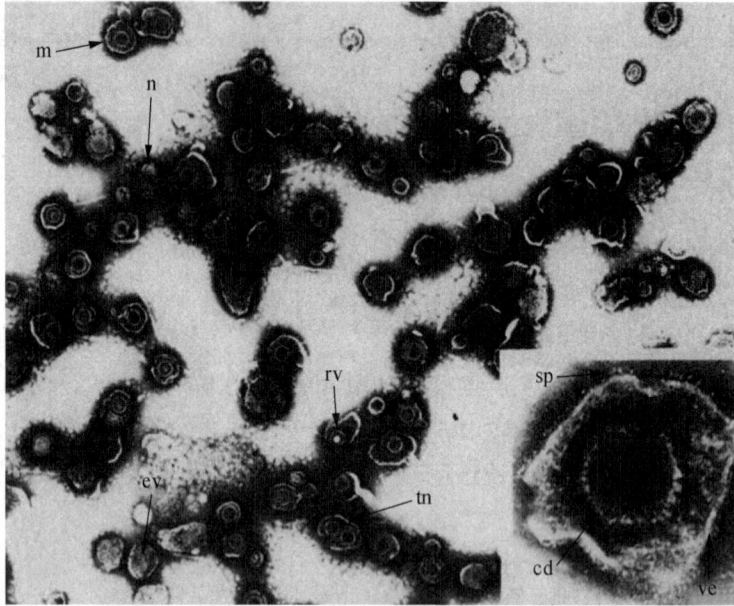

图 26-1　负染的牛传染性鼻气管炎病毒

n：核衣壳；ev：囊膜；rv：囊膜病毒；tn：双核衣壳。17 000×。插图：成熟病毒囊膜细微突起。sp：纤突；cd：芯髓；ve：病毒囊膜。100 000×。图像来源见英文。

Figure 26-1　Negatively stained Infectious bovine rhinotracheitis virus

n: nucleocapsid; ev: envelope; rv: enveloped virus; tn: twin nucleocapsids. 17 000 × . Inset: minute projections on the envelope of a matured virus. sp: virus spikes; cd: virus core; ve: virus envelope. 100 000 × . Source: Talens LT, Zee YC. 1976. Purification and buoyant density of infectious bovine rhinotracheitis virus. Proc Soc Exp Biol Med, 151 (1): 132-135.

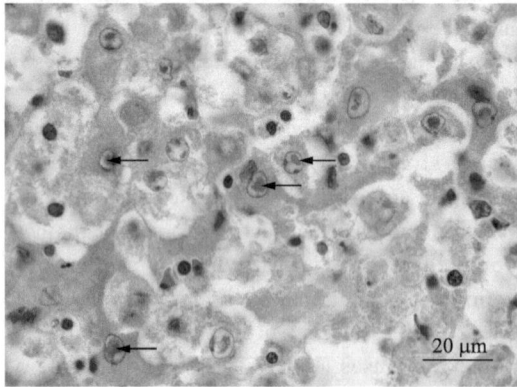

彩图

图 26-2　HE 染色的肝脏显微照片（100×）

箭头所示为疱疹病毒核内嗜酸性包涵体。图像来源见英文。

Figure 26-2　liver stained with hematoxylin and eosin (100 ×)

Intranuclear eosinophilic inclusion bodies of herpesviruses in liver are indicated by arrows. Source: McVey DS, Kennedy M, Chengappa MM. 2013. Veterinary Microbiology. 3rd ed. Hoboken: Wiley-Blackwell.

疱疹病毒在外界环境中不易存活，一般需密切接触才能传染，尤其是交配、舔舐等行为导致的黏膜感染，在集约化牛、猪、鸡养殖场中最易发生。疱疹病毒在阴凉潮湿的环境中可以存活较长时间，潜伏感染的动物为传染源。

Herpesviruses do not survive well outside of the host. Transmission usually requires close contact, especially mucosal infection caused by mating, licking, *etc*. It is most likely to occur in intensive cattle, pig and chicken farms. The viruses may survive longer in cool and moist environments. Latently infected animals serve as infectious source.

第2节　牛传染性鼻气管炎病毒
Section 2　Infectious bovine rhinotracheitis virus

牛传染性鼻气管炎病毒学名为牛疱疹病毒1型（BHV-1），感染牛，可表现为眼、生殖道和呼吸道疾病，是世界动物卫生组织（WOAH）规定的通报疫病。

致病机理： 结膜炎在BHV-1感染中常见，呼吸道型通常表现为鼻气管炎，重则引发致命的支气管肺炎。BHV-1能够感染生殖器，导致龟头包皮炎和外阴阴道炎（**图26-3**）。

疱疹病毒易在源自其天然宿主的细胞

The official name of infectious bovine rhinotracheitis virus is bovine herpesvirus 1 (BHV-1). BHV-1 infection in cattle may lead to the pathological occurrence in eyes, genital and respiratory tracts. Bovine rhinotracheitis is a notifiable disease by World Organisation for Animal Health (WOAH).

Pathogenesis: conjunctivitis is common in BHV-1 infection. Respiratory disease typically presents as rhinotracheitis (infectious bovine rhinotracheitis, IBR), which may lead to severe and often fatal bronchopneumonia. BHV-1 can infect genitalia, resulting in balanoposthitis and vulvovaginitis (infectious pustular vulvovaginitis, IPV) (**Figure 26-3**).

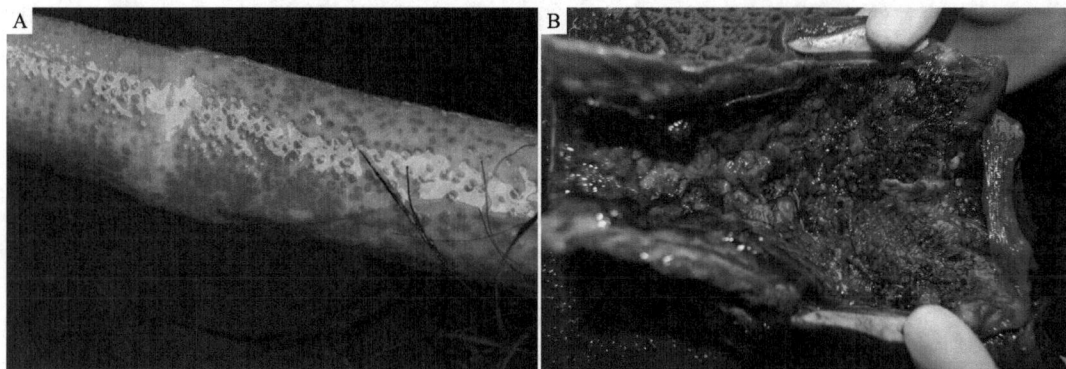

图26-3　牛传染性鼻气管炎的临床症状
A．阴茎黏膜发炎；B．尸检可见坏死性喉气管炎。图像来源见英文。

Figure 26-3　Clinical signs of infectious bovine rhinotracheitis
A. Inflammation on the penile mucosa; B. Post-mortem examinations can reveal lesions of necrotizing laryngotracheitis.
Source: Buchan C. 2020. A practitioner's guide to infectious bovine rhinotracheitis. Veterinary Practice. https://www. veterinary-practice. com/article/a-practitioners-guide-to-infectious-bovine-rhinotracheitis.

彩图

中生长。BHV-1在犊牛肺、睾丸或肾细胞培养生长良好，很快产生细胞病变，形成合胞体和特征性核内包涵体。

实验室诊断： 对该病毒进行检测，可采取水疱液或刮取物进行PCR检测和电镜观察，制作涂片或组织切片进行荧光抗体染色。在许多国家的参考实验室常用PCR检测病毒核酸，酶联免疫技术检测血清抗体，病毒分离鉴定是确诊的依据。

Herpesviruses grow most readily in the cells derived from their natural host. BHV-1 grows well in lung, testis or kidney cells of calves, and there is a rapid cytopathic effect, with syncytia and typical intranuclear inclusion bodies.

Laboratory diagnosis: the methods for detection of bovine herpesvirus 1 include virus-specific PCR and electron microscopic visualization of the virus particles from vesicular fluid or scrapings, and immunofluorescence staining of mucosal smears or tissue sections. Bovine herpesvirus 1 specific PCR for virus nucleic acid detection and specific enzyme immunoassays for antibody detection are now routinely used

防控： 在流行地区可用弱毒活疫苗和灭活疫苗进行预防，疫苗接种虽不能防止感染，但可以降低发病率和患病严重程度。

in reference laboratories in many countries. Virus isolation and characterization provide a definitive diagnosis.

Prevention and control: both live attenuated vaccine and inactivated vaccine are available for BHV-1 in endemic areas. Vaccination does not prevent infection but does reduce the incidence and severity of the disease.

第3节 伪狂犬病毒
Section 3 Pseudorabies virus

伪狂犬病毒（PRV）又名猪疱疹病毒1型或奥耶斯基病毒，只有一个血清型，具有疱疹病毒的典型结构。猪为该病毒的原始宿主，可表现亚临床或潜伏感染，而其他动物感染均可导致死亡。伪狂犬病为WOAH规定的通报疫病，国外称为奥耶斯基病。

致病机理： 幼龄动物感染该病毒最为严重，引起仔猪发热和神经症状，死亡率从5%到100%不等。怀孕母猪感染可能导致流产、死胎或木乃伊胎；成年猪多无明显症状。伪狂犬病发生在其他物种，包括牛、羊、犬、猫和兔，患畜通常出现神经症状，表现为强烈的皮肤瘙痒。病毒最初定位于扁桃体，然后传至头部神经节、脊髓等部位，感染24 h后可从脑中分离到病毒。病毒可以在鼻分泌物中持续14 d，但不随尿液或粪便排毒。病毒可通过摄入、吸入和交配传播。犬通过食用生肉、猪内脏，或通过接触感染的猪或猪尸体而感染PRV，并出现临床症状（抑郁、厌食、瘙痒和吼叫），最后心脏衰竭而死（**图26-4**）。核内包涵体常存在于退化的肝细胞、肾上腺皮质细胞，偶尔存在于脾脏和淋巴结的单核巨噬细胞。病毒容易在许多动物的组织细胞中生长。

实验室诊断： 伪狂犬病毒可用PCR、免疫荧光染色法或ELISA进行检测。ELISA可区分接种基因缺失疫苗和野毒感染产生的抗体，已用于根除计划。

防控： 在流行地区减毒活疫苗可有效降低流行区的死亡，灭活疫苗接种于易感

Pseudorabies virus (PRV) is also known as suid herpesvirus 1 or Aujeszky's disease virus. Only one serotype has been identified. Pseudorabies virus has the typical architecture of virions of the herpesviruses. The original host of PRV is pigs. Pigs may have subclinical and latent infections, whereas infections in all other animals may be fatal. Pseudorabies, initially called Aujeszky's disease in other countries, is a notifiable disease by WOAH.

Pathogenesis: the virus has more severe pathogenicity in younger animals and causes fever and neurological symptoms in piglets and the mortality rate varies from 5% to 100%. Infection of pregnant sow can result in abortion, stillbirth or mummification. Adult pigs usually present no obvious symptoms. Pseudorabies also occurs in other species, including cattle, sheep, dogs, cats, and rabbits in which the clinical signs are usually neurologic and manifested by intense pruritus. The virus is initially localized in the tonsil and then passed to the head ganglion and spinal cord. Viruses can be isolated from the brain 24 h post infection. Viral shedding may persist in nasal secretions for up to 14 d. The virus is not shed in urine or feces. The virus is transmitted by ingestion, inhalation and coitus. Dogs are infected by PRV either through consumption of raw meat or offal from infected swine, or by contact with infected swine or swine carcasses, and they may exhibit symptoms, such as depression, anorexia, pruritus, vocalization, *etc.*, and die of cardiac failures (**Figure 26-4**). Intranuclear inclusion bodies are often present in degenerating hepatocytes, adrenal cortical cells, and occasionally mononuclear phagocytic cells in the spleen and lymph nodes. The virus readily replicates in cells from many species and tissues, including cat, dog, cattle, chicken, and goose.

Laboratory diagnosis: pseudorabies virus infection can be detected by PCR, immunofluorescent staining or ELISA. ELISA is used to differentiate the antibodies induced by gene-deleted vaccines and field infection and has been used in eradication programs.

Prevention and control: attenuated live vaccines are

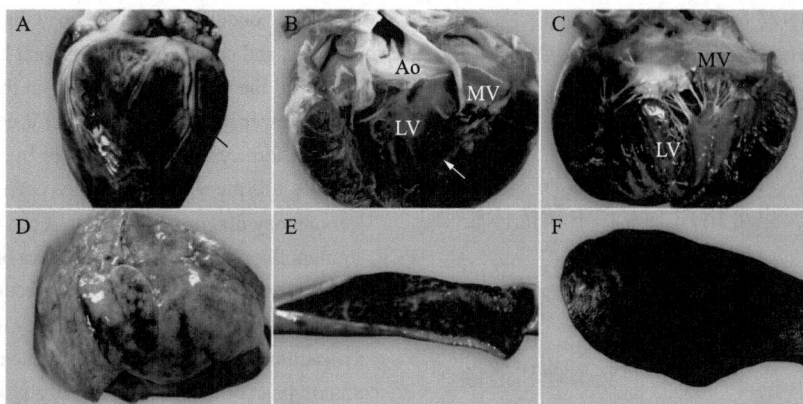

图26-4 实验室感染PRV比格犬的大体病变

A. 心脏，显示心外膜多灶性瘀点出血；B. 心脏，二尖瓣上深红色血栓（白色箭头）；注意左心室内膜（Ao：主动脉；LV：左心室；MV：二尖瓣）中的瘀斑（黑色箭头）；C. 心脏，二尖瓣弥漫性出血（箭头）；D. 肺，因充血和出血呈局部红色；E. 十二指肠，因充血黏膜呈弥漫性发红；F. 脾脏，不完全收缩，边缘有大量暗红色至黑色、柔软、充满血液的突起。图像来源见英文。

Figure 26-4 Gross lesions in experimentally PRV-infected beagle dogs

A. Heart, showing multifocal petechial hemorrhage in the epicardium; B. Heart, a dark-red thrombus attached to the mitral valves (white arrow); note the ecchymosis (black arrow) present in the endocardium of the left ventricle (Ao: aorta; LV: left ventricle; MV: mitral valves); C. Heart, showing extensive hemorrhages in the mitral valves (arrow); D. Lung, showing focal redness arising from congestion and hemorrhage; E. Duodenum, diffuse redness of the mucosa resulting from hyperemia and congestion; F. Spleen, incompletely contracted areas characterized by numerous dark-red to black, soft, blood-filled projections at the margins. Source: Zhang L, Zhong C, Wang J, et al. 2015. Pathogenesis of natural and experimental pseudorabies virus infections in dogs. Virol J, 12: 44.

母猪，产生母源抗体对几周内的新生仔猪提供保护。无伪狂犬病的地区，禁用疫苗。

available and have been successful in reducing death losses in endemic areas. Inactivated vaccines have been used in susceptible sows in endemic areas, producing the antibodies into the colostrum for the protection of newborn pigs during the first several weeks after birth. In the pseudorabies-free areas, vaccination is prohibited.

第4节 禽传染性喉气管炎病毒
Section 4 Avian infectious laryngotracheitis virus

禽传染性喉气管炎病毒又名禽疱疹病毒1型，只有一个血清型，但不同地区的毒株有差异。

致病机理：禽疱疹病毒1型对于所有鸡来说均易感，多感染4～18月龄鸡，导致咳嗽、气喘、流泪、咯血及气管有黄白色干酪物等。轻度感染常表现为产蛋下降和结膜炎，在现代家禽饲养条件下常见。自然条件下该病毒通过上呼吸道和眼感

Avian infectious laryngotracheitis virus (ILTV) is also designated as gallid herpesvirus 1 (GaHV-1). There is only one serotype in the virus, but genetic variation occurs among strains from different regions.

Pathogenesis: all chickens are susceptible to gallid herpesvirus 1, and those at 4-18 months old age are more susceptible. The infection may lead to cough, asthma, tears, hemoptysis, yellow and white cheese-like secretions in the trachea. Mild infection may result in reduced egg production and conjunctivitis and is common in modern poultry industry.

染，在气管中载量最大，仅在鼻腔、气管和下呼吸道中复制。该病毒可用鸡肾细胞或绒毛尿囊膜接种鸡胚进行培养。

实验室诊断： 可通过PCR检测病毒DNA，免疫荧光染色检测气管或其触片中的病毒抗原，也可用ELISA和中和抗体法进行检测。

防控： 种鸡和蛋鸡接种弱毒疫苗，虽然不能防止感染和改变带毒状态，但可不发病。另外，感染场所的彻底清理和消毒对控制该病很重要。

ILTV in natural conditions enters the body through the upper respiratory tract and ocular contact. High viral load is found in the trachea, and the virus replicates only in the nasal cavity, trachea, and lower respiratory tract. ILTV can be cultured using chicken kidney monolayer cells or by inoculation on the chorioallantoic membrane of chicken embryo.

Laboratory diagnosis: ILTV DNA can be detected by PCR. Immunofluorescent staining of trachea or its smear can detect the presence of viral antigen. Its infection can also be detected by ELISA and neutralizing antibody method.

Prevention and control: the use of attenuated vaccines is a common practice for breeding and egg production flocks. Though vaccination does not protect against infection with virulent virus or the development of latency, outbreaks can be avoided. In addition, complete depopulation and disinfection of infected premises is very important to control the disease.

第5节 马立克病病毒
Section 5 Marek's disease virus

马立克病病毒又名禽疱疹病毒2型，是感染鸡的一种重要病原，属于甲型疱疹病毒亚科马立克病病毒属，引起鸡的T淋巴细胞肿瘤。不同毒株间毒力差异较大，有温和型毒株、强毒株、超强毒株和特超强毒株。

致病机理： 马立克病（MD）常发生于2～5月龄的鸡，是一种淋巴增生性疾病，淋巴瘤最为常见，全身脏器长满肿瘤，另外可以使鸡神经麻痹，无法站立而成典型"劈叉"姿势（**图26-5**）。具有感染性的游离病毒只在羽毛囊上皮细胞产生并排出体外，吸入污染的粉尘和皮屑可导致感染。感染的严重程度受病毒毒株、剂量、感染途径，以及鸡的年龄、性别、免疫状况和遗传易感性的影响，通常表现亚临床感染并排毒，隐性感染鸡可终生带毒并排毒。

该病毒可用全血白细胞、4日龄鸡胚卵黄囊或绒毛尿囊膜接种培养，也可用鸡胚和鸭胚成纤维细胞或鸡肾细胞进行培养。

Marek's disease virus (MDV) is also known as gallid herpesvirus 2 (GaHV-2), belonging to the genus *Mardivirus* of the subfamily *Alphaherpesvirinae*. The virus causes tumors of T lymphocytes in chickens, and its virulence varies greatly among strains: mild, virulent, very virulent (vv) or even very virulent plus (vv+).

Pathogenesis: Marek's disease (MD) usually occurs in chickens between 2 and 5 months old, and it is a lymphoproliferative disease, lymphoma being the most common. The visceral organs are full of tumors. In addition, MDV can paralyze the nerves of chickens. Chickens are in a split posture and unable to stand (**Figure 26-5**). The infectious free virus particles are only produced in the feather follicle epithelium, from which they are shed into the environment. Infection is acquired by inhalation of contaminated dust and dander. The outcome of MDV infection in chickens is influenced by the virus strain, dose, and route of infection and also by the age, sex, immune status, and genetic susceptibility of the chickens. Subclinical infection with virus shedding is quite common. Chickens of inapparent infection are lifelong carriers and shedders of viruses.

The virus can be cultured in whole blood leukocytes, yolk sac or chorioallantoic membrane of 4 days old chicken embryo fibroblast cells of chicken or duck embryo, and chicken kidney cells.

图26-5 疑似马立克病病毒感染的病鸡
病鸡向后伸展双腿，无法行走（对马立克病疑似病例进行剖检前拍摄）。图像来源见英文。
Figure 26-5 Clinically diseased chickens suspected of MD virus infection
Pictures taken before conducting post-mortem examination of MD suspected cases where the chickens stretched their legs backwards and unable to walk. Source: Demeke B, Jenberie S, Tesfaye B, et al. 2017. Investigation of Marek's disease virus from chickens in central Ethiopia. Trop Anim Health Prod, 49 (2): 403-408.

彩图

实验室诊断： 确诊可通过病毒分离、采用荧光抗体检测抗原或通过PCR检测病毒DNA，也可用琼脂扩散、病毒中和、间接免疫荧光和ELISA检测抗体。

防控： 疫苗接种是防控该病的主要方法，常规方法是对1日龄雏鸡进行免疫接种。

Laboratory diagnosis: confirmatory diagnosis is made by viral isolation or antigen detection using fluorescent antibodies or by detection of viral DNA using PCR. Antibodies can be detected by agar gel immunodiffusion, viral neutralization, indirect immunofluorescence and ELISA.

Prevention and control: vaccination is the principal method of control. The recommended method is to vaccinate 1 day old chicklings.

第6节 鸭瘟病毒
Section 6 Duck plague virus

鸭瘟病毒又名鸭疱疹病毒1型或鸭肠炎病毒，属于甲型疱疹病毒亚科马立克病病毒属，鸭瘟病毒可以引起鸭瘟和鸭病毒性肠炎（**图26-6**），主要危害鸭、鹅和天鹅，迁徙性水禽是该病传播的主要因素。

致病机理： 本病毒只有一个血清型，可以在鸭胚绒毛尿囊膜或鸭胚成纤维细胞中进行培养，形成核内包涵体。

实验室诊断： 检测可以采用组织切片荧光抗体染色法。

防控： 可接种弱毒疫苗预防该病。

Duck plague virus is also designated as anatid herpesvirus 1 or duck enteritis virus (DEV), belonging to the genus *Mardivirus* of the subfamily *Alphaherpesvirinae*. Anatid herpesvirus 1 causes duck plague and viral enteritis (**Figure 26-6**), and mainly infects ducks, geese and swans. Migratory waterfowls are a major factor in the spread of this disease.

Pathogenesis: the virus has only one serotype and can be cultured in the chorioallantoic membrane of duck embryos or in duck fibroblasts, forming intranuclear inclusion bodies.

Laboratory diagnosis: duck plague virus can be detected by fluorescent antibody staining.

Prevention and control: vaccination with attenuated vaccines can prevent the disease.

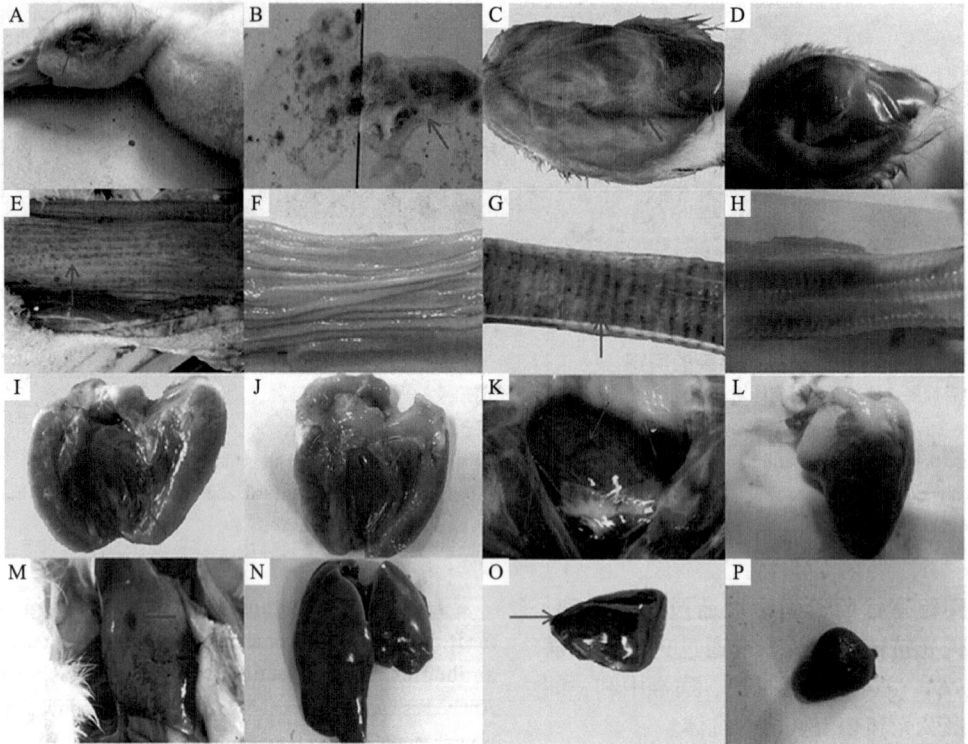

图 26-6 感染鸭瘟病毒鸭的临床症状和大体病变

A. 头部和颈部肿胀；B. 绿色腹泻；C. 头部的淡黄色透明液体；E. 食管黏膜弥漫性出血；G. 气管环出血；I. 心内膜斑点出血；K. 心外膜瘀点出血；M. 肝脏肿大，伴有血斑；O. 脾肿大出血。D、F、H、J、L、N和P 分别代表未感染病毒对照组鸭的头部、食管黏膜、气管、心内膜、心外膜、肝脏和脾脏。图像来源见英文。

Figure 26-6 Clinical symptoms and gross lesions of the DPV-infected ducks

A. Swollen head and neck; B. Greenish diarrhea; C. A light-yellow and transparent liquid of the head; E. Diffuse hemorrhage of the esophageal mucosa; G. Hemorrhage of the annulus trachealis; I. Spotted hemorrhage of the endocardium; K. Petechial hemorrhaging of the epicardium; M. Liver is enlarged with blood spots; O. Splenomegaly and hemorrhage. D, F, H, J, L, N and P represent the head, esophagus mucosa, trachea, endocardium, epicardium, liver, and spleen of ducks from the control group, respectively. Source: Li N, Hong T, Li R, et al. 2016. Pathogenicity of duck plague and innate immune responses of the Cherry Valley ducks to duck plague virus. Sci Rep, 6: 32183.

思考题　Questions

1. 引起狂犬病和伪狂犬病的病原有何不同？

2. 疱疹病毒目成员中有哪些是WOAH规定通报疫病的病原？

3. 马立克病病毒和鸭瘟病毒都属于马立克病病毒属，它们的致病特点有何不同？

1. What's the difference between the pathogens of rabies and pseudorabies?

2. Which members of the *Herpesviridae* are the pathogens of notifiable diseases by WOAH?

3. Both Marek's disease virus and duck plague virus belong to the genus *Mardivirus*. What is the difference between their pathogenic characteristics?

第27章 单链DNA病毒
Chapter 27 Single-stranded DNA viruses

内容提要 单链DNA病毒包括细小病毒科、圆环病毒科和细环病毒科，包含多种对动物有致病性的病毒。其中，细小病毒无囊膜，基因组为单链线状DNA，重要病毒包括猫泛白细胞减少症病毒、犬细小病毒和猪细小病毒等。圆环病毒科的基因组为单链环状，代表病毒为猪圆环病毒。细环病毒科的基因组为单链负链环状DNA，代表病毒为鸡贫血病毒。

Introduction Single-stranded DNA viruses consist of the families *Parvoviridae*, *Circoviridae* and *Anelloviridae*, which contain a variety of viruses that are pathogenic to animals. The parvoviruses are non-enveloped viruses with a genome of linear single-stranded DNA. The important viruses in this family include feline panleukopenia virus, canine parvovirus and porcine parvovirus. The genome of *Circoviridae* is circular single-stranded DNA, and porcine circovirus is the representative. The genome of *Anelloviridae* is circular negative single-stranded DNA, and chicken anemia virus is the representative.

第1节 细小病毒科
Section 1 *Parvoviridae*

细小病毒科是无囊膜、二十面体对称的病毒。病毒粒子含有单链、线状的基因组DNA。细小病毒科的成员可以导致许多动物发病，例如，猫泛白细胞减少症病毒、犬细小病毒、猪细小病毒等。

猫泛白细胞减少症病毒

猫泛白细胞减少症又称猫传染性肠炎，是猫的一种高度传染性的急性病毒病。以高热、厌食、沉郁、呕吐并伴有脱水、腹泻和死亡为特征（**图27-1**）。白细

Members of the *Parvoviridae* are non-enveloped icosahedral viruses that contain a linear single-stranded DNA genome that cause diseases in many animal species, such as feline panleukopenia virus, canine parvovirus and porcine parvovirus.

Feline panleukopenia virus

Feline panleukopenia (FP), also called as feline infectious enteritis, is a highly contagious and acute viral disease of cats, characterized by high fever, anorexia, depression, vomiting, followed by dehydration, diarrhea, and death (**Figure 27-1**). Leukopenia is characteristic of the disease, and the severity of the clinical symptoms is

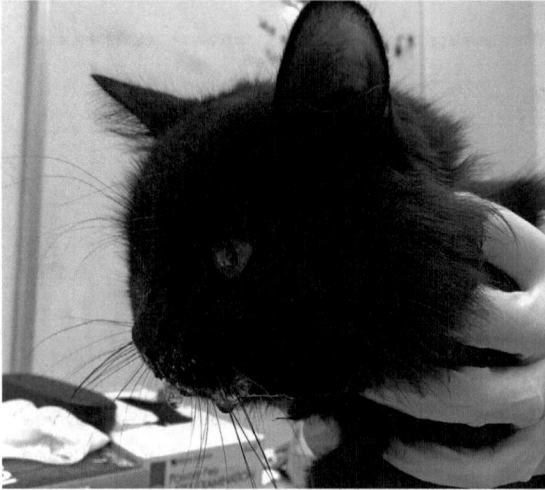

彩图

图27-1 感染猫泛白细胞减少症病毒的病猫唾液分泌过多
图像来源见英文。

Figure 27-1　Excessive secretion of saliva in FPV infected cat
Source: Barrs VR. 2019. Feline panleukopenia: a re-emergent disease. Vet Clin North Am Small Anim Pract, 49 (4): 651-670.

胞减少是猫泛白细胞减少症的特征。临床症状的严重程度与白细胞减少的严重程度直接相关。各个年龄段的猫均易感，但对小猫的致死率最高。猫可以通过消化道和呼吸道感染（**图27-2**），且潜伏期短。子宫内感染猫泛白细胞减少症病毒可导致新生猫死亡或产生以运动失调为特征的中枢神经系统先天性异常。

病毒特征：猫泛白细胞减少症病毒（FPV）是一种典型的细小病毒。病毒粒子无囊膜，呈二十面体对称。基因组为单链DNA分子，包含两个开放阅读框，其中一个至少编码四种蛋白质，参与病毒RNA转录和基因组DNA复制，另一个开放阅读框编码病毒衣壳蛋白。VP3构成主要的衣壳蛋白，与猫泛白细胞减少症病毒的组织嗜性有关。该病毒的复制在宿主细胞的细胞核内完成，由于病毒缺乏自身的DNA聚合酶，细小病毒的复制需要利用宿主细胞的酶，因此在宿主细胞周期的S期晚期或G2期早期进行病毒复制。猫泛白细胞减少症病毒对环境因素和消毒剂的抵抗力很强。0.175%次氯酸钠溶液（1∶30）是最有效和实用的消毒剂。猫科的所有成员都可感染猫泛白细胞减少症病毒。病毒在原代或连续传代的猫肾细胞培养中生长，在犬细胞培养中不生长。

often directly associated with the degree of leukopenia. Cats at all ages are susceptible to the infection, but highest mortality occurs in infected kittens. Cats can be infected by either the oral or respiratory route (**Figure27-2**) with a short incubation period. Intrauterine infection with feline panleukopenia virus may lead to death of neonatal kittens or congenital abnormalities of the central nervous system (CNS) characterized by cerebellar ataxia.

Virus characteristics: feline panleuko-penia virus (FPV) is a typical parvovirus. The virion is unenveloped with icosahedral symmetry. The genome is a single-stranded DNA molecule that includes two open reading frames, one encoding at least four proteins that mediate RNA transcription and DNA replication, and the other encoding the capsid proteins. VP3 constitutes the major capsid protein and is related to tissue/cell tropism of the virus. Virus replication occurs within the nucleus of host cells. Parvoviruses need to utilize host cell enzymes for their own replication due to lacking DNA polymerase. Thus, viral replication requires cells that are proliferating (late S phase or early G2 phase of the cell cycle). FPV is strongly resistant to environmental factors and many commercial disinfectants. Sodium hypochlorite solution at 0.175% (1∶30) is the most effective and practical virucidal disinfectant. All members of the family *Felidae* are likely to be susceptible to FPV infection. FPV grows in primary or continuous feline renal cells, but not in canine cells.

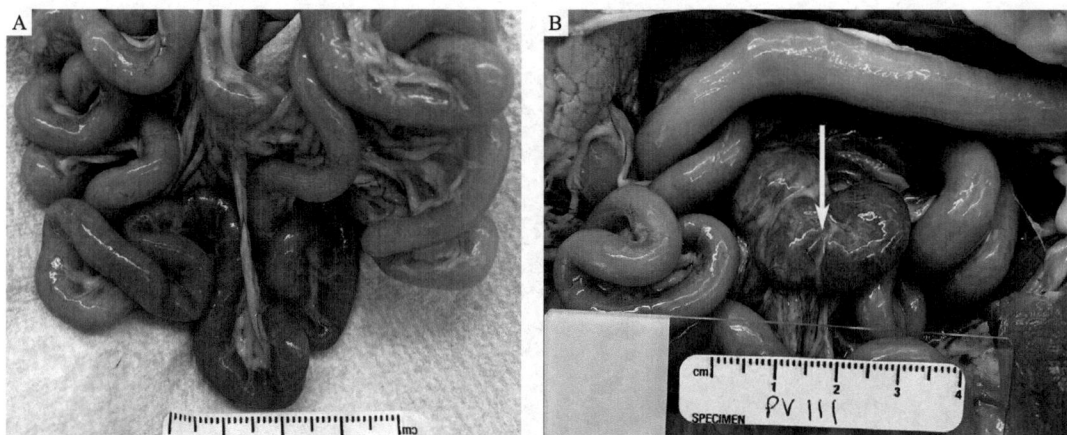

图 27-2 猫泛白细胞减少症病毒引起的猫肠道病理变化

A. 出血性肠炎（空肠）；B. 肠系膜淋巴结肿大、增厚。图像来源见英文。

Figure 27-2 FPV induced intestinal pathologic changes in cats

A. Hemorrhagic enteritis (jejunum); B. Enlarged and thickened mesenteric lymph nodes. Source: Barrs VR. 2019. Feline panleukopenia: a re-emergent disease. Vet Clin North Am Small Anim Pract, 49 (4): 651-670.

彩图

实验室诊断： 结合临床症状、血液中白细胞检测和组织病理学检查是诊断猫泛白细胞减少症的常用手段。

防控： 截至目前，尚无有效的猫泛白细胞减少症的治疗方法。因此，一般通过接种疫苗、隔离感染猫，以及对猫舍进行严格消毒从而实现对该病的防控。市面上已经有猫泛白细胞减少症的灭活苗和减毒活疫苗。母猫抗体会干扰疫苗对幼龄猫的免疫效果。

犬细小病毒

犬细小病毒病以突然发生腹泻、呕吐、厌食、发热、精神沉郁、淋巴细胞减少、脱水和肠出血为特征（**图 27-3**）。幼犬的死亡率高于成年犬，幼犬有时在未出现肠炎的临床症状时发生心肌炎。

病毒特征： 犬细小病毒病是由犬细小病毒 2 型（CPV-2）引起的，它是猫泛白细胞减少症病毒的一个变种。CPV-2 自在犬中出现以来一直在演化，出现了新的变种，称为 CPV-2a 型和 CPV-2b 型。另有一个基因型的 CPV，可以感染犬只，但没有

Laboratory diagnosis: clinical symptoms, the presence of leukopenia, and histopathological examination are useful for presumptive diagnosis of FP.

Prevention and control: there are no effective treatments for feline panleukopenia, thus its control is achieved by vaccination, quarantining infected cats, and rigorous sterilization of cat's lodges. Effective inactivated and modified live vaccines against feline panleukopenia are commercially available. Maternal antibodies can interfere with the immune response of young kittens to vaccines.

Canine parvovirus

A parvoviral disease in dogs is characterized by a sudden onset of diarrhea, vomiting, anorexia, fever, depression, lymphopenia, dehydration and serous hemorrhage (**Figure 27-3**). Mortality is higher in puppies than adults. Very young puppies sometimes develop myocarditis without clinical signs of enteritis.

Virus characteristics: canine parvoviral disease is caused by CPV-2, which is a variant of feline panleukopenia virus. CPV-2 has continued to evolve since its emergence in dogs, with the appearance of new variants that have been designated as CPV-2a and CPV-2b. Another genotype, designated as CPV-1, infects dogs but is not pathogenic. Similar to feline panleukopenia virus, CPV is very resistant to environmental factors,

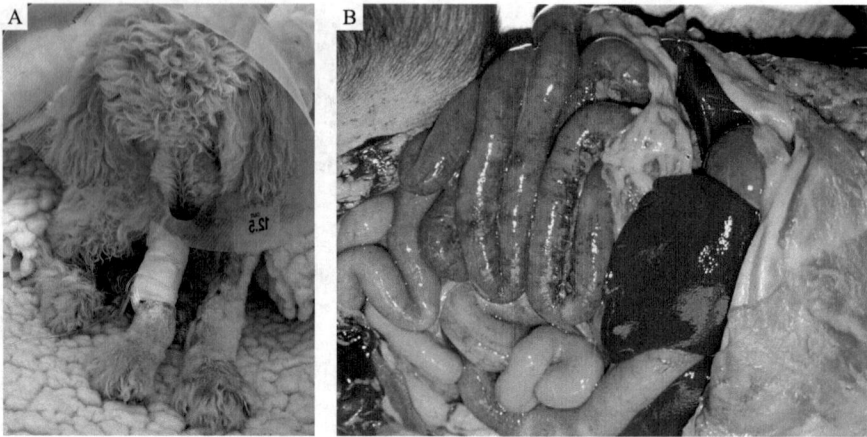

图 27-3 感染犬细小病毒的贵宾犬（A）小肠壁变色及浆膜出血（B）
图像来源见英文。

Figure 27-3 CPV infected poodles (A) discoloration of small intestinal wall and serosal hemorrhage (B)
Source: Sykes JE. 2013. Canine and Feline Infectious Diseases. Amsterdam: Elsevier Inc, 915.

致病性，被称为 CPV-1。与猫泛白细胞减少症病毒类似，CPV 对环境有很强的抵抗力，如极端温度、pH 和常规消毒剂，次氯酸钠溶液（1:30）可将其灭活。CPV-2可在猫胎肺或犬肾原代细胞及连续传代细胞系（如犬细胞系 A72、猫细胞系 NLFK、CRFK）中培养。

实验室诊断：临床体征、病史、病原检测和组织病理学检查常用于犬细小病毒病的初步诊断。

防控：犬细小病毒病重症病例的典型特征是显著脱水和代谢性酸中毒，因此支持疗法是 CPV 治疗的主要手段，通过补充失去的体液和纠正紊乱的电解质平衡和酸中毒，控制感染犬的症状，并通过广谱抗生素预防继发性细菌感染。疫苗接种是预防 CPV-2 感染的最佳方法。犬细小病毒抗体水平与保护程度直接相关。目前，市场上有灭活苗和减毒活疫苗可供选择。CPV 对环境因素具有很强的抵抗力，可长期存在。因此，饲养场所的消毒和幼犬接种疫苗对预防该病非常重要。

such as extreme temperatures, pH and some disinfectants. It can be inactivated by common bleach such as Clorox (1:30). CPV-2 can be propagated in primary cell culture of feline fetal lung or canine kidney as well as cell lines, such as canine cell line A72, feline cell lines NLFK and CRFK.

Laboratory diagnosis: clinical symptoms, medical history, antigen test and histopathological examinations are useful in presumptive diagnosis of canine parvoviral disease.

Prevention and control: severe cases of the canine parvoviral disease are characterized by marked dehydration and metabolic acidosis, thus supportive treatment is important to make up the loss of body fluids and restoring disturbed electrolyte balance and acidosis. Broad-spectrum antibiotics are used to prevent secondary bacterial infections. Vaccination of susceptible canine populations remains the best way to prevent canine parvoviral infection. Antibody levels directly correlate with the degree of protection. Effective inactivated and modified live parvoviral vaccines are commercially available. Canine parvovirus is highly resistant to environmental factors and can persist under adverse conditions for a long period. Therefore, prompt disinfection of premises where infected animals are kept, and vaccination of puppies are important in preventing this disease.

猪细小病毒

猪细小病毒病在世界各地都有发生，导致母猪繁殖障碍，尤其是初产母猪。胎儿经胎盘感染导致死产、木乃伊化和胚胎死亡。

病毒特征：猪细小病毒（PPV）与猫泛白细胞减少症病毒和犬细小病毒相似。PPV只有一种血清型，在抗原上与其他细小病毒不同。猪细小病毒对热、酶和大多数消毒剂都有很强的抵抗力。在下列条件下，73 ℃作用30 min或70 ℃作用1 h，0.5%次氯酸钠作用5 min，0.06%二氯异氰脲酸钾作用5 min，3%甲醛作用1 h，猪细小病毒可被灭活。PPV只感染猪，可在猪胎肾细胞和猪睾丸细胞的原代或继代细胞上繁殖。

实验室诊断：PPV抗原可通过采集流产胎儿组织进行免疫荧光、免疫组化染色或夹心ELISA检测。此外，PCR法也适用于检测PPV。

防控：截至目前，对该病尚无有效药物进行治疗。疫苗接种是确保初产母猪繁殖前产生主动免疫力的最佳方法。灭活疫苗和活疫苗均可用于预防PPV，但疫苗接种的时间点非常重要，确保免疫在初产母猪抗体消失后及在母猪配种之前进行。

鹅细小病毒

鹅细小病毒（GPV）常对8～30日龄的雏鹅致病，也可感染番鸭。临床以精神委顿，离群独偶，鼻孔流出浆液性鼻液，拉灰黄色或黄绿色稀粪，神经紊乱，小肠中后段黏膜坏死脱落、形如腊肠状为特征（**图27-4**），发病率和死亡率很高，对养鹅业生产危害极大。病毒无血凝性，只有一个血清型。可在12～14日龄的鹅胚和番鸭胚中增殖，在肝细胞中常出现包涵体。该病可通过接种灭活苗和弱毒苗进行防控。

Porcine parvovirus

Infection of porcine parvovirus occurs worldwide and can lead to reproductive failure in sows, especially those in gilts. Transplacental infection leads to stillbirth, mummification, embryonic death of fetuses.

Virus characteristics: porcine parvovirus (PPV) is similar to feline and canine parvoviruses. There is only one serotype. PPV is antigenically different than other parvoviruses. PPV is very resistant to heat, enzymes, and most commercial disinfectants. The virus can be inactivated by heat at 73 ℃ for 30 min or at 70 ℃ for 1 h or by 0.5% sodium hypochlorite for 5 min, 0.06% potassium dichloroisocyanurate for 5 min or 3% formaldehyde for 1 h. PPV only infects pigs. The virus can grow in primary fetal porcine renal cells and primary and passaged swine testicle cells.

Laboratory diagnosis: PPV antigens can be detected using the tissues of infected fetuses by immunofluorescent or immunohistochemical staining or sandwich ELISA. PCR can also be used for specific detection of PPV nucleic acid.

Prevention and control: so far, there is no effective treatment for this disease. Vaccination is the best method to ensure active immunity in gilts prior to their pregnancy. Both inactivated vaccines and modified live vaccines are available for the prevention of PPV, but the timing of vaccination is very important, which has to be done after the disappearance of maternal antibodies in gilts and before mating.

Goose parvovirus

Goose parvovirus (GPV), also known as gosling plague virus, affects 8-30 days old goslings and Muscovy ducks. The clinical features are depression, serous nasal fluid, grayish-yellow or yellow-green feces, necrosis in the small intestines with high morbidity and mortality (**Figure 27-4**). The infection of goose parvovirus has a deep impact on the goose industry. This virus has no hemagglutination ability and only one serotype. It replicates in 12-14 days old goose and Muscovy duck embryos. Inclusion bodies in hepatocytes are the features of viral infection. The disease can be controlled by inactivated and attenuated vaccines. Recent studies have reported a trend of co-infection of goose parvovirus and goose circovirus among other pathogens.

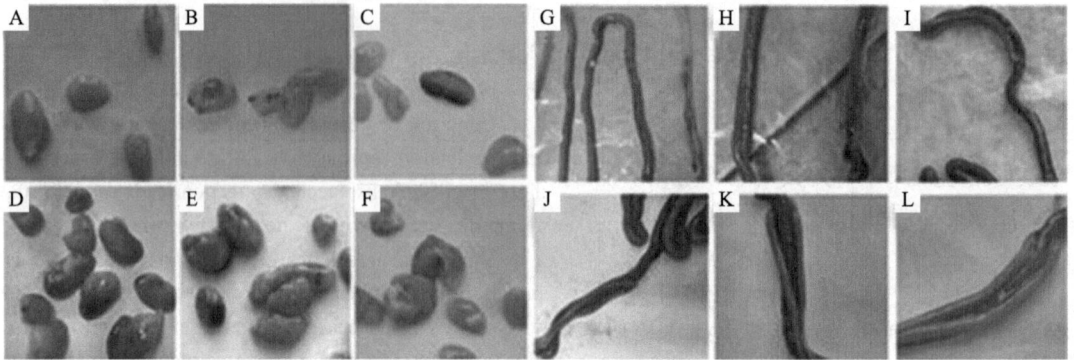

图27-4 感染GPV番鸭的大体病变

A、G为对照组胸腺和盲肠；B为感染GPV第1天，胸腺有出血点；C、D、E分别为感染GPV第2、4和6天，胸腺出现瘀血点；F为感染GPV第8天，胸腺无明显病变；H和J为感染GPV第1、4天，盲肠肿胀、末端充满绿色粪便；I、K和L分别为感染GPV第2、6、8天，盲肠肿胀。图像来源见英文。

Figure 27-4 Gross lesions of GPV-infected ducks

A, G. Thymus and cecum of control groups, respectively; B. Hemorrhagic spots on thymus at 1 days PI; C, D and E. Petechial hemorrhage of thymus at 2, 4, and 6 days PI, respectively; F. No gross lesion in thymus at 8 days PI; H and J. Swelling and full of green feces at end of cecum at 1 and 4 days PI; I, K and L. Swelling in entire cecum at 2, 6, and 8 days PI. Source: Niu Y, Zhao L, Liu B, et al. 2018. Comparative genetic analysis and pathological characteristics of goose parvovirus isolated in Heilongjiang, China. Virol J, 15 (1): 27.

近期有研究报道鹅细小病毒与鹅圆环病毒等病原有共感染的趋势。

第2节 圆环病毒科
Section 2 *Circoviridae*

圆环病毒是动物的重要病原体，类似于细小病毒，微小、无囊膜、二十面体对称，基因组为单链DNA，呈环状。圆环病毒科中的代表病毒为猪圆环病毒。

Circoviruses have been recently shown to be important pathogens of animals. Similar to parvoviruses, they are small non-enveloped icosahedral viruses with a genome of circular single-stranded DNA. Porcine circovirus is the representative of *Circoviridae*.

猪圆环病毒

猪圆环病毒（PCV）包含PCV1和PCV2两个类型，分别是非致病性和致病性病毒。2016年以来，世界各地陆续有发生PCV3和PCV4的报道，虽在猪群中流行率较低，但已引起关注。迄今为止，PCV2感染已与多种综合征联系在一起，统称为猪圆环病毒相关疾病（PCVAD）。PCVAD是一种症状复杂的综合征疾病类型，涉及养猪生产

Porcine circovirus

Porcine circoviruses (PCV) have been divided into nonpathogenic PCV1 and pathogenic PCV2. Since 2016, PCV3 and PCV4 have been reported all over the world and have garnered attention although their prevalence in pigs is low. To date, PCV2 infection has been linked with a variety of syndromes, collectively termed as porcine circovirus-associated disease (PCVAD). PCVAD describes a group of complex syndromes that occur during all stages of pork production. Manifestations of PCVAD

的所有阶段。PCVAD的临床表现从无症状的亚临床感染到急性死亡病例均有分布。典型的临床症状是消瘦、腹泻、呼吸窘迫、皮炎或繁殖障碍（**图27-5**）。

病毒特征：PCV2感染是PCVAD发生的主要原因。PCV2有四种基因型，称为PCV2a、PCV2b、PCV2c和PCV2d。家猪和野猪是PCV2的主要宿主，可持续感染。在PCVAD流行的猪场，PCV2在猪的口鼻液、

can range from asymptomatic subclinical infection to acute death. Clinical signs typically present with wasting, diarrhea, respiratory distress, dermatitis, or reproductive failure (**Figure 27-5**).

Virus characteristics: PCVAD mainly results from the infection by PCV2. PCV2 isolates are divided into four main genotypes, known as PCV2a, PCV2b, PCV2c and PCV2d. Domestic ated and feral pigs are the primary hosts for PCV2. The virus can establish long-term infection and is shed via the oronasal fluid, urine, blood, and feces in pigs for up to 28 weeks on PCVAD affected

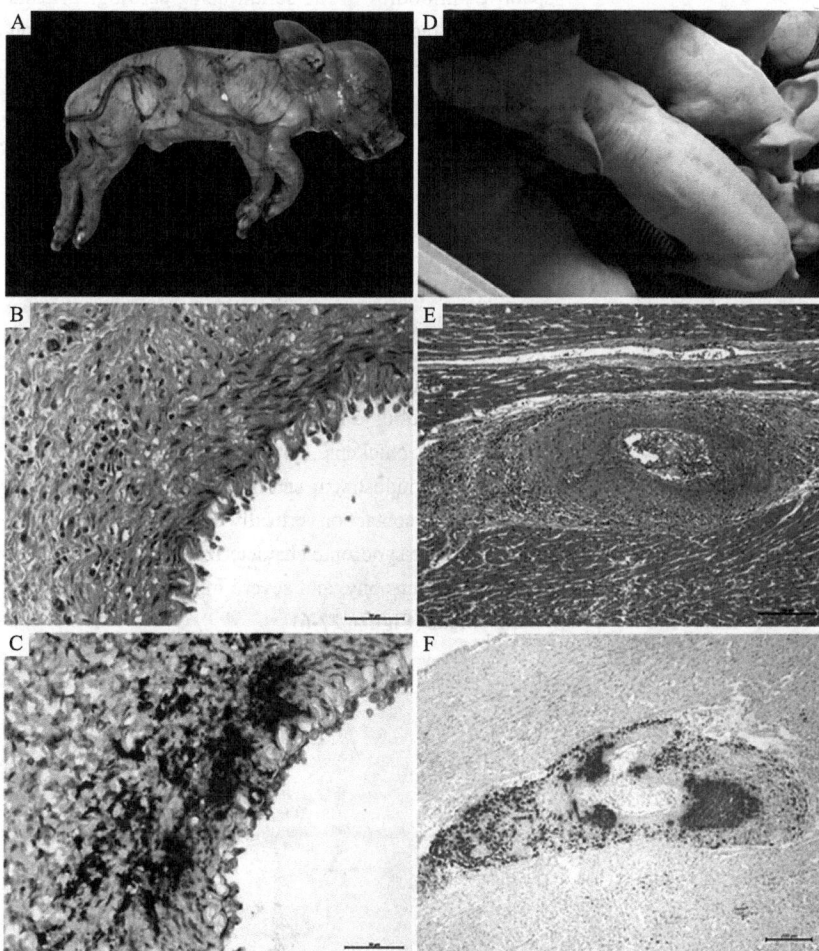

图27-5　PCV3相关的繁殖性疾病（A、B、C）和系统性疾病（D、E、F）个体病例建议诊断标准

PCV3相关的繁殖性疾病：A. 怀孕后期出现繁殖障碍，其特征是死胎和弱胎的比例增加；B. 胎儿脾动脉壁有轻度至中度单核细胞炎性浸润；C. 受损动脉中含有中等至大量PCV3核酸。PCV3相关的系统性疾病：D. 临床表现上猪出现消瘦；E. 心脏出现中度至重度非化脓性动脉炎；F. 受损动脉中含有大量PCV3核酸。图像来源见英文。

Figure 27-5　Proposed diagnostic criteria for the individual case definition of PCV3-associated reproductive (A, B, C) and systemic (D, E, F) disease

PCV3-reproductive disease: A. Stillborn piglet from a litter with a late reproductive problem characterized by increased percentage of stillborn and weak-born piglets; B. Mild-to-moderate mononuclear inflammatory infiltrates in the arterial wall of the fetal spleen; and C. Moderate to high amount of PCV3 nucleic acid in the damaged arterial area. PCV3-systemic disease: D. Clinical picture of a pig showing wasting; E. Moderate-to-severe non-suppurative arteritis in the heart; and F. High amount of PCV3 genome in the damaged artery. Source: Saporiti V, Franzo G, Sibila M, et al. 2021. Porcine circovirus 3 (PCV3) as a causal agent of disease in swine and a proposal of PCV3 associated disease case definition. Transbound Emerg Dis, 68 (6): 2936-2948.

彩图

尿液、血液和粪便中传播，持续时间长达28周龄。水平传播是最常见的传播方式。PCV2在体外可以在猪肾和睾丸上皮细胞中增殖。

　　实验室诊断：实验室诊断方法包括：①通过易感细胞培养，从感染动物的血清或组织中分离PCV2，或通过PCR鉴定感染组织中的PCV2核酸；②用病毒特异性抗体进行免疫荧光或免疫组化染色，检测肺或淋巴病变组织切片中的PCV2抗原；③采用间接荧光抗体、病毒中和试验或ELISA等血清学检测血清中PCV2特异性抗体。

　　防控：目前控制PCV2最有效的方法是接种疫苗。

farms. Transmission most often occurs from one pig to another through the horizontal route. Porcine epithelial cells from kidney and testicle are used for PCV2 replication.

Laboratory diagnosis: laboratory procedures used to confirm PCV2 infection include: ①isolation of PCV2 from the serum or tissues of infected animals using susceptible cell culture, or by identification of PCV2 nucleic acids in infected tissues by PCR; ② detection of PCV2 antigens in histological sections of lung or lymphoid lesions by immunofluorescent or immunohistochemical staining with virus specific antibodies; and ③ detection of PCV2-specific antibodies in the serum using serological tests such as indirect fluorescent antibody, virus neutralization assay or ELISA.

Prevention and control: the most effective method for controlling PCV2 is vaccination.

第3节　细环病毒科
Section 3　*Anelloviridae*

　　鸡贫血病毒（CAV），曾称为鸡传染性贫血因子，除鸡之外不感染其他家禽，是养禽业的世界性难题，可通过直接接触水平传播或经卵垂直传播。CAV可引起雏鸡严重贫血、骨髓萎缩和严重免疫抑制综合征（**图27-6**）。

Chicken anemia virus (CAV), once known as chicken infectious anemia factor, does not infect other poultry except chickens. Now it is a worldwide problem in the poultry industry. It can be transmitted horizontally through direct contact or vertically through eggs. CAV can cause a disease syndrome characterized by severe anemia, bone marrow atrophy, and severe immunosuppression in young chicks (**Figure 27-6**).

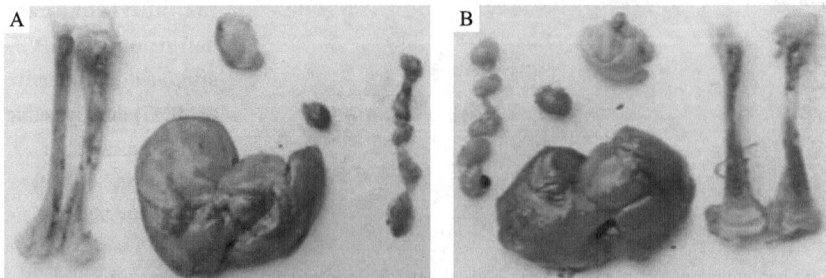

图27-6　感染鸡贫血病毒的雏鸡肝脏、骨髓、脾脏、法氏囊和胸腺的大体病变

A. 感染组；B. 对照组。图像来源见英文。

Figure 27-6　Gross lesions in liver, bone marrow, spleen, bursa and thymus in chicks infected with chicken anemia virus

A. Infected group; B. Control group. Source: Wani MY, Dhama K, Latheef SK, et al. 2014. Experimental pathological studies of an Indian chicken anaemia virus isolate and its detection by PCR and FAT. Pak J Biol Sci, 17 (6): 802-811.

思考题　Questions

1. 细小病毒的基因组及病毒颗粒有何结构特点。

1. What are the structural characteristics of parvovirus genome and virus particles?

2. 试述圆环病毒科的分类及主要特征。

2. Please describe the classification and main characteristics of *Circoviridae*.

第28章 逆转录病毒
Chapter 28 / Retroviruses

内容提要 逆转录病毒科的病毒在复制过程中有逆转录过程，其基因组为单链正链RNA，包括甲型、乙型、丙型、丁型和戊型逆转录病毒属，以及慢病毒属。

Introduction The members of *Retroviridae* have a reverse transcription process during replication. Their genomes are positive single-stranded RNA, including the genera *Alpha-, Beta-, Gamma-, Delta-* and *Epsilonretrovirus*, and *Lentivirus*.

第1节 逆转录病毒科的主要特征
Section 1 Main characteristics of *Retroviridae*

逆转录病毒科是有囊膜包裹的单链正链RNA病毒，与肿瘤的形成及获得性免疫缺陷综合征有关，包含正逆转录病毒亚科和泡沫逆转录病毒亚科，前者有6个属（甲型逆转录病毒属、乙型逆转录病毒属、丙型逆转录病毒属、丁型逆转录病毒属和戊型逆转录病毒属，以及慢病毒属），后者包括泡沫病毒属等5个属。

逆转录病毒含逆转录酶和源于宿主细胞的囊膜，平均直径在80～100 nm，为球形颗粒，具有独特的三层结构，最内层为二倍体RNA、螺旋对称的核衣壳蛋白和逆转录酶形成的复合物。中层为二十面体对称的衣壳，直径约60 nm。外层为源于宿主细胞膜的囊膜，由脂质以及env基因编码的糖蛋白组成。

逆转录病毒感染细胞时，病毒与细胞膜融合有两种机制：一是病毒与细胞膜受体结合发生膜融合，不依赖于pH；二是病毒与细胞器膜融合，需低pH环境。病

Retroviridae are positive single-stranded RNA viruses with a capsid envelope that are associated with tumor formation and acquired immune deficiency syndromes. *Retroviridae* includes *Orthoretroviridae* and *Spumaretroviridae*. The former consists of 6 genera (*Alpharetrovirus, Betaretrovirus, Gammaretrovirus, Delta-retrovirus, Epsilonretrovirus* and *Lentivirus*) and the latter of 5 genera (*e.g. Spumavirus*).

Retroviruses contain reverse transcriptase and a vesicle membrane derived from the host cell, with an average diameter of 80-100 nm and spherical particles with a unique three-layer structure, the innermost layer being a complex formed by diploid RNA, helix-symmetric nucleocapsid protein and reverse transcriptase. The middle layer is an icosahedrally symmetric capsid with a diameter of about 60 nm. The outer layer is a vesicle membrane derived from the host cell membrane and consists of lipids as well as a glycoprotein encoded by the *env* gene.

When a retrovirus infects a cell, there are two mechanisms by which the virus fuses with the cell membrane: first, the virus binds to the cell membrane receptor to undergo membrane fusion, which is not pH dependent; second, the virus fuses with the organelle

毒进入细胞后，释放病毒RNA，在逆转录酶的作用下逆转录成双链DNA，双链DNA转移至细胞核，形成前病毒。前病毒与宿主基因组一起复制，并传递给后代细胞。前病毒DNA在细胞核内转录成病毒mRNA并释放入细胞质中，最后病毒在宿主细胞质中组装，在细胞膜处通过出芽方式释放，同时形成囊膜。当逆转录病毒将自己的基因组整合到细胞基因组中后，它们的基因组就会传给下一代。目前这些内源性逆转录病毒占人类基因组的5%～8%。大多数内源性逆转录病毒的功能是未知的，通常被称为"垃圾DNA"。但是，许多内源性逆转录病毒在宿主生物学中起着重要作用，例如，控制基因转录、胎盘发育过程中的细胞融合，以及抵抗外源性逆转录病毒的感染。

脂溶剂、去垢剂及56 ℃加热30 min均能灭活逆转录病毒，但其对紫外线及X射线照射的抵抗力较强，可能与其具有二倍体RNA基因组有关。

to undergo membrane fusion, which requires a low pH environment. Upon entry into the cell, the virus releases viral RNA, which is reverse transcribed into double-stranded DNA by reverse transcriptase, and the double-stranded DNA is transferred to the nucleus to form the provirus. The provirus replicates with the host genome and is passed on to the progeny cells. The proviral DNA is transcribed into viral mRNA in the nucleus and released into the cytoplasm, and finally the virus assembles in the host cytoplasm and is released by budding through the cell membrane while acquiring a capsid. When retroviruses integrate their genome into the cellular genome, their genome is then passed on to the next generation. Currently, these endogenous retroviruses account for 5% to 8% of the human genome. The function of most endogenous retroviruses is unknown, and they are often referred to as "junk DNA". However, many endogenous retroviruses play important roles in host biology, such as control of gene transcription, cell fusion during placental development, and resistance to infection by exogenous retroviruses.

The retroviruses can be inactivated by lipid solvents, detergents, and heating at 56 ℃ for 30 min. They are more resistant to ultraviolet and X-ray radiation compared with other viruses, probably as a result of their diploid RNA genome.

第2节 甲型逆转录病毒属
Section 2 *Alpharetrovirus*

甲型逆转录病毒属于逆转录病毒科，正逆转录病毒亚科。

分类：根据病毒囊膜蛋白，禽白血病病毒（ALV）可以分为10个亚群，分别为A～J。其中A～E和J亚群宿主为鸡，F、G亚群宿主为雉，Ⅰ亚群宿主为鹌鹑。根据传播模式的不同，禽白血病病毒又可分为外源性和内源性逆转录病毒。其中E亚群为内源性病毒，此类病毒一般不具有致病性；其他亚群为外源性病毒，可诱发不同类型肿瘤及其他繁殖障碍性疾病（图28-1）。

基因组：ALV病毒粒子直径80～100 nm，

The genus *Alpharetrovirus* belongs to the subfamily *Orthoretrovirinae* of the family *Retroviridae*.

Classification: based on the viral capsid protein, avian leukosis viruses are grouped into 10 subgroups designated as A to J. Subgroups A to E and J infect chicken, subgroups F and G infect pheasant, and subgroup I infects quail. Based on the transmission mode, they can be grouped into "exogenous" and "endogenous" retroviruses. Subgroup E viruses are endogenous that generally not pathogenic, while viruses in the other subgroups are exogenous that induce different types of tumors and other reproductive disorders (**Figure 28-1**).

Genome: ALV is a single-stranded positive-sense RNA virus with a genome of 7.2 kb. It is spherical, enveloped and 80-100 nm in diameter. There are 2 long terminal repeats (LTRs) of about 600 bp at the 5′ and

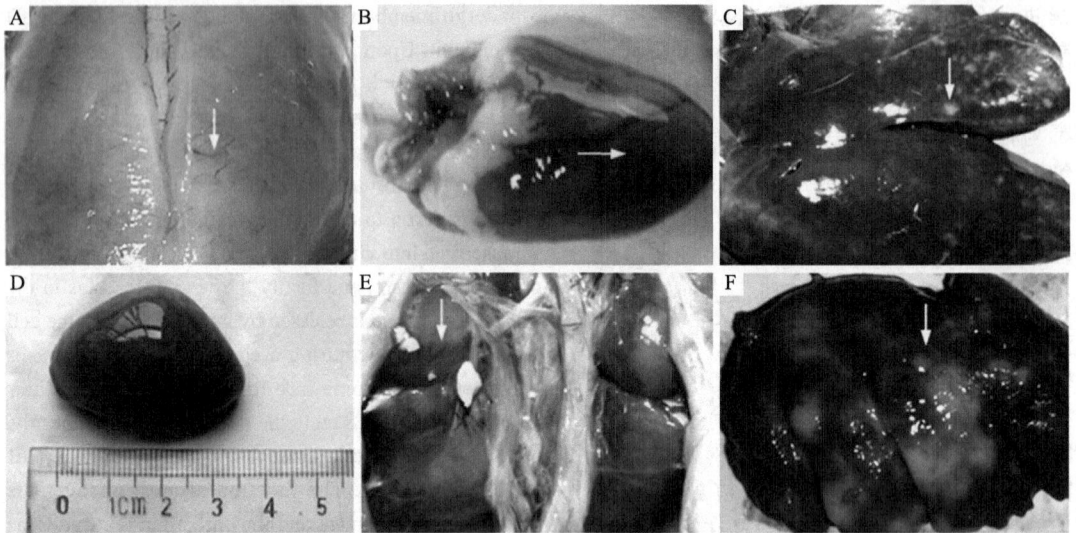

图 28-1　ALV 感染鸡的大体病变

A. 胸肌苍白、出血；B. 心肌出血；C. 肝肿大，多处可见白色隆起灶；D. 脾肿大；E. 花斑肾；F. 肺表面有弥漫性白色隆起灶。图像来源见英文。

Figure 28-1　Gross lesion of ALV infected chickens

A. Hemorrhage on pale pectoral muscle; B. Hemorrhage on myocardium; C. Hepatomegaly with multifocal white raised foci; D. Splenomeagly; E. Kidney with piebald; F. Multifocal to diffuse white raised foci on the lung surface. Source: Wang G, Jiang Y, Yu L, et al. 2013. Avian leukosis virus subgroup J associated with the outbreak of erythroblastosis in chickens in China. Virol J, 10: 92.

彩图

有囊膜，球形。ALV是单链正义RNA病毒，基因组全长约7.2 kb。在5′端和3′端有2个约600 bp的长末端重复（LTR）。ALV有3个主要基因：*gag/pro*、*pol*、*env*，分别编码群特异性抗原（Gag）、逆转录酶/整合酶和囊膜蛋白。不同亚群的*gag*基因序列高度相似，可编码6种蛋白（p19、p10、p27、p2、p15和p12），其中p27为衣壳蛋白，其保守性最高，可以作为抗原，用于检测鸡群是否感染ALV。

流行病学： ALV于20世纪60年代首次发现，随后迅速在全球传播，对养禽业造成重大损失。自2004年起，ALV-J在我国鸡群中发现，其发病率和致死率高达50%。

禽白血病一年四季均可发生。在自然情况ALV下主要感染鸡，其中白羽肉鸡最易感，也可以感染野鸡、鸭、鹌鹑、鸽子、火鸡等其他禽类。发病鸡、带毒鸡和受感

3′ ends. ALV genome has 3 major genes: *gag/pro*, *pol*, *env*, which encode group-specific antigen(Gag), reverse transcriptase/integrase and envelope proteins, respectively. The *gag* gene of different subgroups is of high similarity and encodes 6 proteins (p19, p10, p27, p2, p15 and p12), of which p27, a capsid protein, is most conserved and can be used as a target antigen to diagnose whether chickens are infected with ALV.

Epidemiology: ALV was first discovered in the 1960s and then rapidly spread around the world, causing great losses to the poultry industry. Since 2004, ALV-J was found in China with high morbidity and mortality over 50%.

Avian leukemia can occur throughout the year. Under natural conditions, ALV mainly infects chickens, especially white-feathered broilers. It can also infect pheasants, ducks, quails, pigeons, turkeys and other birds. Sick chickens, subclinically infected chickens and infected breeding eggs are the sources of transmission. ALV can be transmitted horizontally, vertically and genetically. ALV is most harmful to chicklings, followed by hens and young chickens. Infected layers will stop

染的种蛋均是传染源。ALV可通过水平传播、垂直传播和遗传方式进行传播。ALV对雏鸡危害最大，其次为母鸡和幼龄鸡。感染的蛋鸡会停止产蛋，死亡率达20%以上。

自2008年起，随着我国ALV清除计划的实施，其在规模化肉鸡和蛋鸡养殖场中的流行率显著减少，得到了有效控制。但近些年ALV在小规模本地鸡中的流行率却呈上升趋势。

鸡发病后表现为渐行性消瘦，呈企鹅状行走；鸡冠和肉髯苍白，后期萎缩；皮肤多处有出血点且止血困难，个别出现血泡，破溃后血流不止，最终失血过多而死；个别鸡也会形成血痂，反复脱落和出血；有的表现为腹水症；有的病鸡突然死亡。随着病情发展，可在病鸡许多内脏器官（常见于肝脏、脾脏）产生大小和数量各异、灰白色或淡灰黄色的肿瘤，病鸡最后因极度消耗衰竭而亡。

laying, with a mortality of more than 20%.

Since 2008, with the implementation of ALV eradication program in China, its prevalence in the large-scale broiler and layer farms has been significantly reduced and effectively controlled. However, in recent years, its prevalence on small-scale local chickens is increasing.

Since the onset of the clinical signs, the chickens show progressive emaciation and penguin-like walking. The crowns and beards are pale and atrophied at the later stage. There are many hemorrhagic spots on the skin that are difficult to clot. Some infected chickens suffer from bloody vesicles on the skin and unceasing bleeding from ruptured vesicles, and eventually die of excessive blood loss. Some vesicles become blood scabs and may fall off and bleed repeatedly. Some infected chickens show ascites while others die suddenly. With the progression of the disease, many internal organs (commonly liver and spleen) can produce gray-white or light gray-yellow tumors in different sizes and numbers, and finally the infected chicken die of cachexia.

第3节 乙型逆转录病毒属
Section 3 *Betaretrovirus*

乙型逆转录病毒有B型和D型两种形态，该病毒的包装通常在细胞质中进行，首先形成A型的病毒颗粒，然后从细胞膜出芽获得囊膜后形成B型或D型粒子。

绵羊肺腺瘤病毒： 又称绵羊肺癌病毒，是绵羊最常见的逆转录病毒，属于单链负链RNA病毒，为D型或B/D嵌合型病毒。难以对其进行体外培养，只能通过接种易感绵羊来获得病毒抗原，因此对其生物学特性了解甚少。

绵羊肺腺瘤分布广泛，除澳大利亚和新西兰以外，在世界各地可呈散发、地方性流行或暴发等。各品种的绵羊均为易感动物，没有性别、年龄差异，但在羔羊中潜伏期较短。

临床上该病毒可引起绵羊和山羊的慢

The genus *Betaretrovirus* includes betaviruses of type B and type D in terms of morphotype. Their packaging is usually carried out in the cytoplasm. First, type A virus particles are formed, and then type B or D particles are formed after the type A virus particles obtain the cytoplasmic membrane by budding.

Ovine pulmonary adenocarcinoma (OPA) virus: known as sheep lung cancer virus, it is the most common retrovirus in sheep. It is a single-stranded negative RNA virus that is either a type D or type B/D chimera. It is difficult to culture *in vitro*, so virus antigens can be only obtained by inoculating susceptible sheep. Therefore, its biological characteristics are poorly understood.

OPA is widespread. Besides Australia and New Zealand, the disease can be sporadic, endemic or pandemic all over the world. All breeds of sheep are susceptible. No particular gender or age of sheep appears to be predisposed to OPA, but the latent period is shorter in lambs.

Clinically, the virus can cause a chronic and con-

性感染性肺脏疾病。该病主要临床特征包括咳嗽，呼吸困难，消瘦，鼻腔内有水样分泌物，肺部癌变等，能引起绵羊肺部出现实变（固化），肺泡上皮细胞和支气管上皮细胞进行性肿瘤增生（**图28-2**）。羊最后因缺氧和继发细菌性肺炎而死亡。该病临床上容易和维士纳/梅迪病相混淆。但绵羊肺腺癌导致鼻分泌物较多，强迫病羊低头或者抬高病羊后躯时，有大量的带有泡沫的液体从鼻孔内流出。这一现象具有示病特征。

tagious lung disease in sheep and goats. The main clinical signs include coughing, dyspnea, weight loss, watery nasal discharge, and lung cancer. The infection can cause lung consolidation (solidification) in sheep, and progressive tumor from proliferation of alveolar and bronchial epithelial cells (**Figure 28-2**). The host eventually dies of hypoxia and secondary bacterial pneumonia. It can be difficult to differentiate OPA from Visna/Maedi disease clinically. However, sheep infected with OPA have increased nasal secretions. When infected sheep are forced to lower their heads or lift the hindquarters, a large amount of foamy liquid flows out from the nostrils, which is the specific characteristic of this disease.

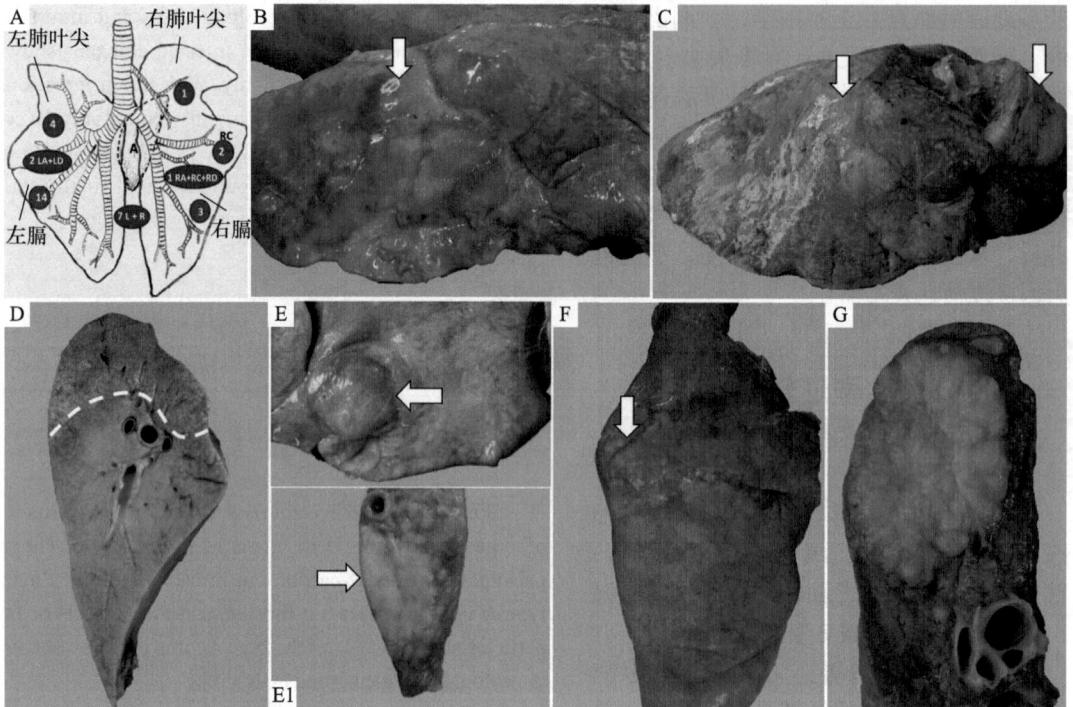

图 28-2 绵羊肺腺瘤的大体病变

A. 肺部病变分布；B. 经典型绵羊肺腺瘤（OPA），累及膈叶（箭头）；C. 经典型OPA，累及多个肺叶（箭头）；D. 经典型OPA，肺切面显示肿瘤和正常组织之间的连接（中断线）；E、E1. 非典型OPA，胸膜下白灰色结节（箭头）。在切面上，肿瘤呈珍珠色且干燥（箭头）；F. 肺黏液瘤样结节的大体特征（箭头）；G. 肿瘤切面呈多叶、白灰色、凝胶状。图像来源见英文。

Figure 28-2　Gross features of spontaneous OPA in sheep

A. Distribution of pulmonary lesions; B. Classical form of OPA, involving the diaphragmatic lobe (arrow) ; C. Classical form of OPA, affecting multiple pulmonary lobes (arrows) ; D. Classical form of OPA on cut surface showing the junction (dot line) between the tumor and normal tissue; E and E1. Atypical form of OPA showing a subpleural white-grey nodule (arrow). On cut surface, the tumor is pearly white and dried (arrow) ; F. Gross features of pulmonary myxoma-like nodule (arrow) ; G. The myxomatous mass shows a multilobular, white-gray, gelatinous feature on cut surface. Source: Toma C, Bâlteanu VA, Tripon S, et al. 2020. Exogenous Jaagsiekte sheep retrovirus type 2 (exJSRV2) related to ovine pulmonary adenocarcinoma (OPA) in Romania: prevalence, anatomical forms, pathological description, immunophenotyping and virus identification. BMC Vet Res, 16 (1): 296.

第4节 丙型逆转录病毒属
Section 4 *Gammaretrovirus*

本属病毒包含4组：复制完全型病毒、复制缺陷型病毒、爬行类病毒，以及禽类病毒，分别有猫白血病病毒、猫肉瘤病毒、蝮蛇逆转录病毒及禽网状内皮组织增殖病毒。但是实际情况往往更复杂，如猫白血病病毒的不同毒株，可能是复制完全型，也可能是复制缺陷型。

The genus *Gammaretrovirus* contains 4 groups of viruses: replication-competent viruses, replication-defective viruses, reptilian viruses and avian viruses. These groups include feline leukemia virus, feline sarcoma virus, pit viper retrovirus and avian reticuloendotheliosis virus. However, the reality is often more complicate. For example, different strains of feline leukemia virus can either be replication-competent or replication-defective.

猫白血病病毒

猫白血病病毒为复制完全型病毒，遗传信息存在于病毒RNA，可不依赖于其他病毒完成自身复制过程。该病毒可能无致病性，也可导致白血病和免疫抑制（**图28-3**），其分布全球。

内源性猫白血病病毒是一种复制缺陷型的前病毒，可不依赖于其他病毒完成自

Feline leukemia virus

Feline leukemia virus (FeLV) is a replication-competent virus. The genetic information exists in the viral RNA and can complete its own replication process without reliance on other viruses. It may not be pathogenic but may also lead to leukemia and immunodeficiency (**Figure 28-3**), with occurence all over the world.

Endogenous feline leukemia virus is a replication-defective provirus that can complete its own replication process independent of other viruses. It has high sequence

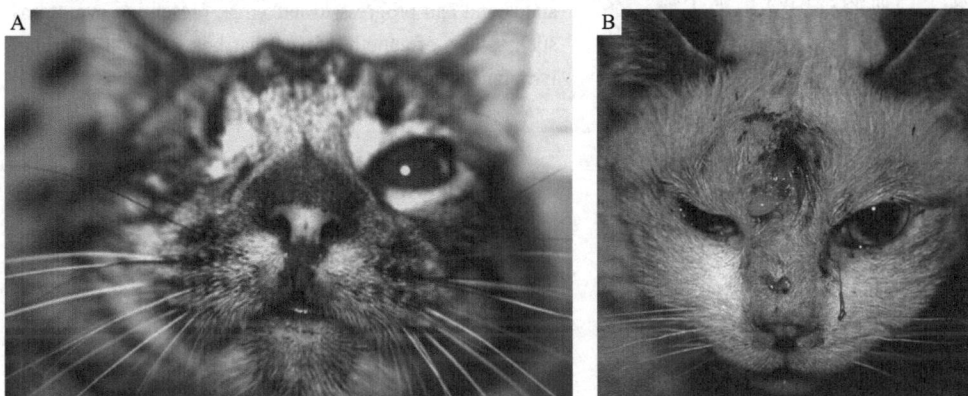

图28-3 进行性猫白血病病毒感染猫的条件感染

A. 一只6月龄的雄性已去势家猫的鼻平面溃疡，伴有进行性FeLV感染；B. 患有地方性FeLV感染的一只暹罗猫出现了严重的鼻隐球菌病。图像来源见英文。

Figure 28-3　Opportunistic infections in cats with progressive FeLV infection

A. Ulceration of the nasal planum in a 6-month-old male neutered domestic cat with progressive FeLV infection; B. Severe nasal cryptococcosis in a Siamese cat with endemic FeLV infection. Source: Sykes JE, Hartmann K. 2016. Feline leukemia virus infection. Veterinary Key. https://veteriankey.com/feline-leukemia-virus-infection.

身复制过程，与外源性猫白血病病毒有很高的序列相似性，但由于突变，不能产生感染性的病毒粒子。

外源性猫白血病病毒同样不依赖于其他病毒完成自身复制过程，可产生感染性的病毒颗粒，引起猫患病。

猫肉瘤病毒

猫肉瘤病毒为外源性猫白血病病毒缺陷型，缺失 *env* 基因，因此在复制和增殖过程中，肉瘤病毒依赖于外源性猫白血病病毒才能产生成熟的病毒粒子。

受感染猫的鼻腔分泌物、尿液、粪便和乳汁都可以传播病毒。猫与猫之间可能会通过相互之间的咬伤、互相梳理毛发，在少数情况下通过共用猫砂盆和餐盘进行传播。大部分猫出现自限性感染，无病毒血症，产生中和抗体和猫肿瘤病毒膜相关抗原（FOCMA）抗体，不排毒，不发生白血病。部分感染猫表现为病毒血症，通过分泌物排毒；产生中和抗体，但FOCMA抗体下降，属于病情不稳定状况，继而发生白血病，出现食欲不振、发热、腹泻、牙龈出血、淋巴结肿大、肿瘤等。

similarity with exogenous feline leukemia virus. However, they cannot produce infectious viral particles due to mutations.

Exogenous feline leukemia virus is also independent of other viruses to complete its own replication process and can produce infectious viral particles that can cause diseases in cats.

Feline sarcoma virus

Feline sarcoma virus is an exogenous feline leukemia virus-defective type, lacking the *env* gene. Therefore, during replication and proliferation, sarcoma virus depends on exogenous feline leukemia virus in order to produce mature viral particles.

The virus can be transmitted by nasal secretions, urine, feces and milk of infected cats. It may be transmitted from cat to cat through mutual biting, mutual grooming and, in rare cases, by sharing litter boxes and food bowls. Most cats experience self-limiting infection without viral viremia. They produce neutralizing antibodies and antibodies against feline oncornavirus membrane antigen (FOCMA), and they neither shed the virus nor develop leukemia. Some infected cats develop viremia and would shed the virus through secretions. They produce neutralizing antibodies, but FOCMA antibody levels decline. This condition is characterized by unstable disease status and can progress to leukemia, presenting symptoms such as loss of appetite, fever, diarrhea, gingival bleeding, lymph nodes enlargement, and tumors, *etc.*

第5节 丁型逆转录病毒属
Section 5 *Deltaretrovirus*

丁型逆转录病毒属包含牛白血病病毒，导致牛、绵羊、山羊等产生淋巴肉瘤。该病在大多数国家均有发现，每年的发病率在14/100 000～165/100 000之间。

致病机理：肿瘤的发生可能与其前病毒的 *v-onc* 基因整合到细胞的DNA有关。主要靶细胞为B细胞。感染牛最初无症状，然后发展为持续性淋巴细胞增多，最终产生肿瘤。肿瘤组织尤其是疾病晚期的肿瘤中并不含病毒或病毒抗原，但如将肿瘤组

Bovine leukemia virus is a type D retrovirus that causes lymphosarcoma in cattle, sheep and goats. The disease is found in most countries, and the annual incidence is 14-165 per 100 000.

Pathogenesis: occurrence of the tumor may be related to integration of the *v-onc* gene of provirus into the host cell DNA. The main target cells are B cells. Infected cattle are initially asymptomatic, then develop persistent lymphocytosis and eventually tumors. Tumor tissues, especially those at the late stage, do not contain viruses or viral antigens, but if lymphocytes and sensitive cells are co-cultured with the tumor cells, infectious virus

织用淋巴细胞与敏感细胞共同培养，可获得有感染性的病毒。

与其他逆转录病毒一样，该病毒并无很高的传染性，病毒水平传播不易波及相邻牛群，除非密切或长时期接触。损伤、直肠检查的手套、注射用针头、外科器械等可传播病毒。病毒可经胎盘感染胎儿，一般新生犊牛的感染率小于10%，且已证实，通过正常接触病毒不能由牛传给绵羊，反之亦然。

实验室诊断： 国际贸易组织指定的方法是用琼脂凝胶免疫扩散试验或ELISA，检测血清中病毒gp51和p24的抗体，gp51的抗体出现较早，而且稳定。

防控： 牛白血病是WOAH规定通报的疫病。每隔2~3个月对牛群做血清学检查，淘汰阳性牛，直至建立无病牛群。可试用灭活疫苗进行免疫预防；并可加强饲养管理，切断血液、牛奶、分泌物、排泄物、注射器或手术器械等传播途径，以防止牛白血病病毒感染的扩散。

particles can be obtained.

As with other retroviruses, the virus is not highly infectious and horizontal transmission of the virus is not likely to spread to adjacent herds unless there is close or prolonged contact. However, injuries, gloves for rectal examination, needles for injection, and surgical instruments can transmit the virus. The virus can infect the fetus via the placenta, generally less than 10% of newborn calves, and it has been confirmed that the virus cannot be transmitted from cattle to sheep through normal contact and vice versa.

Laboratory diagnosis: the method designated by World Trade Organization is to use an agar gel immuno-diffusion test or ELISA to detect antibodies against the virus gp51 and p24. The antibodies against gp51 appear earlier and are stable.

Prevention and control: bovine leukemia is a notifiable disease listed by WOAH. Serological diagnosis should be conducted on the cattle herd every 2 to 3 months. Positive cattle should be culled until a disease-free herd is established. Inactivated vaccines can be used for prevention. Bovine leukemia virus could be preventable by strengthening feeding management, and blocking transmission involving blood, milk, secretions, excrement, syringes, surgical instruments and other factors.

第6节 慢病毒属
Section 6 *Lentivirus*

慢病毒属的特征是有较长的潜伏期，如人类免疫缺陷病毒、猴免疫缺陷病毒、马传染性贫血、猫免疫缺陷病毒、维士纳/梅迪病毒等。

维士纳/梅迪病毒

维士纳/梅迪病毒为逆转录病毒科，慢病毒属代表成员，病毒核酸类型为正链单链RNA。病毒粒子呈球形或卵圆形，有囊膜，大小为90~100 nm，含有直径约为40 nm的芯髓。

维士纳/梅迪病毒来自冰岛，绵羊最为易感，潜伏期可达2~6年，可导致一种消耗性、全身性疾病，具有终生感染的

Members in the genus *Lentivirus* are characteristic of a long latent period, such as human immunodeficiency virus, simian immunodeficiency virus, equine infectious anemia virus, feline immunodeficiency virus, and Visna/Maedi virus.

Visna/Maedi virus

Visna/Maedi virus is a representative member of the genus *Lentivirus* in the family *Retroviridae* and has a positive single-stranded RNA genome. The viral particles are spherical or ovoid in shape, with an enveloped membrane, measuring approximately 90-100 nm in size. They contain a core with a diameter of approximately 40 nm.

Visna/Maedi virus originates from Iceland and affects sheep, which are highly susceptible. The incubation period can last up to 2-6 years and can lead to a debilitating, systemic disease with lifelong infection capability.

能力，因为引发的病程慢，又被称为慢病毒。病羊衰弱、消瘦，最后终归死亡。该病毒可导致两种病症：维士纳病引起绵羊脑膜炎，使羊日渐消瘦，距关节运动不灵活，休息时经常跖骨后段着地；而梅迪病会引起间质性肺炎，即肺实质和间质出现纤维化病变，从而使羊呼吸困难，导致渐进性消瘦。该病毒可以水平传播和垂直传播，主要通过呼吸道、奶、精液、昆虫叮咬等方式传播给其他绵羊。该病呈世界性分布，主要分布在欧洲、非洲、美洲、亚洲，因引种于1966年由澳大利亚传入我国。目前广泛存在于我国的云南、陕西、甘肃等12个地区。

该病毒可在绵羊的多种细胞中增殖，如羊的脉络丛、肾和唾液腺等细胞，并常在2～3周后产生特征性的细胞病变，多数细胞可变为多核巨细胞。

实验室诊断：可以通过琼脂免疫扩散试验对病毒进行检测，或逆转录聚合酶链反应（RT-PCR）检测病毒RNA。通过酶联免疫吸附试验检测血清中的gp135和p28抗体。

防控：该病毒危害大，尚无特效疫苗和药物防治，是WOAH通报疾病。死亡的绵羊应该焚烧掩埋，所有器具无害化处理，用2%氢氧化钠对圈舍进行全面消毒。

马传染性贫血病毒

马传染性贫血病毒，是慢病毒属的另外一个代表成员。只感染马属动物，马最易感，骡、驴次之。对外部环境（冷、热、干燥等）有强抵抗力，多呈地方性流行或散发。能凝集鸡、蛙、豚鼠和人O型红细胞，可在马白细胞中生长，出现细胞病变。

该病毒可通过吸血昆虫（虻、厩蝇、蚊、蠓等）的叮咬机械性传播，也可通过胎盘、病马的血液及其各种分泌排泄物传播，同时污染的诊疗器械也是一重要的传

Due to its slow progression, it is referred to as a slow virus. Infected sheep become weak and emaciated, ultimately resulting in death. There are two distinct diseases caused by this virus: Visna disease, which causes meningoencephalitis in sheep, leading to gradual emaciation, impaired joint movement, and frequent resting with the rear part of the metatarsus touching the ground; and Maedi disease, which causes interstitial pneumonia, characterized by fibrotic changes in the lung parenchyma and interstitium, resulting in breathing difficulties and progressive wasting. The virus can be transmitted horizontally and vertically, primarily through respiratory routes, milk, semen, and insect bites to other sheep. The disease has a worldwide distribution, mainly found in Europe, Africa, the Americas, and Asia, and was brought into China from Australia in 1966 by importing breeding sheep. Currently, it is widespread in 12 regions of China, including Yunnan, Shaanxi, and Gansu.

Visna/Maedi virus can proliferate in a variety of sheep cells, such as the choroid plexus cells, renal, salivary gland cells, *etc*. The characteristic cytopathic effect usually occurs after 2 to 3 weeks of infection and most cells become multinucleated giant cells.

Laboratory diagnosis: Visna/Maedi virus can be detected by a variety of tests, such as agar diffusion or reverse transcription polymerase chain reaction (RT-PCR) targeting specific fragment of the virus RNA. ELISA can be used to detect antibodies to the gp135 or p28 protein in the serum.

Prevention and control: the disease can cause serious economic losses. There are no specific vaccines and drugs to prevent or control it. It is a notifiable disease by WOAH. Dead sheep should be burned and buried. All instruments should be treated as biohazard wastes, and the folds should be thoroughly disinfected with 2% sodium hydroxide.

Equine infectious anemia virus

Equine infectious anemia virus (EIAV) is another representative member of the genus *Lentivirus*. EIAV only infects equines with the horses most susceptible, followed by mules and donkeys. The virus has strong resistance to the external environment (cold, hot, dry, *etc*.), and can be either endemic or sporadic. EIAV can agglutinate erythrocytes of chicken, frog, guinea pig and human type-O blood, and grow in horse leukocytes with cytopathic effect.

The virus can be mechanically transmitted through

播媒介。

病毒先感染巨噬细胞，接着感染淋巴细胞，导致病马终身具有细胞结合的病毒血症。临床症状分为急性、亚急性、慢性和亚临床性4种，其中急性型最典型。表现为发热、贫血、黄疸等，常在四肢下端、胸前、腹下等处出现无热无痛的水肿。约80%病马死亡，而耐过的马终身持续感染。

马传染性贫血为WOAH规定通报疫病，在我国列为二类传染病。国际通用检测方法是琼脂凝胶免疫扩散（AGID）试验（**图28-4**）。可采用疫苗进行预防，无治疗意义。国外通常采用检测加淘汰的防控手段。

biting of blood-sucking insects (horse flies, stable flies, mosquitoes, midges, *etc.*), as well as through the placenta and the blood of infected horses and their excretions. Contaminated medical instruments are also important fomites.

The virus first infects macrophages and then lymphocytes, resulting in a lifelong cell-bound viremia in the infected horses. The clinical progression of the disease can be divided into acute, subacute, chronic and subclinical types, of which the acute type is most typical. The clinical signs include fever, anemia, jaundice, etc., and often feverless and painless edema in the lower extremities, chest, and cranial thorax. Approximately 80% of infected horses die, while those survivors have a lifelong infection.

EIA is a notifiable disease by WOAH but is classified as a second-class infectious disease in China. The international diagnostic standard is the agar gel immuno-diffusion (AGID) test (**Figure 28-4**). Vaccines can be used for prevention. Treatment of infected horses is useless. It is mainly prevented and controlled through detection and culling of infected animals.

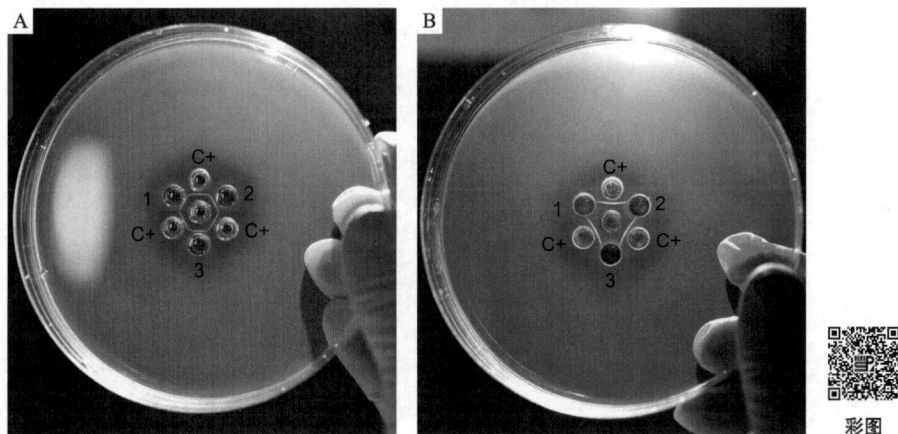

图28-4　琼脂凝胶免疫扩散试验
A．在阳性琼脂凝胶免疫扩散试验中，由于抗原（中心孔）与阳性对照（C+）和样本（1、2和3）中的抗体结合，出现免疫沉淀。在图中，三个样本（1、2和3）对EIA呈阳性，因此在所有孔中都可以看到一条线，呈现出六边形的外观；B．在阴性琼脂凝胶免疫扩散试验中，由于抗原（中心孔）和阳性对照（C+）中的抗体结合，出现免疫沉淀，但在样本（1、2和3）中不产生免疫沉淀。在图中，三个样本（1、2和3）对EIA呈阴性，因此只在含有阳性对照的孔中看到一条线，呈现三角形外观。图像来源见英文。

Figure 28-4　AGID test
A. In a positive AGID test, an immunoprecipitation occurs due to the binding of the antigen (central well) and the antibodies in the positive controls (C+) and the samples (1, 2 and 3). In the picture, the three samples (1, 2 and 3) are positive to EIA and therefore a line can be seen in all wells, giving the appearance of a hexagon; B. In a negative AGID test, an immunoprecipitation occurs due to the binding of the antigen (central well) and the antibodies in the positive controls (C+), but the immunoprecipitation does not happen in the samples (1, 2 and 3). In the picture, the three samples (1, 2 and 3) are negative to EIA and therefore a line can only be seen in the wells containing positive controls, giving the appearance of a triangle. Source: Camino E, Cruz F. 2017. EIA: equine infectious anemia. VISAVET Outreach Journal. https://www.visavet.es/en/articles/eia-equine-infectious-anemia.php.

思考题　Questions

1. 逆转录病毒具有什么样的结构？

2. 禽白血病病毒中，哪个亚群是内源性逆转录病毒？

3. 乙型逆转录病毒的包装通常在哪里完成？

4. 如何区别绵羊肺腺瘤病与维士纳/梅迪病？

1. What's the structure of retroviruses?

2. Which ALV subgroup is an endogenous virus?

3. Where does the packaging of the *Betaretrovirus* usually occur?

4. How to distinguish ovine pulmonary adenomatosis from Visna/Maedi disease?

第29章 双链RNA病毒
Chapter 29 Double-stranded RNA viruses

内容提要 双链RNA病毒包括呼肠孤病毒科和双RNA病毒科的众多成员，均具有分节段的双链RNA。不同属病毒基因组的节段数不同，病毒呈二十面体对称，无囊膜。

Introduction Double-stranded RNA viruses can be divided into two families of *Reoviridae* and *Birnaviridae*. All the viruses share the same feature of segmented RNA genomes, while the number of the segments varies. They are in icosahedral symmetry and have no envelope.

第1节 呼肠孤病毒科
Section 1 *Reoviridae*

正呼肠孤病毒属

呼肠孤病毒无处不在，宿主范围很广，可以感染多种动物。最初从人类和动物的呼吸道和肠道分离出具有分节段的双链RNA基因组的二十面体病毒，称之为呼肠孤病毒。

环状病毒属

环状病毒属是呼肠孤病毒科的成员，可感染昆虫，脊椎动物的感染多由昆虫传播。环状病毒属血清型约有100余种。大多数环状病毒不会引起人和动物的严重疾病，仅引起轻微发热。仅有少量的环状病毒，如蓝舌病毒和非洲马瘟病毒，较为重要。环状病毒基因组由10段双链RNA组

Orthoreovirus

Reoviruses are ubiquitous, with a very wide host range. Reovirus can infect a variety of animals. Icosahedral viruses with double-stranded RNA genomes were first isolated from the respiratory and intestinal tracts of humans and animals, hence the term reovirus.

Orbivirus

Orbivirus is a member within *Reoviridae*. They commonly infect insects that may transmit the viruses to vertebrates. About 100 serotypes are known in the genus *Orbivirus*. Majority of these viruses do not cause serious diseases in humans and animals, but may cause mild fever. Important animal pathogens include bluetongue virus and African horse sickness virus. Their genomes consist of 10 segments of double-stranded RNA, with

成，总基因组大小为18 kb，其复制周期与呼肠孤病毒相似。与呼肠孤病毒科的其他成员相比，环状病毒对低pH较敏感。

轮状病毒属

轮状病毒首次发现于1963年，直径约为75 nm，结构类似于车轮的辐条，因此病毒以拉丁词rota（车轮）命名。轮状病毒感染可破坏肠道绒毛上皮细胞，导致肠道对水、盐、糖等营养物质吸收减少，引起腹泻。研究发现，肠上皮细胞间的紧密连接被轮状病毒的NSP4蛋白破坏，使组织液渗漏到肠道中加重腹泻，进一步导致动物脱水（图29-1）。

病毒特征：轮状病毒粒子呈二十面体对称，无囊膜，衣壳分三层，每一层都由不同的病毒蛋白（VP）构成，也被称为三层粒子，因形似"轮子"而得名。外层由VP7和VP4两种蛋白组成；具有免疫原性的VP7糖蛋白是该层的主要结构蛋白，780个VP7分子以260个三聚体形式构成正二十面体结构；120个VP4分子以二聚体方式在病毒表面形成60个刺突。中间层包含"轮子"的"辐条"，亦由780个VP6分子组成260个三聚体构成的正二十面体结构。内部层由120个VP2分子构成，包裹病毒双链RNA基因组。十二个拷贝的VP1和VP3转录酶复合物在二十面体晶格的五重顶点处附着到VP2的内表面。VP1是依赖于RNA的RNA聚合酶，而VP3具有鸟酰转移酶和甲基转移酶活性。

实验室诊断：在世界范围内，轮状病毒是婴幼儿及其他1～8周龄的幼龄动物腹泻的主要病因，成年动物一般不感染。潜伏期1～3 d；典型症状包括水样腹泻、发热、腹痛和呕吐，导致脱水；感染后3～4 d通过粪便大量排毒（可高达每克粪便 10^{11} 个病毒粒子）。实验室诊断可用疾病早期收集的粪便样品，用电子显微镜观察病毒粒子，或用乳胶凝集试验或夹

a total size of 18 kb. The replication cycle is similar to reoviruses. Orbiviruses are sensitive to low pH compared with other *Reoviridae*.

Rotavirus

Rotavirus was first described in 1963. Its virion is about 75 nm in diameter and looks like the spokes of a wheel. Therefore, the virus was named with the Latin word *rota* (= wheel). Enterocytes on the villi are destroyed due to rotavirus infection, and diarrhea occurs as a result of reduced absorption of water, salts and sugars from the gut. There is evidence that the tight junctions between cells are damaged by the non-structural protein NSP4, allowing leakage of fluid into the gut with aggravated diarrhea, or even dehydration (**Figure 29-1**).

Virus characteristics: the virion is in icosahedral symmetry and has no envelope. The name rotavirus was given for its capsid appearance of triple-layered wheel-like particle composed of three layers of distinct virus proteins. The outer layer is composed of VP7 and VP4 proteins. VP7 is a viral glycoprotein of high immunogenicity and the key component of the outer layer. A total of 780 copies of VP7 are grouped as 260 trimers in icosahedral symmetry. The outer layer is decorated by 60 spikes, each of which is formed by a dimer of VP4. The intermediate layer, looking like "spokes" of the "wheel", is formed by 780 copies of VP6 arranged as 260 trimers with the same icosahedral symmetry. The innermost protein layer is composed of 120 copies of VP2 and houses the dsRNA genome. Twelve copies of the VP1/VP3 transcription enzyme complexes are attached to the inner surface of the VP2 layer at the five-fold vertices of the icosahedral lattice. VP1 is an RNA-dependent RNA polymerase, and VP3 has guanylyltransferase and methyltransferase activities.

Laboratory diagnosis: rotaviruses mainly cause diarrhea in infants, children and many kinds of young animals at 1-8 weeks of age worldwide but not in adult animals. There is a latent period of 1-3 days. Typical symptoms include watery diarrhea, fever, abdominal pain, vomiting, and even dehydration. The virus can be secreted in the feces of infected animals in high titers (10^{11} viral particles per gram) and maximum shedding occurs 3-4 days post infection. Laboratory diagnosis rests on demonstration of the virus in the stool collected early in the illness by electron microscope, detection of viral antigens by latex agglutination tests or sandwich ELISA,

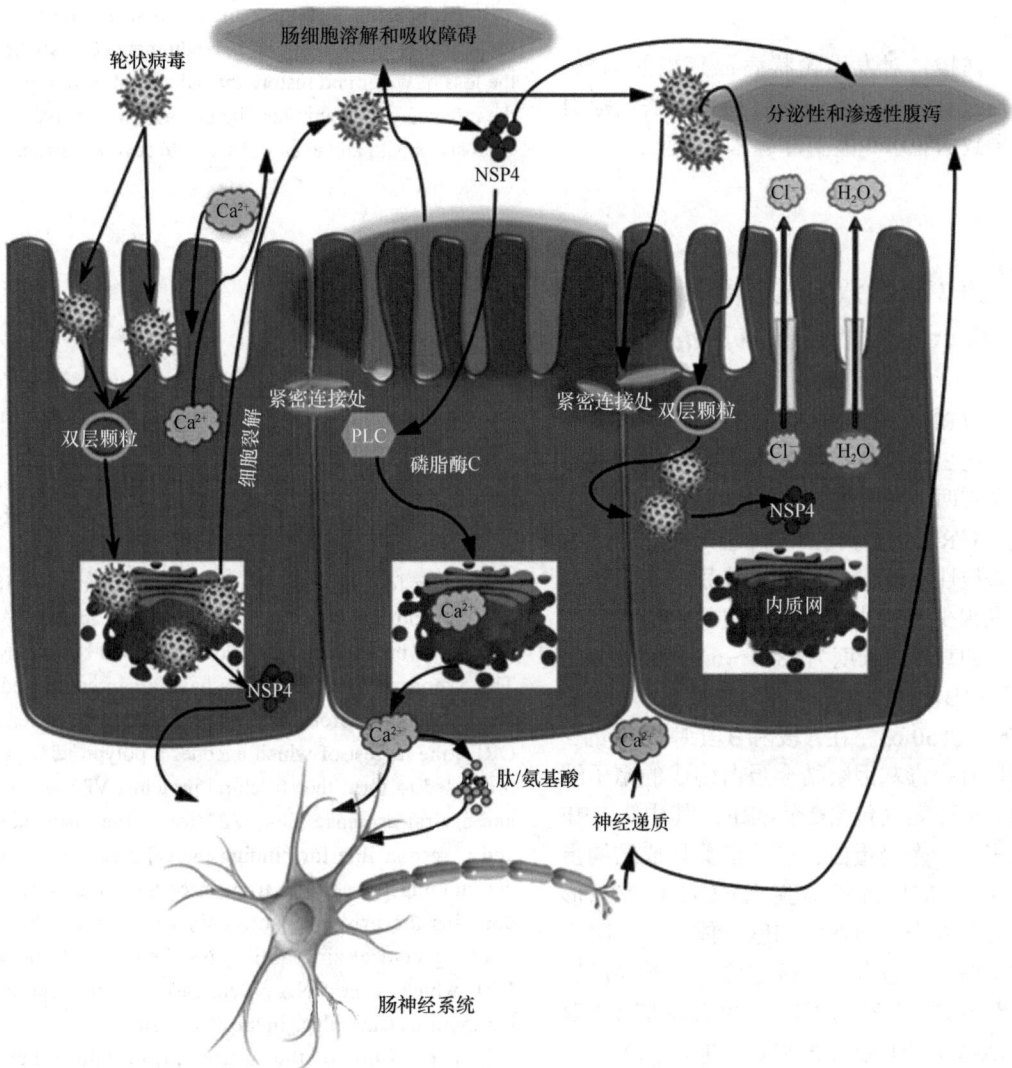

图 29-1 轮状病毒（RV）的致病机制

RV在肠细胞内复制可诱导渗透性腹泻。RV能增加细胞内钙（Ca^{2+}）的浓度，破坏细胞骨架和紧密连接，使细胞间通透性增加。此外，RV产生非结构蛋白4（NSP4），这是一种肠毒素，通过磷脂酶C依赖（PLC）机制诱导Ca^{2+}从内质网流出，进一步导致电解质失衡和分泌性腹泻。RV还可以刺激肠神经系统（ENS，通过NSP4依赖机制），进一步促进分泌性腹泻和肠蠕动。图像来源见英文。

Figure 29-1 Potential mechanisms of rotavirus (RV) pathogenesis

RV replication inside enterocytes induces osmotic diarrhea. RV also increases the concentration of intracellular calcium (Ca^{2+}), disrupting the cytoskeleton and the tight junctions, increasing paracellular permeability. In addition, RV produces non-structural protein 4 (NSP4), an enterotoxin that induces Ca^{2+} efflux from endoplasmic reticulum via the phospholipase C dependent (PLC) mechanism further contributing to electrolyte imbalance and secretory diarrhea. RV can also stimulate the enteric nervous system (ENS, via NSP4 dependent mechanism), further contributing to secretory diarrhea and increasing intestinal motility. Source: Vlasova AN, Amimo JO, Saif LJ. 2017. Porcine rotaviruses: epidemiology, immune responses and control strategies. Viruses, 9 (3): 48.

心 ELISA 检测病毒抗原，或 RT-PCR 检测病毒核酸；ELISA 可用于检测血清特异性抗体。

防控：对发生轮状病毒感染的动物，可通过补充水和电解质支持性疗法治疗腹泻导致的脱水、酸中毒、休克和死亡。

or demonstration of rotavirus nucleic acids by RT-PCR. ELISA can be used to detect antibodies specific to the virus.

Prevention and control: a supportive therapy by replenishing saline solution is recommended to make up the loss of water and restore the balance of electrolytes of the infected young animals, thereby avoiding dehydration, acidosis, shock, and even death due to severe diarrhea.

第 2 节　双 RNA 病毒科
Section 2　*Birnaviridae*

双 RNA 病毒科成员可以感染脊椎动物、昆虫、软体动物和甲壳类动物，基因组含有两个双链 RNA。

双 RNA 病毒科的基因组是分段的，由两条线性双链 RNA 组成。个别病毒粒子中也发现少量的非病毒基因组的核酸。完整病毒的基因组长度为 5880～6400 bp，分为 A 段和 B 段。A 段长 3100～3200 bp，B 段长 2750～2850 bp。在 A 段和 B 段基因组的 5′端均有病毒基因组结合蛋白构成的帽子结构。A 段至少包含 2 个 ORF，其中大 ORF 编码一个多聚蛋白，进一步剪切成结构蛋白 VP2、VP3 和病毒蛋白酶 VP4。VP2 形成病毒衣壳，与宿主细胞受体结合，决定细胞嗜性，包含多个抗原位点，诱导产生中和抗体。B 段编码 RNA 聚合酶 VP1。双 RNA 病毒的复制在细胞质中进行。

双 RNA 病毒科的病毒粒子结构简单，大小约 65 nm，无囊膜，六边形，由单层二十面体对称的壳粒蛋白组成。

双 RNA 病毒耐热（60 ℃，60 min）、pH 3～9 环境、乙醚和氯仿。3% 福尔马林、2% 氢氧化钠、紫外线及 γ 射线可迅速灭活病毒。

Members of the family *Birnaviridae* infect vertebrates, insects, mollusks and crustaceans. Their genomes are composed of two segments of RNA.

The genomes of the viruses in the family are segmented and consist of two segments of linear double-stranded RNA. Non-genomic nucleic acids are also found in a few virions. The complete genome is 5880-6400 bp and is divided into two segments: segment A of 3100-3200 bp and segment B of 2750-2850 bp in length. The 5′ ends of both segments have a cap composed of genome-binding proteins. Segment A contains at least 2 ORFs, the largest of which encodes a polyprotein that is processed to form the structural proteins, VP2 and VP3, and a viral protease VP4. VP2 forms the virus capsid and is responsible for binding the cellular receptor and determining the cellular tropism of the virus. VP2 also contains the principal antigenic sites responsible for eliciting neutralizing antibodies. Segment B encodes VP1, which is the RNA polymerase. The replication of birnaviruses takes place in the cytoplasm.

The virions of the viruses in the family have a simple structure, 65 nm in diametere, non-enveloped and hexagonal with a single capsid shell of having icosahedral symmetry.

The viruses are relatively heat-stable (60 ℃ for 60 min), and are stable in pH 3-9, ether and chloroform. Formalin at 3%, sodium hydroxide at 2%, or irradiation by ultraviolet light or γ-ray could inactivate the viruses rapidly.

传染性法氏囊病病毒

传染性法氏囊病是由传染性法氏囊病病毒（IBDV）引起的一种雏鸡高传染性

Infectious bursal disease virus

Infectious bursal disease (IBD) is a highly contagious disease of young chickens caused by infectious bursal disease virus (IBDV), characterized

疾病，其主要特点是病毒在鸡法氏囊中选择性复制，导致淋巴滤泡坏死，淋巴样细胞减少或消失，导致免疫抑制。3～6周龄雏鸡发病最为严重，发病率100%，死亡率可高达90%（**图29-2**）。1962年，该病首次在美国特拉华州的甘伯勒被发现。由于感染IBDV后，雏鸡对其他传染性疾病的易感性增加，并且干扰其他疫苗的免疫效果，所以该病对家禽业影响重

by depletion of lymphoid cells and immunosuppression as a result of selective viral replication in the bursa of Fabricius, and necrosis or apoptosis of lymphoid follicles. Chicklings at 3-6 weeks of age are most susceptible with 100% morbidity and up to 90% mortality (**Figure 29-2**). The disease was first discovered in Gumboro, Delaware, in 1962. It is economically important to the poultry industry worldwide due to increased susceptibility of infected chicklings to other diseases and interference with the infectiveness of vaccines against other infections. In

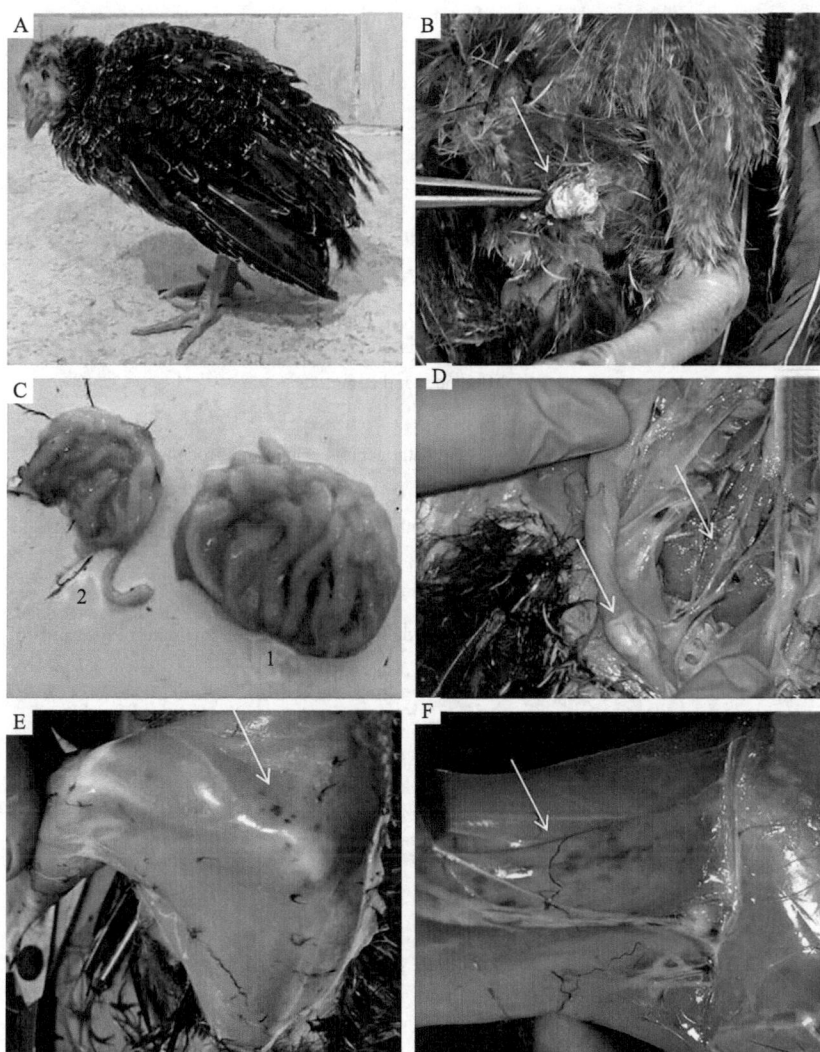

图29-2　自然感染传染性法氏囊病病毒（IBDV）雏火鸡的临床症状和死后变化

A. 火鸡昏睡，羽毛凌乱；B. 火鸡白肛门处有白色腹泻粪便；C. 5周龄雏火鸡的法氏囊，健康火鸡正常大小的法氏囊（1）；感染IBDV火鸡的萎缩法氏囊（2）；D. 轻度肾炎（箭头）和直肠内可见白色腹泻物（箭头）；E、F. 火鸡大腿肌有瘀点出血（箭头）。图像来源见英文。

Figure 29-2　Clinical signs and post-mortem changes of IBDV in naturally infected turkey poults

A. Turkey poult with dullness and ruffled feathers; B. Turkey poult with whitish diarrhea and soiled vent; C. Bursae of 5-week-old turkey poults, normal-sized bursa of healthy turkey poult (1) and atrophied bursa of turkey poult suspected to be infected with IBDV (2); D. Lesion of mild nephritis (arrow) and whitish diarrhea inside rectum (arrow); E, F. Petechial hemorrhages on thigh muscles of turkey poult (arrow). Source: Mosad SM, Eladl AH, El-Tholoth M, et al. 2020. Molecular characterization and pathogenicity of very virulent infectious bursal disease virus isolated from naturally infected turkey poults in Egypt. Trop Anim Health Prod, 52 (6): 3819-3831.

彩图

大。更值得引起注意的是，近年来，在欧洲、拉丁美洲、东南亚、非洲和中东均出现了导致雏鸡严重死亡的超强毒株IBDV（vvIBDV）。

实验室诊断： RT-PCR或实时荧光定量PCR等可用于检测病毒核酸。

防控： 接种灭活苗或活疫苗是确保鸡产生主动免疫力的最佳方法，可用于预防鸡传染性法氏囊病。母鸡免疫可为雏鸡提供母源抗体，使后者获得被动免疫。

recent years, very virulent strains of IBDV (vvIBDV) causing severe mortality in chicken have emerged in Europe, Latin America, Southeast Asia, Africa and the Middle East.

Laboratory diagnosis: RT-PCR or real-time fluorescence PCR can be used for specific detection of viral RNA.

Prevention and control: immunization, either with inactivated or live vaccines, is the best way to ensure that chickens develop active immunity against infectious bursal disease in chickens. Vaccination of the hens can provide the chicklings with maternal antibodies to promote passive immunity.

思考题　Questions

1. 列举具有重要致病作用的双链RNA病毒。

2. 呼肠孤病毒科有哪些重要的属？比较其异同。

3. 试述轮状病毒的致病机理。

4. 试述传染性法氏囊病病毒的致病机理。

5. 试述传染性法氏囊病的诊断与防控措施。

1. List some of the important pathogenic double-stranded RNA viruses.

2. What are the important genera in the family *Reoviridae*? Tell their differences and commons.

3. What is the pathogenic mechanism of rotaviruses?

4. What is the pathogenic mechanism of infectious bursal disease virus?

5. What are the diagnosis, prevention and control measures of infectious bursal disease?

第30章 单负链病毒目
Chapter 30 / *Mononegavirales*

内容提要 单负链病毒目主要包括副黏病毒科、弹状病毒科和波纳病毒科，其共同特征是单链负链RNA基因组、相似的病毒复制机制、相似的基因序列构成，以及具有囊膜的病毒粒子。

Introduction The members of *Mononegavirales* mainly include *Paramyxoviridae*, *Rhabdoviridae*, and *Bornaviridae*. These viruses share common characteristics: negative single-stranded RNA genomes, similar viral replication mechanisms, genome composition, and enveloped virions.

第1节 副黏病毒科
Section 1 *Paramyxoviridae*

副黏病毒科的病毒有囊膜，具有多形性，病毒粒子直径150～500 nm。病毒基因组为单链负链RNA。核衣壳呈螺旋形对称。

Viruses in the family *Paramyxoviridae* are enveloped, pleomorphic and 150-500 nm in diameter, and contain linear single-stranded negative RNA genomes. Their nucleocapsids are in helical symmetry.

犬瘟热病毒

犬瘟热是犬类中最严重的病毒性疾病。急性犬瘟热表现为双相热、眼鼻分泌物增加、厌食、精神沉郁、呕吐腹泻、脱水、白细胞减少、呼吸困难、皮疹、脚垫过度角质化和神经症状（图30-1）。临床症状严重程度存在明显的个体差异，从无症状到明显的临床症状均可发生。该病的潜伏期为3～5 d。死亡率在很大程度上依赖于犬的免疫状态，幼犬致死率最高。

病毒特征：犬瘟热病毒（CDV）具有多形性，粒子常呈球形，直径150～300 nm（图30-2）。核衣壳呈螺旋状对称，外面被

Canine distemper virus

Canine distemper (CD) is the most serious viral disease of dogs. It is characterized by biphasic fever, ocular and nasal discharges, anorexia, depression, vomiting, diarrhea, dehydration, leukopenia, respiratory distress, skin rash, hyperkeratosis of the foot pads and signs of the central nervous system (**Figure30-1**). Severity of the disease varies greatly among the infected individuals, ranging from asymptomatic or subclinical infection to severe symptoms. The disease has a latent period of 3 to 5 days. The mortality rate of the disease largely depends on the immune status of infected dogs, being the highest in puppies.

Virus characteristics: canine distemper virus (CDV) is polymorphic and variable in shape and size. Viral particles are often spherical and vary from 150 nm to

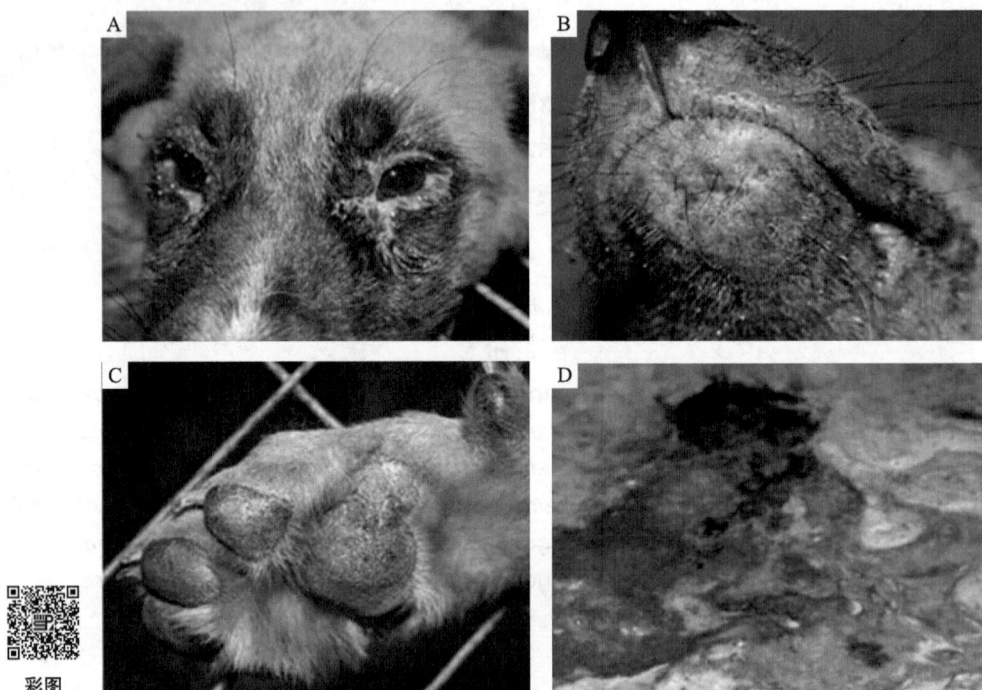

彩图

图 30-1　犬瘟热病毒感染犬的典型临床症状

A. 眼部有分泌物；B. 面部出现皮疹；C. 脚垫变硬；D. 血便。图像来源见英文。

Figure 30-1　The typical clinical signs of CDV infected dogs

A. The ocular discharge from the eyes; B. Red rashes on face; C. Hardened footpat; D. Bloody diarrhea. Source: Tan B, Wen YJ, Wang FX, et al. 2011. Pathogenesis and phylogenetic analyses of canine distemper virus strain ZJ7 isolate from domestic dogs in China. Virol J, 8: 520.

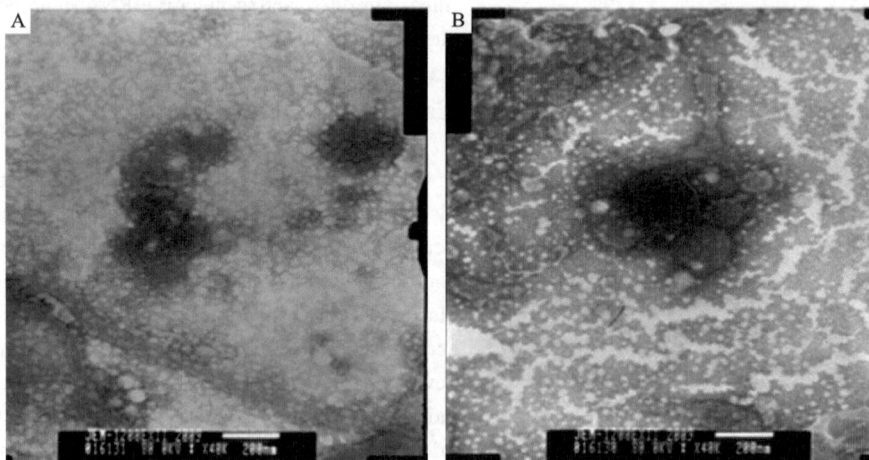

图 30-2　犬瘟热病毒的电镜图片

A. 犬瘟热病毒粒子——负染；B. 犬瘟热病毒粒子的组装。图像来源见英文。

Figure 30-2　The morphology of CDV under electron microscope

A. The morphology of CDV isolate under electron microscope (negative staining) ; B. The particles of assembling viruses. Source: Tan B, Wen YJ, Wang FX, et al. 2011. Pathogenesis and phylogenetic analyses of canine distemper virus strain ZJ7 isolate from domestic dogs in China. Virol J, 8: 520.

具有纤突的囊膜包围。病毒基因组为线性单链负链RNA。病毒表面糖基化的H蛋白介导与易感细胞表面受体的结合，另一种糖基化蛋白F介导CDV与宿主细胞融合。CDV只有一种抗原类型，对紫外线、干燥及50℃以上的温度、常用消毒剂敏感。CDV可以感染多种动物，除了犬，犬科的其他成员，鼬科和浣熊科的大部分成员均易感。CDV可在犬和雪貂的原代肾细胞中繁殖。

实验室诊断：①采集病料进行CDV的分离和鉴定；②免疫荧光、免疫组化或犬瘟热病毒胶体金检测试纸条检测疑似犬临床样品或细胞培养物中的病毒抗原；③酶联免疫吸附试验（ELISA）检测血清中的特异性抗体水平。

防控：犬瘟热的治疗措施多是辅助性的支持疗法，如使用抗生素用于预防细菌继发性感染，输液补充电解质，使用抗惊厥药缓解神经症状。疫苗接种是控制该病的最好方法。

新城疫病毒

从家禽和野鸟中分离到的禽副黏病毒可分为9个血清型。其中，新城疫病毒（NDV）是典型的禽副黏病毒血清1型病毒，也是禽副黏病毒中致病性最强的病原。新城疫是鸡的一种高度传染性疾病，以呼吸窘迫、腹泻和神经症状为主要特征（**图30-3**）。新城疫的严重程度取决于引起感染毒株的毒力。毒性强的毒株被称为强毒型或者速发型，可以导致禽类高达90%以上的致死率。中等毒力的NDV引起不严重的疾病类型，死亡率往往低于25%。弱毒株对鸡无致病性，可用作疫苗。新城疫的潜伏期为4～11 d。

病毒特征：新城疫病毒（NDV）具有多形性，病毒粒子直径100～300 nm，属于有囊膜病毒。NDV基因组RNA编码了6个蛋白质。囊膜表面的HN蛋白具有血凝素

300 nm in diameter (**Figure 30-2**). The viral nucleocapsid is in helical symmetry and enclosed by a glycoprotein envelope that contains spikes. The viral genome is a linear single-stranded negative RNA. The glycoprotein H is responsible for viral binding to the receptor of susceptible cells. The glycoprotein F is involved in fusing with the infected cells. CDV has only one antigenic type and is sensitive to ultraviolet rays, dryness, temperature above 50 ℃, and commonly used disinfectants. CDV infects a wide range of animal species. In addition to dogs, other members of the *Canidae*, many members of the families *Mustelidae* and *Procyonidae* are susceptible. CDV can be propagated in primary canine and ferret kidney cells.

Laboratory diagnosis: ① isolation and identification of CDV from clinical samples; ② detection of canine distemper viral antigens in clinical samples by immuno-fluorescent technique, immunohistochemistry or sandwich immunochromatographic strip; ③ detection of specific antibodies by enzyme-linked immunosorbent assay (ELISA).

Prevention and control: supportive therapy is largely used for CD. Antibiotics are used to prevent secondary bacterial infections. Fluid therapy is recommended to replenish electrolytes. Anticonvulsants can be employed for symptomatic relief of neurological disorders. Vaccination is the best way for controlling the disease.

Newcastle disease virus

Avian paramyxoviruses isolated from domestic and wild birds can be divided into 9 serologic groups. Newcastle disease virus (NDV) is the prototype virus of the avian paramyxovirus serotype 1 that is most pathogenic to chickens, causing Newcastle disease, a highly infectious disease characterized by respiratory distress, diarrhea, and neurological signs (**Figure 30-3**). The severity of the disease is dependent upon the virulence of the strain that is responsible for the infection. The most virulent strains are referred to as velogenic types and have mortality rates as high as 90% or even higher in infected birds. The disease caused by mild strains is less severe with the mortality rate often less than 25%. The low virulent strains are almost avirulent to chickens and are employed as vaccines. The latent period of the Newcastle disease varies from 4 d to 11 d.

Virus characteristics: Newcastle disease virus is pleomorphic and enveloped with a diameter varying from 100 nm to 300 nm. The viral RNA genome encodes 6 major polypeptides. The glycoprotein HN is responsible for

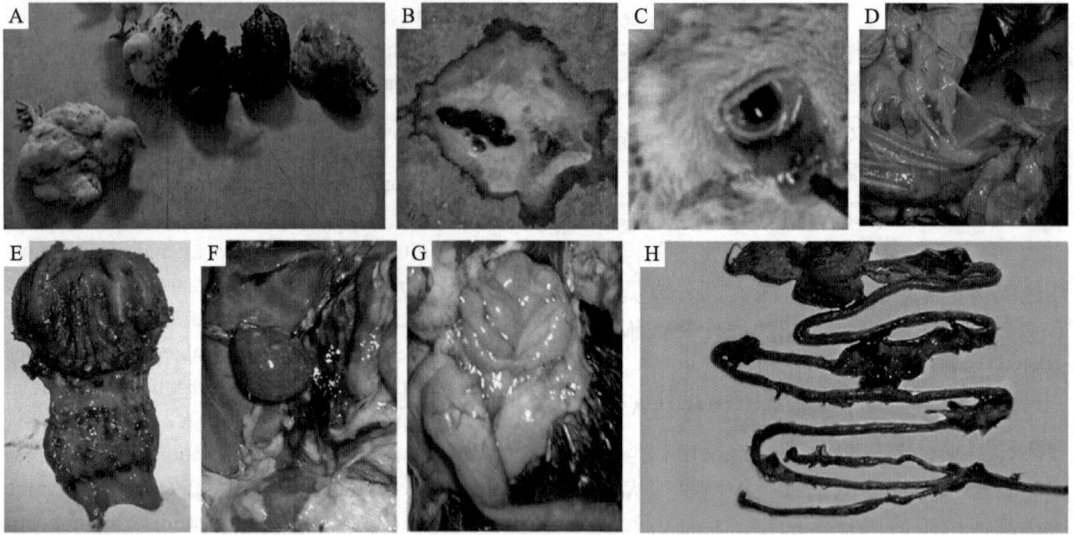

图 30-3 感染新城疫后鸡只临床症状

A. 感染雏鸡群：联合麻痹、斜颈、四肢和翅膀麻痹；B. 绿色稀粪；C. 下眼睑出血；D. 胸腺出血；E. 腺胃乳头出血；F. 斑驳脾；G. 法氏囊水肿；H. 肠腔和盲肠扁桃体出血。图像来源见英文。

Figure 30-3 The clinical symptoms of chickens with Newcastle disease

A. Group of infected chicks: unileg paresis, torticolis, paralysis of legs and wings; B. Greenish diarrheic feces; C. Hemorrhage in the lower eyelid; D. Thymus showing mild hemorrhages; E. Hemorrhage at the tip of the proventricular glands; F. Mottled spleen; G. Severe edema cloacal bursa; H. Intestinal lumen and cecal tonsils showing severe hemorrhage. Source: Desingu PA, Singh SD, Dhama K, et al. 2017. Clinicopathological characterization of experimental infection in chickens with sub-genotype VIIi Newcastle disease virus isolated from peafowl. Microb Pathog, 105: 8-12.

和神经氨酸酶活性，糖蛋白F介导被感染细胞之间的细胞融合。NDV可被热、紫外线、氧化剂和化学消毒剂灭活。

实验室诊断：①将疑似发病鸡的呼吸道分泌物或组织悬液（脾、肺或脑）等病料通过接种鸡胚或细胞培养物来分离和鉴定病毒。②通过免疫荧光技术检测感染组织或细胞培养中的新城疫病毒抗原。③通过血凝抑制试验、病毒中和试验或ELISA检测血清中的NDV特异性抗体。

防控：由于新城疫病毒只有一种血清型，不同毒株间微小的抗原差异不足以影响疫苗的总体效果，疫苗接种是预防新城疫的首要方法。NDV弱毒活疫苗可采用饮水免疫或者气雾免疫。

hemagglutination and has neuraminidase activities, while the glycoprotein F causes fusion among infected cells. Newcastle disease virus can be inactivated by physical and chemical agents, such as heat, ultraviolet light, oxidants, and commonly used chemical disinfectants.

Laboratory diagnosis: ① isolation and identification of the virus by inoculating embryonated eggs or cell culture with respiratory exudate or tissue suspensions (spleen, lung, or brain); ② detecting NDV antigens in affected tissues or cell cultures by the immunofluorescence technique; ③ detecting the level of anti-NDV antibodies by hemagglutination inhibition test, virus neutralization test, or ELISA.

Prevention and control: since there is only one serotype in NDV and minor antigenic differences among different strains are not a concern of compromising the vaccine efficacy, vaccination is of prime importance in preventing ND. The avirulent strain-based NDV live vaccines can be administered in drinking water or by aerosol.

第2节 弹状病毒科
Section 2 *Rhabdoviridae*

弹状病毒呈子弹状，有囊膜，基因组为单链负链RNA，宿主广泛。弹状病毒科包括狂犬病毒属、水疱病毒属、牛暂时热病毒属和粒外弹状病毒属等，具有相同的形态，复制需要依赖于RNA依赖的RNA聚合酶，成熟的病毒粒子从细胞膜出芽或进入胞质空泡中。

Viruses in the family *Rhabdoviridae* are bullet-shaped and enveloped with a single-stranded negative RNA genome and have a wide range of hosts. The family consists of *Lyssavirus, Vesiculovirus, Ephemerovirus, Novirhabdovirus, etc*. The viruses have similar morphology. The replication relies on RNA-dependent RNA polymerase. Mature virions bud from the cell membrane or enter cytoplasmic vacuoles.

狂犬病毒

狂犬病毒（RV）是一种弹状病毒，属于狂犬病毒属，病毒颗粒呈杆状或子弹状（**图30-4**）。该病毒在4 ℃可以保存数周，在−70 ℃可以保存数年，可被紫外线照射、热处理（50 ℃，1 h）、脂类溶剂、胰蛋白酶、洗涤剂和极端pH迅速灭活。狂犬病毒的宿主范围很广，所有的温血动物

Rabies virus

Rabies virus (RV) belongs to the genus *Lyssavirus* in the family *Rhabdoviridae*. The virions are rod- or bullet-shaped (**Figure 30-4**). It can survive at 4 ℃ for weeks or −70 ℃ for years, and can be killed rapidly by exposure to ultraviolet radiation, heat (1 h at 50 ℃), lipid solvents, trypsin, detergents and by the extremes of pH. The virus has a wide host range, infecting all warm-blooded animals, including humans.

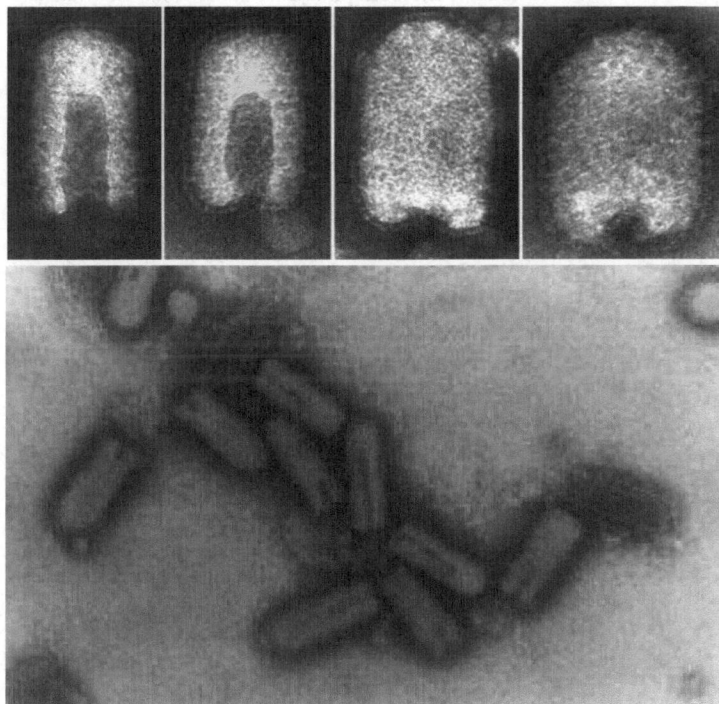

图30-4 电镜下狂犬病毒的形态特征
图像来源见英文。

Figure 30-4 The morphological characteristics of rabies viruses under electron microscope
Source: Matsumoto S. 1963. Electron microscope studies of rabies virus in mouse brain. J Cell Biol, 19 (3): 565-591.

包括人类都能被感染。

实验室诊断：① 抗原或核酸：通过脑组织切片的直接免疫荧光染色或者免疫组织化学染色。脑组织中形成的内基氏小体中含有RV抗原，可通过免疫荧光法进行检测。内基氏小体在人和动物感染中较为常见。此外，亦可通过RT-PCR检测脑组织中的RV基因组片段。② 病毒分离：可将疑似动物的组织液接种到乳鼠脑内。感染小鼠可出现脑炎和死亡。此外，将病料接种到仓鼠和小鼠细胞系细胞可以快速分离RV。

防控：目前还没有治疗狂犬病的方法，主要依赖疫苗接种进行预防。

Laboratory diagnosis: ① antigens or nucleic acids: rabies virus infection can be identified by immunofluorescence or immunohistochemistry staining using anti-rabies monoclonal antibodies. Negri bodies contain RV antigens and can be demonstrated by immunofluorescence. Both Negri bodies and rabies antigens can usually be found in the infected animals or humans. RT-PCR can be used to amplify the parts of a RV genome from brain tissues; ② viral isolation: brain tissues from the suspected cases are intracerebrally inoculated into suckling mice. Infection in mice results in encephalitis and death. Hamster and mouse cell lines can be inoculated for isolation of rabies virus.

Prevention and control: there is no treatment for rabies. Vaccination is the best way to prevent this disease.

第3节 波纳病毒科
Section 3 *Bornaviridae*

波纳病毒（BDV）为球形、有囊膜的病毒粒子，直径大约90 nm，单链负链RNA。主要感染马，但感染也出现在包括人、牛、兔、羊、犬、猫、某些驯化的鸟等多种动物。BDV具有嗜神经性，已被认为是人的行为异常和动物脑炎的病原。波纳病毒可引起马神经异常性疾病，主要局限于中欧地区。

Viruses in the family *Bornaviridae* are spherical, enveloped and approximately 90 nm in diameter, and have a single-stranded negative RNA genome. Horses are susceptible hosts, but infections also occur in a variety of animals including humans, cattle, rabbits, goats and sheep, dogs, cats, some domesticated birds, *etc.* BDV is neurotropic and has been considered to be the cause of human behavioral abnormalities and animal encephalitis. BDV causes a severe neurological disorder in horses, mainly seen in the Central Europe.

思考题 Questions

1. 单负链病毒目的成员有哪些？

1. What are the members of *Mononegavirales*?

2. 副黏病毒科中具有重要致病作用的病毒有哪些？

2. What are the important pathogenic viruses in *Paramyxoviridae*?

3. 副黏病毒的形态结构与基因组的特征有哪些？

3. What are the morphological structures and genomic characteristics of paramyxoviruses?

4. 新城疫病毒的致病机理、检测方法和防控措施有哪些?

5. 犬瘟热病毒的致病机理、检测方法和防控措施有哪些?

6. 弹状病毒的形态结构与基因组的特征有哪些?

4. What are the pathogenic mechanisms, diagnostic methods, and prevention and control measures of Newcastle disease virus?

5. What are the pathogenic mechanisms, detection methods, and prevention and control measures of canine distemper virus?

6. What are the morphological structures and genomic characteristics of rhabdoviruses?

第31章 分节段的负链RNA病毒
Chapter 31 Segmented negative-stranded RNA viruses

内容提要 分节段的负链RNA病毒有正黏病毒科、近布尼亚病毒科、砂粒病毒科白纤病毒科、内罗病毒科和汉坦病毒科等，其特征是通过不同毒株基因组节段之间的重配快速演化出新毒株。正黏病毒科以流感病毒为代表，其基因组含6～8个基因片段；对兽医学有重要意义的有马流感病毒、禽流感病毒、猪流感病毒等，后两者还具有重要的公共卫生意义。

Introduction Segmented negative-stranded RNA viruses, including *Orthomyxoviridae, Peribunyaviridae, Arenaviridae, Phenuiviridae, Nairoviridae* and *Hantaviridae, etc.*, have an evolutionary advantage in that they can rapidly evolve through reassortment among the genome segments of different strains. Influenza virus is the prototype of *Orthomyxoviridae*, whose genome contains 6-8 segments. Equine influenza virus, swine influenza virus and avian influenza virus are of great importance to veterinary medicine, and the last two are of great public health significance.

第1节 正黏病毒科
Section 1 *Orthomyxoviridae*

正黏病毒科有7个病毒属：甲型（A型）流感病毒属、乙型（B型）流感病毒属、丙型（C型）流感病毒属、丁型流感病毒属、鲑传贫病毒属、索戈托病毒属和夸兰扎病毒属。甲型流感病毒是马、猪、人类、貂、海豹、鲸和家禽的病原体；乙型流感病毒是人的病原体，但也有感染海豹的报道；丙型流感病毒是人和猪的病原体，极少引起严重疾病；丁型流感病毒在2018年被正式确认，与丙型流感病毒最为相似，但细胞嗜性更广，除了感染牛之外，还可以感染雪貂、猪和豚鼠。

分类： 根据表面蛋白血凝素（HA）和神经氨酸酶（NA）的抗原性，可进一步将

Orthomyxoviridae includes 7 genera: *Influenza A virus, Influenza B virus, Influenza C virus, Influenza D virus, Isavirus, Thogotovirus* and *Quaranjavirus*. Influenza A virus is a common pathogen of horses, pigs, humans, mink, seals, whales, and domestic poultry. Influenza B virus is a pathogen of humans, but its infection has also been reported in seals. Influenza C virus infects humans and pigs, but rarely causes serious diseases. Influenza D virus was officially confirmed in 2018, which is most similar to influenza C virus, but has a wider cell tropism. In addition to infecting cattle, it can also infect ferrets, pigs and guinea pigs.

Classification: influenza A virus can be further divided into several different subtypes based on the antigenicity of the surface proteins hemagglutinin (HA)

甲型流感病毒分为不同亚型，目前共有18个HA和11个NA亚型（**图31-1**）。水禽是流感病毒的天然贮存库。流感病毒株的命名包括：型（A、B或C）/宿主（猪、马、鸡、火鸡等）/分离地区（省或州）/毒株序号/分离年份（亚型）。例如：甲型流感病毒/鸡/苏格兰/1959（H5N1），是1959年分离获得的第一个H5亚型高致病性禽流感（HPAI）毒株。

and neuraminidase (NA). There are 18 HA and 11 NA subtypes in influenza A virus (**Figure 31-1**). Wild aquatic birds are the natural reservoirs. In naming specific influenza virus strains, type (A, B, or C)/ host (swine, equine, chicken, turkey, *etc.*)/ geographic origin (at a province or state level)/ strain number/year of sampling (subtype) are included in order. For example: A/chicken/ Scotland/1959 (H5N1), the first highly pathogenic avian influenza (HPAI) virus strain (H5 subtype) isolated in 1959.

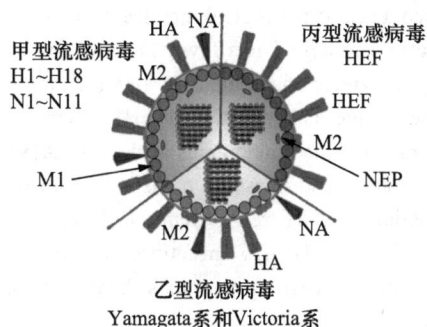

图31-1 甲型、乙型和丙型流感病毒结构示意图

甲型流感病毒各型由其表面蛋白血凝素（HA）和神经氨酸酶（NA）决定，目前共有18个HA和11个NA。乙型流感病毒分两个谱系（B/Yamagata系和B/Victoria系）。甲型和乙型流感病毒表面均含有HA、NA和基质蛋白M2，内部有8个基因组片段。丙型流感病毒表面只有纤突蛋白（HEF）和M2，其中HEF在病毒感染中起作用。丙型流感病毒有7个基因组片段。图像来源见英文。

Figure 31-1 Schematic of Influenza A, B and C virus structure

Influenza A viruses are defined by their surface proteins hemagglutinin (HA) and neuraminidase (NA) of which there are 18 HA and 11 NA. Influenza B viruses are categorized into two lineages (B/Yamagata and B/Victoria). The surface of both A and B viruses contains HA, NA, and the matrix protein M2. Internally, A and B viruses both have 8 genomic segments. Influenza C viruses have only one external spike protein (HEF) which functions both in viral entry and egress and M2 protein. Type C viruses have 7 internal genomic segments. Source: Francis ME, King ML, Kelvin AA. 2019. Back to the future for influenza preimmunity-looking back at Influenza virus history to infer the outcome of future infections. Viruses, 11 (2): 122.

主要特性：病毒颗粒呈多形性，一般呈球形，直径为80～120 nm；或为丝状，直径20 nm，长200～300 nm。病毒有囊膜，其表面有两种糖蛋白纤突：一种为棒状的血凝素蛋白（HA），另一种为蘑菇状的神经氨酸酶（NA），具有神经氨酸酶活性，HA与NA的比例为（4～5）：1。囊膜与基质蛋白M1和M2相连，M1在脂质双层下形成一内层，另一种基质蛋白M2的四聚体插入M1，构成少量的离子通道。甲型流感病毒的基因组包括8个单链负链的RNA片段，片段长度为890～2341个核苷酸，编码10种蛋白。第1～3节段分别编码聚合酶蛋白PB2、PB1和PA，节段4和6分别编码表面糖蛋白HA和NA，节段5编码与病毒RNA结合在一起的核蛋白NP，第7、8节段分别编码M1/M2和NS1/NS2。

Main features: influenza virus particles are pleomorphic, normally spherical in a size of 80-120 nm in diameter, or filamentous in a size of 20 nm in diameter and 200-300 nm long. The virus has an envelope with two distinct types of glycoprotein spikes: one is the HA with a rod-shaped form, and the other is the NA with a mushroom-shaped form and neuraminidase activity. The ratio of HA to NA is about (4-5) : 1. The envelopes are linked to the matrix proteins M1 and M2. Of them, M1 forms the inner layer under the lipid bilayer, while M2 forms tetramers and inserts into M1, thus forming a certain numbers of ion channels. The genome of influenza A virus contains eight segments of negative-stranded RNA, varying between 890 and 2341 nucleotides in length, which encode 10 proteins. Segments 1 to 3 encode RNA-dependent polymerases PB2, PB1 and PA, respectively. Segments 4 and 6 encode the surface glycoprotein HA and NA, respectively. The segment 5 encodes nucleoprotein (NP) that binds the viral RNA (vRNA). The segment 7 codes for M1 and M2, and the segment 8 for proteins NS1 and NS2.

禽流感病毒

禽流感病毒（AIV）分高致病性和低致病性两类，由高致病力毒株引起的高致病性禽流感（HPAI），是WOAH通报疫病，旧称"真性鸡瘟"，引起鸡和火鸡突发性死亡，无前驱症状。禽流感病毒，特别是H5和H7病毒的一个关键致病标志是HA裂解位点的氨基酸序列，大多数低致病性禽流感（LPAI）病毒在裂解位点处只有一个碱性氨基酸（精氨酸），而高致病性禽流感病毒在裂解位点含有多种碱性氨基酸。AIV主要感染呼吸道和肠道，并在上皮细胞内复制。在大多HPAI毒株感染中，存在病毒血症和全身扩散，导致胰腺、心脏、脑、骨骼肌、皮肤等器官或组织炎症和坏死。

禽流感病毒亚型特异性RT-PCR和实时RT-PCR已广泛应用，ELISA也可用于病毒的快速检测。可采取气管、肺、气囊或泄殖腔拭子接种鸡胚或鸭、牛和猴子的肾细胞进行病毒的分离，尿囊液或细胞培养物可通过HA试验、夹心ELISA或测序进行鉴定。

猪流感病毒

H1N1、H3N2和H1N2流感病毒是引起猪流感的主要亚型；症状包括发热、呼吸困难、喷嚏、咳嗽和流涕。有些猪会发生肺水肿或支气管肺炎。猪流感为WOAH通报疫病。人流感病毒对呼吸道上皮唾液酸α-2,6-半乳糖苷受体具有亲嗜性；禽流感和马流感病毒对唾液酸α-2,3-半乳糖苷受体具有亲嗜性。猪呼吸道上皮同时具有上述两种受体，对人流感和禽流感都具有易感性，因此猪被认为是流感病毒重配的混合器。该病毒可通过RT-PCR、血凝或血凝抑制、免疫荧光等方法进行检测，可

Avian influenza virus

Avian influenza virus (AIV) is categorized into two groups, one with high pathogenicity and the other with low pathogenicity. The disease caused by high pathogenic avian influenza (HPAI) viruses has historically been called as "fowl plague". HPAI is a notifiable disease by WOAH and causes sudden death without prodromal symptoms in chickens and turkeys. In AIV, particularly H5 and H7 viruses, a key pathogenic marker is the amino acid sequence at the proteolytic cleavage site of the hemagglutinin protein. Most low pathogenic avian influenza (LPAI) viruses have a single alkaline amino acid (arginine) at the cleavage site, but HPAI viruses contain multiple alkaline amino acids at the site. AIVs infect the respiratory and intestinal tracts and replicate in their epithelial cells. There is viremia and systemic dissemination in most HPAIV infections, leading to inflammation and necrosis in pancreas, heart, brain, skeletal muscle, skin, *etc.*

Subtype-specific RT-PCR and real-time RT-PCR for AIV have been developed and utilized widely. Sandwich ELISA can also be used for rapid viral detection. AIV can be cultivated and isolated in chick embryos or kidney cells of duck, calf and monkey using the samples of trachea, lung, air sac, or cloacal swabs from infected chickens. The presence of AIV in allantoic fluids or cell culture can be confirmed by HA assay, sandwich ELISA, or sequencing.

Swine influenza virus

Influenza virus H1N1, H3N2 and H1N2 are the main subtypes responsible for outbreaks in pigs. The symptoms include fever, dyspnea, sneezing, coughing, and nasal discharge. Some animals develop pulmonary edema or bronchopneumonia. Swine influenza is a notifiable disease by WOAH. Human influenza viruses have affinity to α-2,6-sialic acid receptors, while AIV and EIV (equine influenza virus) show affinity to α-2,3-sialic acid receptors. Pigs possess both α-2,3 and α-2,6 receptors and are thus susceptible to both AIV and human influenza virus. Therefore, pigs have been considered as the "mixer" of influenza viruses. The virus can be detected by RT-PCR, HA or HI, immunofluorescence, *etc.* Virus isolation can be achieved using embryonated eggs or MDCK

用鸡胚或MDCK细胞进行分离。

cells.

马流感病毒

马流感病毒属于甲型流感病毒，有H7N7和H3N8两个亚型。两个亚型间没有交叉免疫反应。马流感是WOAH规定通报疫病。近年来的疫情主要由H3N8引起，该病毒已发生抗原漂移。该病死亡率低，但发病率可接近100%。临床主要表现为高热，频繁强烈咳嗽。该病毒可用鸡胚或细胞分离，并可通过血凝抑制或RT-PCR检测。

Equine influenza virus

Equine influenza virus belongs to influenza A virus and has two subtypes H7N7 and H3N8. There is no immunological cross-reactivity between these subtypes. Equine influenza is a notifiable disease by WOAH. H3N8 virus has been identified to be mainly responsible for recent outbreaks and exhibits modest genetic drift. The mortality rate of equine influenza is low, but the morbidity rate can reach 100%. The disease is manifested by high fever, and frequent, strong and dry cough. The virus can be isolated using embryonated chicken eggs or cell culture, and it can be detected by HI or RT-PCR.

第2节 近布尼亚病毒科
Section 2 *Peribunyaviridae*

近布尼亚病毒呈球状或多形态，直径80～120 nm，有囊膜，成熟的病毒粒子表面有糖蛋白纤突。病毒粒子有4种结构蛋白，包括两种囊膜表面的糖蛋白（Gn和Gc）、包裹基因组的核衣壳蛋白（N）和转录酶蛋白（L）（图31-2）。

近布尼亚病毒的基因组11.2～12.5 kb，由3个负链（或双向）单链RNA片段组成，3个片段分别为大（L）、中（M）和小（S）。各个属的节段大小不同，L为6.3～12 kb，编码RNA依赖的RNA聚合酶；M为3.5～6 kb，编码一种多聚蛋白，经过加工形成两种糖蛋白（Gn和Gc）和一种非结构蛋白（NSm）；S为1～2.2 kb，编码核衣壳蛋白（N）和一种非结构蛋白（NSs）。

病毒所含脂质来自宿主细胞膜，糖则构成糖蛋白的侧链。病毒在细胞质复制，通过高尔基体空泡出芽释放。病毒对热、酸极为敏感，去污剂、脂溶剂和常见的消毒剂可将其灭活。

Peribunyaviruses are enveloped, spherical or pleomorphic with a diameter of 80-120 nm. There are surface projections (spikes) emanating from the envelope surface of mature virions. The virions consist of 4 structural proteins, including two external glycoproteins (Gn and Gc) in the envelope, a nucleocapsid protein (N) that encapsulates the genome, and a transcriptase protein (L) (**Figure 31-2**).

The genomes of peribunyaviruses are 11.2-12.5 kb and consist of 3 segments of negative (or bi-sense) single-stranded RNA, designated as large (L), medium (M) and small (S), respectively. The RNA segments differ in size among the genera: the L segment ranges from 6.3 kb to 12 kb in size and encodes a single large protein (L), an RNA-dependent RNA polymerase. The M segment ranges from 3.5 kb to 6 kb and encodes a polyprotein that is processed to form two glycoproteins (Gn and Gc) and a nonstructural protein (NSm). The S segment has a size from 1 kb to 2.2 kb and encodes the nucleoprotein (N) and a nonstructural protein (NSs).

Virions also contain lipids derived from host cell membranes, and sugars act as side chains of the glycoproteins. The viruses replicate in the cytoplasm and release by budding through the vesicles of Golgi. Virions are quite sensitive to heat and acidic conditions, and are inactivated readily by detergents, lipid solvents, and

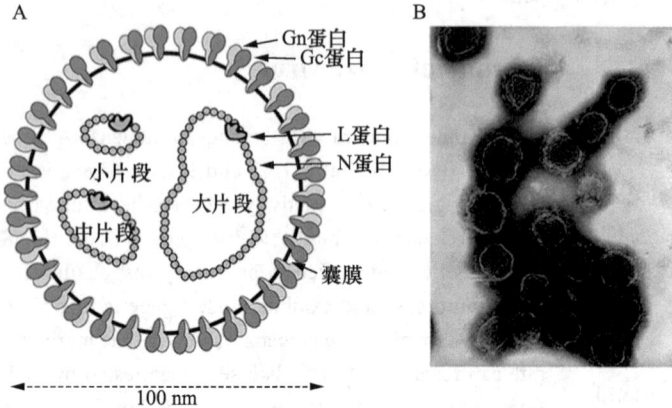

图31-2　近布尼亚病毒病毒粒子结构

A. 病毒粒子切面。表面有Gn和Gc糖蛋白。螺旋状核衣壳内有三个独特的单链RNA片段（大、中、小三个片段），由N蛋白包裹并与L蛋白相关。B. 加州脑炎病毒负染透射电镜照片（图片：由美国疾病预防控制中心 Frederick Murphy 和 Erskine Palmer博士提供）。图像来源见英文。

Figure 31-2　Peribunyavirus virion structure

A. representation of a virion in cross section. The surface spikes comprise the Gn and Gc glycoproteins. The helical nucleocapsids are circular and comprise each of the unique single-stranded RNA segments (L, large; M, medium; S, small) encapsidated by N protein and associated with the L protein. B. negative-stained transmission electron microscopy photograph of California encephalitis virus virions (image: CDC/Drs Frederick Murphy and Erskine Palmer). Source: Hughes HR, Adkins S, Alkhovskiy S, et al. 2020. ICTV virus taxonomy profile: *Peribunyaviridae*. J Gen Virol, 101 (1): 1-2.

赤羽病毒

赤羽病毒因最初在日本的赤羽村发现而得名。该病毒在日本、澳大利亚和以色列，引起反刍动物，特别是牛的季节性胎儿畸形。妊娠期间感染胎儿，引起先天性关节弯曲或积水性无脑综合征，蚊和库蠓为传播媒介。细胞培养的灭活苗安全有效。

common disinfectants.

Akaba virus

Akaba virus is named after the village of Akaba in Japan, where it was first discovered. Akaba virus is the cause of periodic outbreaks of fetal malformation in ruminants, especially cattle, in Japan, Australia, and Israel. The virus is transmitted by mosquitoes and *Culicoides* species and causes arthrogryposis and hydranencephaly syndrome in infected fetuses during pregnancy. Inactivated virus vaccines produced in cell culture have been proven safe and efficacious.

第3节　白纤病毒科
Section 3　*Phenuiviridae*

裂谷热病毒

裂谷热病毒为白纤病毒科白蛉病毒属成员，因最初在肯尼亚的东非大裂谷发现而得名。裂谷热病毒主要流行于非洲撒哈拉沙漠以南地区，有时侵入邻近的北部地区，如埃及、沙特阿拉伯和也门。该病毒可通过蚊虫传播，引起反刍动物尤其是绵

Rift Valley fever virus

Rift Valley fever virus (RVFV) belongs to the member of *Phlebovirus*, *Phenuiviridae* and gets its name because it was first found in the Great Rift Valley, Kenya. RVFV is endemic in sub-Saharan Africa, and periodically occurs in adjacent northern regions such as Egypt, Saudi Arabia and Yemen. It is a zoonotic, mosquito-transmitted virus that causes a severe, frequently fatal, disease in

羊和山羊严重的致死性疾病，人兽共患，幼龄动物死亡率高，可引起怀孕母畜流产（图31-3）。

ruminants, especially sheep and goats. Mortality is high in young animals, and pregnant ruminants often abort following infection (**Figure 31-3**).

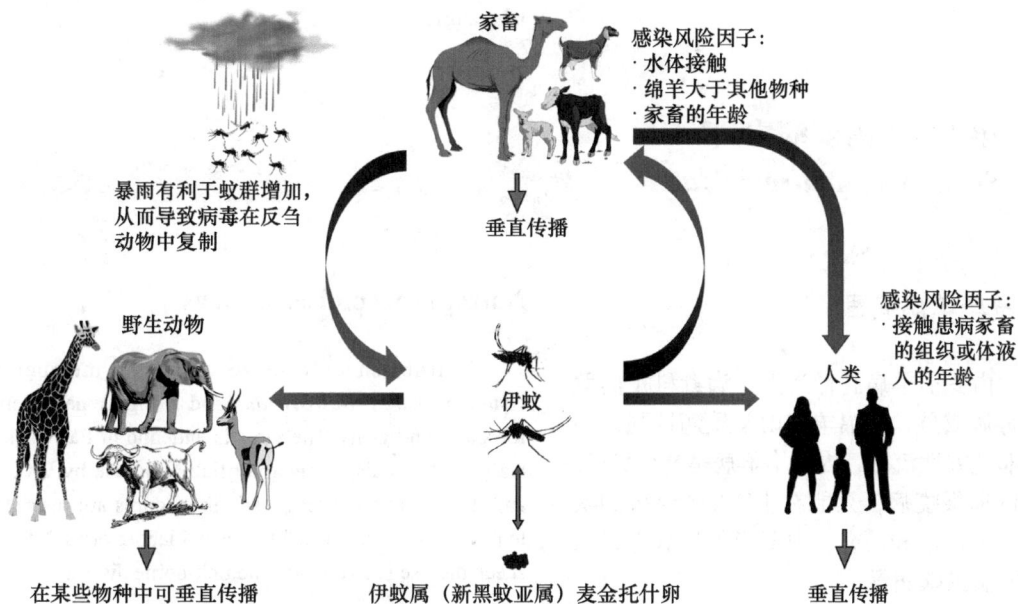

图31-3 裂谷热病毒的传播

裂谷热病毒可通过伊蚊属（新黑蚊亚属）麦金托什卵传播和流行。野生有蹄类动物和家畜也可存在低水平传播。暴雨季，蚊虫数量激增导致牲畜感染增加，虫媒和反刍动物扩增病毒。随着越来越多的牲畜感染，裂谷热病毒传播至人类的机会增加。人类可通过蚊虫叮咬或接触感染动物的组织和体液感染。图像来源见英文。

Figure 31-3 Rift Valley fever virus cycle

Between epidemics RVFV may be maintained through transovarial transmission in *Aedes* (*Neomelaniconion*) *mcintoshi* eggs. Wild ungulates and livestock can also harbour low level infection. During heavy rains, a surge in mosquito populations leads to increased infection of livestock and viral amplification between numerous vector species and ruminants occurs. As more livestock become infected, the chances of spillover into humans increases. Human infection can occur via mosquito bite, or more commonly, via contact with infected animal tissue and fluid. Source: Wright D, Kortekaas J, Bowden TA, et al. 2019. Rift Valley fever: biology and epidemiology. J Gen Virol, 100 (8): 1187-1199.

裂谷热病毒在靶器官复制极快，并能达到很高滴度。经蚊虫叮咬或口咽部进入体内；经30～72 h潜伏期后侵入肝实质及网状内皮器官，常引起感染的绵羊和山羊广泛的肝坏死和出血，有些感染动物因神经元坏死造成脑炎。

裂谷热病毒因其对人和多种动物致病以及可通过多种蚊子迅速传播而受到广泛关注。人感染的死亡率约为1%，但近来有报道高达30%。裂谷热为WOAH通报疫

RVFV rapidly replicates to very high titers in target tissues. After entry by mosquito biting or through the oropharynx via aerosols, there is a latent period of 30-72 h. The virus invades the parenchyma of the liver and reticuloendothelial organs and causes extensive liver necrosis in infected sheep and goats, and widespread hemorrhages. Encephalitis with neuronal necrosis also occurs in some infected animals.

Extensive attention has been paid to RVFV because of its pathogenicity to humans and multiple species of animals, as well as its rapid spread by a variety of mosquito species. The overall human mortality rate is

病，该病毒分离鉴定必须在生物安全4级实验室操作，所有关于该病毒的研究与诊断仅限于某些国家参考实验室。

approximately 1%, but has been reported as high as 30% in recent outbreaks. It is a notifiable disease by WOAH and is listed as one of the Biosafety Level 4 laboratory pathogens. Research and diagnostic procedures related to RVFV are restricted to certain national reference laboratories.

第4节 内罗病毒科
Section 4 *Nairoviridae*

内罗毕羊病病毒

内罗毕羊病病毒为内罗病毒科正内罗病毒属成员，对绵羊和山羊致病性强，在东非地方性流行，内罗毕羊病是WOAH规定的通报疫病，只能在生物安全4级的实验室做诊断和研究。内罗毕羊病病毒是人畜共患虫媒病毒。

克里米亚/刚果出血热病毒

克里米亚/刚果出血热病毒（CCHFV）是一种重要人畜共患病的病原，动物感染后通常无症状，但反刍动物出现高滴度的病毒血症。人类可通过蜱虫叮咬、屠宰感染的牲畜或在护理感染患者期间感染该病毒（图31-4），可引起人严重的出血热，死亡率15%～40%，诊断和研究必须在生物安全4级的实验室进行。

Nairobi sheep disease virus

Nairobi sheep disease virus is a member of *Orthonairovirus*, *Nairoviridae* and is highly pathogenic for sheep and goats. The virus is endemic in East Africa. Nairobi sheep disease is a notifiable disease by WOAH and its research and diagnostic procedures are restricted to be performed in Biosafety Level 4 laboratories. Nairobi sheep disease is a zoonotic and tick-borne disease.

Crimean-Congo hemorrhagic fever virus

Crimean-Congo hemorrhagic fever virus (CCHFV) is the cause of an important zoonotic disease. The infected animals generally show no clinical signs, but infection in ruminants results in high-titer viremia. Humans can become infected with CCHFV via tick bites and butchering of infected livestock and in the health-care setting during the care of infected patients (**Figure 31-4**). The infection in humans causes severe hemorrhagic fever, and the mortality rate is 15%-40%. All laboratory work must be performed in Biosafety Level 4 laboratories.

第5节 汉坦病毒科
Section 5 *Hantaviridae*

汉坦病毒

汉坦病毒可引起人类出血热，包括肾

Hantaan virus

Hantaan virus can cause hemorrhagic fever in

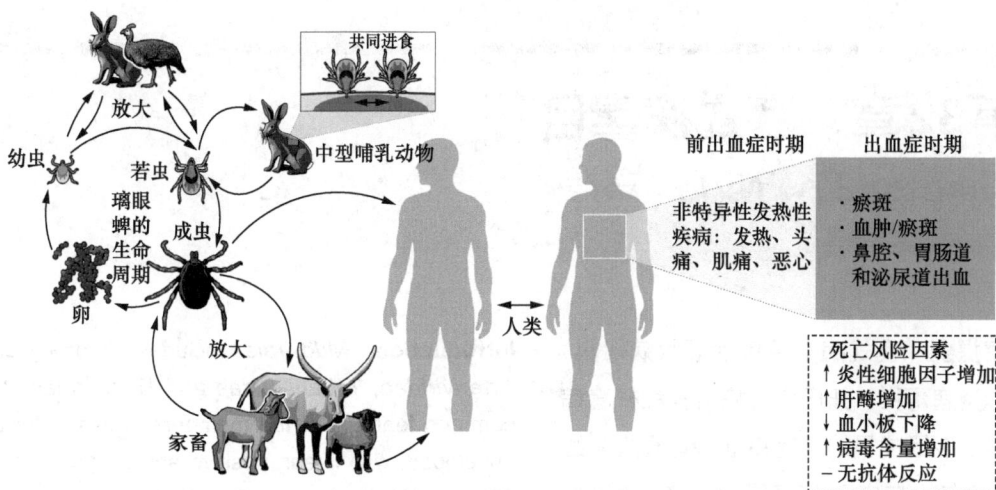

图 31-4 克里米亚/刚果出血热病毒（CCHFV）

CCHFV的自然宿主和媒介为璃眼蜱属的蜱虫。蜱在生命周期任一阶段均可通过叮咬病毒血症动物或与受感染蜱虫共同进食时感染，哺乳动物为病毒的重要放大宿主。图像来源见英文。

Figure 31-4 Crimean-Congo hemorrhagic fever virus (CCHFV)

The natural reservoir and vector for CCHFV are ticks of the *Hyalomma* genus. Ticks can become infected at any life-cycle stage during feeding on a viremic animal or during co-feeding with an infected tick, and mammals likely act as important amplification hosts for the virus. Source: Hawman DW, Feldmann H. 2018. Recent advances in understanding Crimean-Congo hemorrhagic fever virus. F1000Res, 7 (F1000 Faculty Rev): 1715.

综合征出血热和汉坦病毒肺综合征。啮齿动物是该病毒的贮存宿主，人通常在接触啮齿动物粪便后发病，主要控制措施是灭鼠。

humans, including hemorrhagic fever with renal syndrome and hantavirus pulmonary syndrome. These viruses are maintained in rodents. Infections in humans usually occur after direct contact with rodent feces, and the most effective way for control is deratization.

思考题 Questions

1. 为什么说猪是流感病毒重配的混合器？

2. 非洲猪瘟病毒、非洲马瘟病毒、裂谷热病毒等起源于非洲，这些病毒有何共同点？

1. Why do we call pigs as the "mixer" of influenza viruses?

2. African swine fever virus, African horse fever virus and Rift Valley fever virus all originated from Africa. What do these viruses have in common?

第32章 套式病毒目
Chapter 32 Nidovirales

内容提要 套式病毒目包括冠状病毒科、动脉炎病毒科、托巴套式病毒科和杆套病毒科。该目成员的共有特征是有囊膜和线性、正链、单链RNA基因组。这些病毒复制过程中产生的3′共有末端嵌套（巢状）4组或更多组的亚基因组mRNA，但在遗传复杂性和病毒结构上有很大不同。

Introduction *Nidovirales* includes *Coronaviridae*, *Arteriviridae*, *Tobaniviridae* and *Roniviridae*. The common features of the members in this order are enveloped, with linear, positive and single-stranded RNA (ssRNA) genomes. These viruses generate 4 or more sets of nested subgenomic mRNA at the 3′ coterminal ends during replication, but they are very different in the genetic complexity and virus structures.

第1节 冠状病毒科
Section 1 *Coronaviridae*

冠状病毒科由正冠状病毒亚科和勒托病毒亚科组成。冠状病毒颗粒为球状，有囊膜和纤突，核衣壳螺旋状对称。其基因组为线状单链正链RNA，是已知RNA病毒中基因组最大的，其5′端有帽子，3′端有poly（A）。基因组编码纤突糖蛋白（S），主要嵌膜蛋白（M），次要嵌膜蛋白（E），血凝素脂酶（HE）。成员病毒在细胞质内复制，经出芽进入内质网，通过胞吐释放。可在人类和动物中引起亚临床、轻度或致死性的呼吸道和胃肠道疾病。

冠状病毒科中感染猪的6种冠状病毒简述如下。

甲型冠状病毒属

猪传染性胃肠炎病毒（TGEV）仅1

The *Coronaviridae* consists of the subfamilies of *Coronavirus* and *Torovirus*. The virions of coronaviruses are spherical in shape and have envelope and spikes, and their nucleocapsid is spirally symmetrical. Their genomes are a linear positive single-stranded RNA and the largest one among known RNA viruses. They all have a cap at the 5′ end and poly(A) tail at the 3′ end. The genomes encode the spike glycoprotein(S), main membrane protein(M), minor membrane-embedded protein(E), and hemagglutinin lipase (HE). These viruses replicate in the cytoplasm, enter the endoplasmic reticulum through budding and are released through exocytosis. The viruses can cause subclinical, mild or fatal respiratory and gastrointestinal diseases in human and animals.

The 6 coronaviruses that infect pigs in the *Coronaviridae* are shown below.

Alphacoronavirus (α-CoV)

Porcine transmissible gastroenteritis virus

个血清型，引起仔猪腹泻（**图32-1**），与人、狗和猫的冠状病毒有抗原相关性。该病毒的抗体可与同属的猪呼吸道冠状病毒发生抗原性交叉反应。猪肾（PK）和猪睾丸（ST）细胞系是分离病毒的首选。该病毒感染在冬季多发，但具有自限性。免疫组织化学、免疫荧光、实时RT-PCR均可用于确诊，通过接种2～7日龄猪或PK以及ST细胞可分离病毒，ELISA可用于区分其与猪呼吸道冠状病毒感染。目前无治疗药物，但有商品化的灭活疫苗和活疫苗可接种。

(TGEV) has only 1 serotype, causing diarrhea of piglets (**Figure 32-1**). It is antigenically related to coronaviruses in humans, dogs and cats. Antibodies against TGEV can antigenically cross-react with porcine respiratory coronavirus (PRCoV) of the same genus. Porcine kidney (PK) and porcine testis (ST) cell lines are the first choices for virus isolation. This virus infection frequently occurs in winter, but it is self-limiting. Immunohistochemistry, immunofluorescence staining, and real-time RT-PCR can be used to confirm the diagnosis. Viruses can be isolated by inoculating 2-7 days old pigs or PK and ST cells. ELISA method is available to distinguish it from porcine respiratory coronavirus. There are no effective therapeutic drugs at present, but commercial inactivated and live

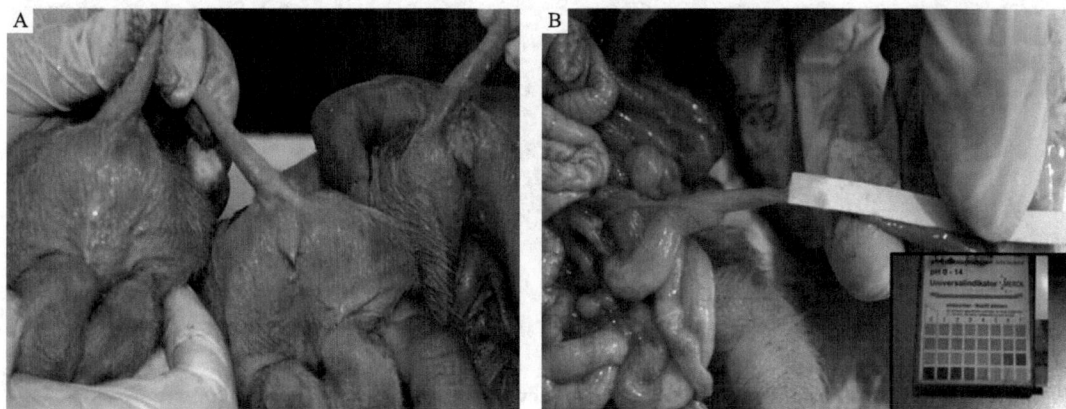

图 32-1　感染 TGEV 新生仔猪的大体病变

A. 仔猪会阴部被粪便污染；B. 肠祥壁薄、充满气体，含有 pH 5～6 的黄色水样腹泻物。图像来源见英文。

Figure 32-1　Gross lesions in neonatal piglets infected by TGEV

A. Piglets presented with stool-stained perineum; B. Intestinal loops are thin walled, distended by gas and contain abundant yellow-watery diarrhea with pH 5–6. Source: Piñeyro PE, Lozada MI, Alarcón LV, et al. 2018. First retrospective studies with etiological confirmation of porcine transmissible gastroenteritis virus infection in Argentina. BMC Vet Res, 14 (1): 292.

猪呼吸道冠状病毒（PRCoV）与TGEV抗体可发生抗原性交叉反应，是TGEV的天然缺失突变体。病猪常表现亚临床感染，也可引起猪的弥漫性支气管间质性肺炎。在感染该病毒的猪群，传染性胃肠炎的严重程度可能会降低。RT-PCR用于诊断以及与TGEV的鉴别；PK和ST细胞可用于分离病毒；ELISA可有效检测该病毒。无治疗药物和疫苗可用。

vaccines are available.

Porcine respiratory coronavirus (PRCoV) can antigenically cross-react with anti-TGEV antibodies and is a natural deletion mutant of TGEV. Infected pigs often show subclinical infections, but can also show symptoms of diffuse bronchial interstitial pneumonia in pigs. In pigs infected with PRCoV, the severity of TGE may be reduced. RT-PCR is used for diagnosis and can distinguish it from TGEV. PK and ST cells are used for virus isolation. ELISA can effectively detect this virus. No therapeutic drugs and vaccines are available.

猪流行性腹泻病毒（PEDV）在抗原性上不同于TGEV和猪血凝性脑脊髓炎病毒，但该病毒和猫传染性腹膜炎病毒之间有抗原交叉反应。该病毒感染的哺乳仔猪100%发病，50%～100%死亡，引起严重水样腹泻（**图32-2**）。常用免疫荧光、免疫组织化学诊断，也可用RT-PCR或实时RT-PCR确诊。有商品化的弱毒和灭活疫苗。

Porcine epidemic diarrhea virus (PEDV) is antigenically different from TGEV and porcine hemagglutinating encephalomyelitis virus (PHEV), but there is antigenic cross-reactivity between this virus and feline infectious peritonitis virus. The morbidity and mortality are 100% and 50%-100%, respectively in suckling piglets infected with this virus, causing severe watery diarrhea (**Figure 32-2**). The most commonly, reliable and accurate detection methods are immunofluorescence and immunohistochemistry. RT-PCR and real-time RT-PCR can be used in diagnosis. Commercial attenuated and inactivated vaccines are available.

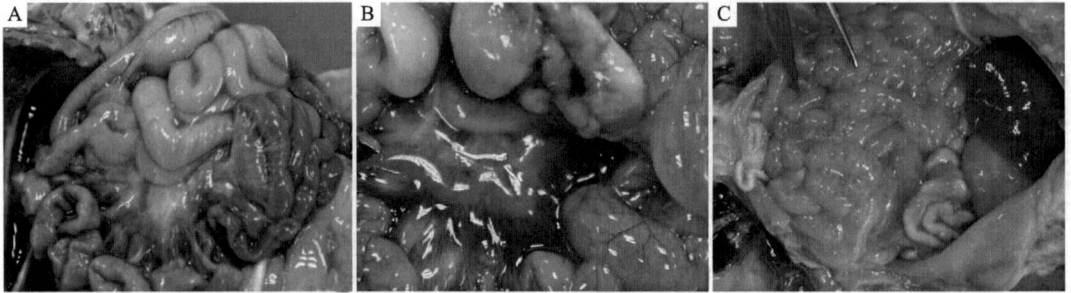

图32-2 实验室感染PEDV的新生仔猪大体病变
A. 未感染仔猪，无肠道损伤；B. 接种PEDV-LY4-98毒株的仔猪，肠系膜严重充血；C. 接种PEDV-LY1毒株的仔猪，小肠壁薄且含水样内容物。图像来源见英文。

Figure 32-2 The gross lesions of PEDV infection in newborn piglets
A. No intestinal lesions in uninfected piglets; B. Severe hyperaemia in the mesentery inoculated with PEDV-LY4-98 strain; C. Thin-walled and watery contents in small intestine inoculated with PEDV-LY1 strain. Source: Sun J, Li Q, Shao C, et al. 2018. Isolation and characterization of Chinese porcine epidemic diarrhea virus with novel mutations and deletions in the S gene. Vet Microbiol, 221: 81-89.

猪肠道阿尔法冠状病毒（SeACoV）于2017年在我国广东首次发现，曾称为猪急性腹泻综合征冠状病毒（SADS-CoV）。该病毒与蝙蝠α-冠状病毒HKU2株具有94.9%的核苷酸序列相似性，提示有共同祖先。SeACoV是TGEV和PEDV之间的一种新型重组病毒，主要在空肠增殖，具有多物种趋向性，存在跨种传播风险，对猪的致病性仍存在争议。常用加有胰蛋白酶的Vero细胞分离病毒。常用免疫荧光、RT-PCR、实时RT-PCR进行临床诊断。无疫苗可用。

Swine enteric alphacoronavirus (SeACoV) was first discovered in Guangdong, China in 2017 and initially called swine acute diarrhea syndrome coronavirus (SADS-CoV). This virus has 94.9% nucleotide sequence identity with the bat α-coronavirus HKU2 strain, indicating the same ancestor. SeACoV is a novel recombinant between TGEV and PEDV. This virus mainly multiplies in the jejunum of affected pigs. There is a risk of cross-species transmission of this virus in multiple animals because of its multi-species tropism. Its pathogenicity to pigs is still controversial. Vero cells with trypsin are commonly used to isolate viruses. Immunofluorescence technology, RT-PCR, and real-time RT-PCR are often used in clinical diagnosis. No vaccines are available.

乙型冠状病毒属

　　猪血凝性脑脊髓炎病毒（PHEV）是引起以脑脊髓炎、呕吐和消瘦为特征的幼猪呕吐和消瘦病（VWD）的原因。猪是其唯一宿主，并隐性带毒，成年猪感染多为自限性。免疫组化、RT-PCR、病毒中和和血凝抑制（HI）可用于诊断PHEV感染。使用原代猪肾或猪甲状腺细胞分离病毒。无可用的疫苗。

丁型冠状病毒属

　　猪德尔塔冠状病毒的抗血清与PEDV和TGEV无交叉反应，病毒可在补充有胰酶或小肠内容物无菌滤液的猪睾丸细胞（ST）中连续繁殖。病毒主要在感染猪小肠复制增殖，组织学病变的特征是急性、多灶性至弥漫性、轻度至重度萎缩的近端空肠至回肠的肠炎（**图32-3**）。RT-PCR为首选检测方法，间接荧光抗体、病毒中和、ELISA也用于诊断。该病毒应与PEDV和TGEV鉴别。目前没有疫苗可用，也无治疗方法。

Betacoronavirus (β-CoV)

　　Porcine hemagglutinating encephalomyelitis virus (PHEV) is the cause of vomiting and wasting disease (VWD) in young pigs, characterized by encephalomyelitis, vomiting and weight loss. Pigs are the only host of PHEV and also act as carriers of the virus without clinical manifestations. Its infection in adult pigs is often self-limiting. PHEV infection can be diagnosed by immunohistochemistry, RT-PCR, virus neutralization and hemagglutination inhibition. Primary porcine kidney or porcine thyroid cells are used to isolate the virus. No vaccines are available.

Deltacoronavirus (δ-CoV)

　　The antiserum against **porcine deltacoronavirus** (PdCoV) has no cross-reactivity with PEDV and TGEV. The virus can continuously propagate in ST cells supplemented with trypsin or sterile filtrates of small intestinal contents. It mainly replicates in the small intestine of infected pigs. The histological changes are characterized by acute, multifocal to diffuse, mild to severe atrophic enteritis from proximal jejunum to ileum (**Figure 32-3**). RT-PCR is the preferred detection method. Indirect fluorescent antibody, virus neutralization and ELISA are also used for diagnosis. This virus should be differentially diagnosed from PEDV and TGEV. There are no vaccines or treatments available at present.

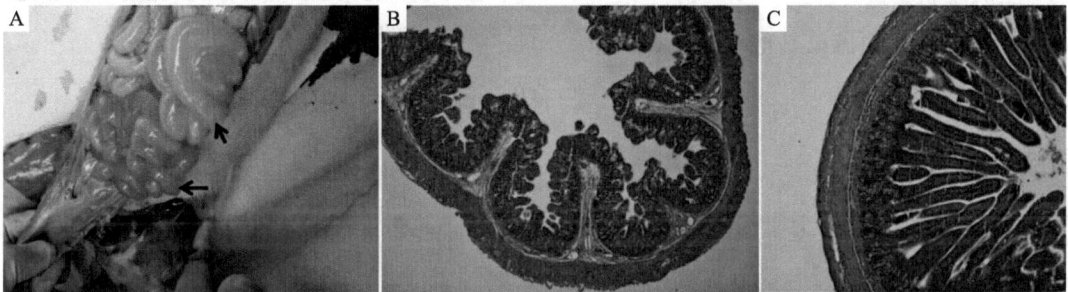

图32-3　实验室感染猪德尔塔冠状病毒的无菌猪肠道变化
A. 接种病毒72 h后的肠道：肠壁薄而透明（十二指肠至结肠），肠腔内汇积大量黄色液体（箭头）；B. 接种病毒72 h后的空肠：出现急性弥漫性和严重萎缩性空肠炎（放大倍数×40）；C. 未接种病毒的猪：空肠上皮绒毛正常（放大倍数×80）。图像来源见英文。

彩图

Figure 32-3　Intestinal changes in gnotobiotic pigs inoculated with Porcine deltacoronavirus
A. Intestine at hour post-inoculation (hpi) 72, showing thin and transparent intestinal walls (duodenum to colon) and accumulation of large amounts of yellow fluid in the intestinal lumen (arrows); B. Jejunum at hpi 72, showing acute diffuse, severe atrophic jejunitis (magnification ×40); C. Jejunum of noninoculated pig, showing normal villous epithelium (magnification ×80). Source: Jung K, Hu H, Eyerly B, et al. 2015. Pathogenicity of 2 porcine deltacoronavirus strains in gnotobiotic pigs. Emerg Infect Dis, 21 (4): 650-654.

冠状病毒科中感染人的7种冠状病毒简述如下。

目前感染人的冠状病毒分别是甲型冠状病毒属（α-CoV）的人冠状病毒229E（HCoV-229E）和人冠状病毒NL63（HCoV-NL63）；乙型冠状病毒属（β-CoV）的人冠状病毒OC43（HCoV-OC43）、人冠状病毒HKU1（HCoV-HKU1）、中东呼吸综合征冠状病毒（MERS-CoV）、严重急性呼吸综合征冠状病毒（SARS-CoV）和新型冠状病毒（SARS-CoV-2）。其中，具有高致病性或致死性的是 MERS-CoV、SARS-CoV 和 SARS-CoV-2。

人类冠状病毒是导致人普通感冒的第二大常见原因。其中 HCoV-NL63、HCoV-HKU1、HCoV-229E 和 HCoV-OC43 的感染通常无症状或有轻度至中度的上呼吸道症状。而 SARS-CoV、MERS-CoV 和 SARS-CoV-2 则引起致死性的严重呼吸综合征。

RT-PCR是用于检测人冠状病毒的"金标准"。ELISA和侧流免疫层析试剂盒均可用于诊断；荧光免疫色谱法可检测抗原，等温扩增方法可用于检测病毒核酸。

预防SARS-CoV-2感染引起的新型冠状病毒肺炎（COVID-19），有多种疫苗可用，如全病毒灭活疫苗、腺病毒载体疫苗、mRNA疫苗等。

冠状病毒科中感染其他动物的6种冠状病毒简述如下。

甲型冠状病毒属

猫传染性腹膜炎病毒（FIPV）引起家养和一些野生猫科动物的高度致死性猫传染性腹膜炎（FIP）。该病毒主要在巨噬细胞中复制，可在猫的单核巨噬细胞中增殖。乳鼠可用于病毒分离。猫肠道冠状病毒抗体可与该病毒发生交叉反应，但并不起到保护作用。FIPV感染临床表现分为渗出型和非渗出型（**图32-4**）。腹腔穿刺术确定腹腔渗出可确诊渗出型 FIP，而RT-

The 7 coronaviruses that infect humans in the *Coronaviridae* are shown below.

The α-coronaviruses that infect humans are human coronavirus 229E (HCoV-229E) and NL63 (HCoV-NL63). The β-coronaviruses that infect human are human coronavirus OC43 (HCoV-OC43) and HKU1 (HCoV-HKU1), Middle East respiratory syndrome coronavirus (MERS-CoV), severe acute respiratory syndrome coronavirus (SARS-CoV) and SARS-CoV-2. Among them, MERS-CoV, SARS-CoV and SARS-CoV-2 are highly pathogenic or lethal to humans.

Human coronaviruses are the second group of pathogens for common cold in humans. The infections of HCoV-NL63, HCoV-HKU1, HCoV-229E and HCoV-OC43 are usually asymptomatic or cause a mild to moderate upper respiratory tract disease. However, SARS-CoV, MERS-CoV and SARS-CoV-2 usually lead to severe respiratory syndrome in humans, fatal in some cases.

RT-PCR is the recommended method for detection of human coronavirus, and ELISA and lateral flow immunochromatographic kits can be also used for diagnosis. Fluorescence immunochromatography can be employed for detection of viral antigen and isothermal amplification methods for detection of viral nucleic acids.

To prevent COVID-19 caused by SARS-CoV-2, a variety of vaccines are available, such as inactivated virus vaccines, adenovirus vectored vaccines and mRNA vaccines.

The 6 coronaviruses that infect other animals in the *Coronaviridae* are shown below.

Alphacoronavirus (α-CoV)

Feline infectious peritonitis virus (FIPV) causes highly lethal feline infectious peritonitis (FIP) in domestic cats and some wild felines. This virus mainly replicates in macrophages and can multiply in feline mononuclear phagocytes. Suckling mice can be used for virus isolation. The antibodies against cat intestinal coronavirus react with the virus, but they do not play a protective role in FIPV infection. FIPV infection can be exudative and non-exudative (**Figure 32-4**). Intraperitoneal puncture can confirm the diagnosis of exudative FIP, while RT-PCR and immunohistochemistry can be used to diagnose non-exudative FIP. Virus neutralization test and ELISA can detect serum antibodies of infected cats for diagnosis.

PCR和免疫组化可诊断非渗出型 FIP。病毒中和试验和ELISA 可检测感染猫血清抗体而用于诊断。无有效治疗方法，有温度敏感突变株FIPV疫苗可用于健康易感猫的预防。

There are no effective treatments. The temperature-sensitive mutant FIPV has been used as vaccines for protection of healthy and susceptible cats.

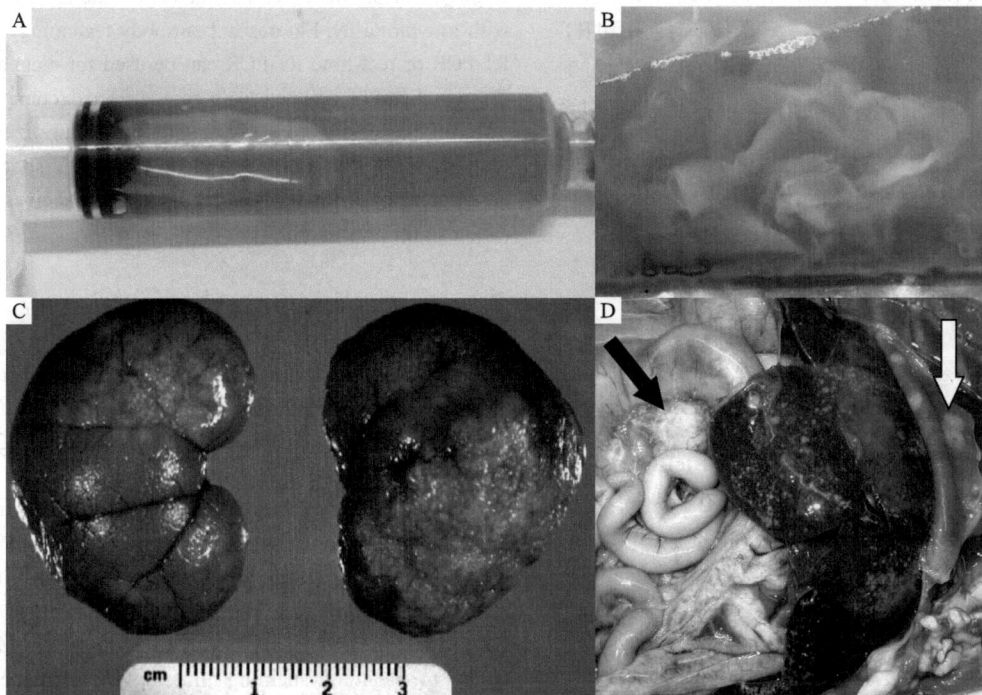

图32-4　FIPV感染猫的大体病变

A. 典型渗出型FIP猫的腹腔积液；B. 腹腔积液内可见大量纤维蛋白；C. 猫肾脏表面沿血管出现多灶性至愈合性肉芽肿性炎症（白色，粗糙外观）；D. 肠、肝、淋巴结、脾脏和膈肌。覆盖腹腔壁和腹膜的表面有白色至黄色软块（白色箭头），大肠淋巴结呈黄色、肿大（黑色箭头）。图像来源见英文。

彩图

Figure 32-4　Gross lesions in cat infected by FIPV

A. Peritoneal effusion from a cat with classic effusive form of FIP; B. Close view of abdominal effusion and large clumps of fibrin; C. Cat kidneys with multifocal to coalescing granulomatous inflammation (white, rough appearance) following the superficial blood vessels; D. Intestine, liver, lymph node, spleen, and diaphragm. White-to-yellow soft plaques covering the parietal and visceral peritoneal surfaces (white arrow). The lymph nodes associated with large intestine are enlarged and yellow (black arrow). Source: Drechsler Y, Alcaraz A, Bossong FJ, et al. 2011. Feline coronavirus in multicat environments. Vet Clin North Am Small Anim Pract, 41 (6): 1133-1169.

　　猫肠道冠状病毒（FECV）的致病性低，单纯的感染常局限于肠道，并引起无临床表现的肠炎，尤其是幼猫，随后猫迅速产生抵抗力。感染猫可持续带毒，是重要的病毒储存库。该病毒和猫传染性腹膜炎病毒在遗传上密切相关，后者可能来自相对无致病性的猫肠道冠状病毒的基因突

　　Feline enteric coronavirus (FECV) has low pathogenicity. Simple infections are often confined to the intestines and cause enteritis without clinical manifestations, especially in kittens. Then the exposed cats quickly develop resistance. Infected cats can continue to carry the virus and are an important virus reservoir. FECV and feline infectious peritonitis virus are genetically closely related, and the latter may come from relatively nonpathogenic feline intestinal coronavirus by

变株，可参考猫传染性腹膜炎病毒的检测方法进行诊断。

犬冠状病毒（CCoV）与包括 TGEV、FECV 和 FIPV 在内的其他冠状病毒抗原有相关性，是家养和野生犬种的一种高度传染性肠道病原，引起腹泻。犬感染后发病率高，但死亡率低。荧光抗体技术和 RT-PCR、实时 RT-PCR 等可用于诊断。通过犬纤维肉瘤细胞株 A-72、犬肾脏原代细胞系分离病毒。病毒中和试验和 ELISA 可用于检测特异的 CCoV 抗体。有灭活和弱毒活疫苗用于预防。

乙型冠状病毒属

牛冠状病毒（BCoV）与其他物种的冠状病毒抗原相关，只有一种血清型，但又分为 EBCoV（肠道型）和 RBCoV（呼吸道型）。BCoV 感染可引起牛和野生反刍动物三种不同的疾病综合征：新生 1~3 周龄牛腹泻、成年牛的冬痢、牛的呼吸道感染。BCoV 可凝集仓鼠、小鼠和大鼠的红细胞，多数毒株在乳鼠、牛肾细胞或 Vero 细胞等内增殖。通过 RT-PCR、实时 RT-PCR 和荧光抗体或免疫组化、病毒分离和中和实验进行诊断。母牛肠道型病毒的自然感染可经初乳保护犊牛，但对呼吸道型病毒尚无疫苗可用。

马冠状病毒（ECoV）与牛冠状病毒（BCoV）、人冠状病毒（HCoV-OC43）和猪冠状病毒（PHEV）的关系最密切。该病毒难以在细胞培养中分离和繁殖，仅可经 HRT-18 细胞分离病毒。ECoV 引起马空肠和结肠炎（图 32-5）。RT-PCR 和实时 RT-PCR 为首选诊断方法；ECoV 感染马的血清抗体与 BCoV 具有交叉反应，配对血清样本中 BCoV 的中和抗体滴度升高 4 倍以上可证实感染。

兔冠状病毒抗血清与 FIPV、CCoV 和 TGEV 有交叉反应。该病毒感染兔表现为

mutations. Diagnosis can be performed by referring to the detection method for feline infectious peritonitis virus.

Canine coronavirus (CCoV) is classified as an α-coronavirus and antigenically associated with other coronaviruses, including TGEV, FECV and FIPV. It is a highly infectious intestinal pathogen in domestic and wild canines, causing diarrhea. The incidence rate is high, but with low mortality. Fluorescent antibody technology and RT-PCR or real-time RT-PCR can be used for diagnosis. The virus can be isolated from canine fibrosarcoma cell line A-72 or primary kidney cells. Virus neutralization test and ELISA can be used to detect CCoV-specific antibodies. Inactivated and attenuated live vaccines are available for prevention.

Betacoronavirus (β-CoV)

Bovine coronavirus (BCoV) belongs to β-coronavirus and is antigenically related to other coronaviruses. The virus has only one serotype but is divided into EBCoV (intestinal type) and RBCoV (respiratory type). BCoV infection can cause three different disease syndromes in cattle and wild ruminants: diarrhea in newborn 1-3 weeks old cattle, winter diarrhea in adult cattle and respiratory tract infection in cattle. BCoV can agglutinate the red blood cells of hamsters, mice and rats. Most strains propagate in suckling mice, bovine kidney cells or Vero cells. RT-PCR, real-time RT-PCR, fluorescent antibody or immunohistochemistry, virus isolation and neutralization experiments are usually used for diagnosis. Natural infection of cows with EBCoV (intestinal type) can protect calves through colostrum, but no vaccines are available for RBCoV.

Equine Coronavirus (ECoV) is most closely related to bovine coronavirus (BCoV), human coronavirus (HCoV-OC43) and porcine coronavirus (PHEV). This virus is difficult to be isolated because of its failure to reproduce in cell culture, except HRT-18 cells. Equine coronavirus causes enteritis (colon and jejunum) (**Figure 32-5**). RT-PCR and real-time RT-PCR are the preferred diagnostic methods. The serum antibodies from ECoV-infected horses cross-react with BCoV. The infection can be confirmed by the increase, 4 times or more, of neutralizing antibody titers of BCoV in paired serum samples of the same infected horse.

The antiserum against **rabbit coronavirus** cross-reacts with FIPV, CCoV and TGEV. Rabbits infected

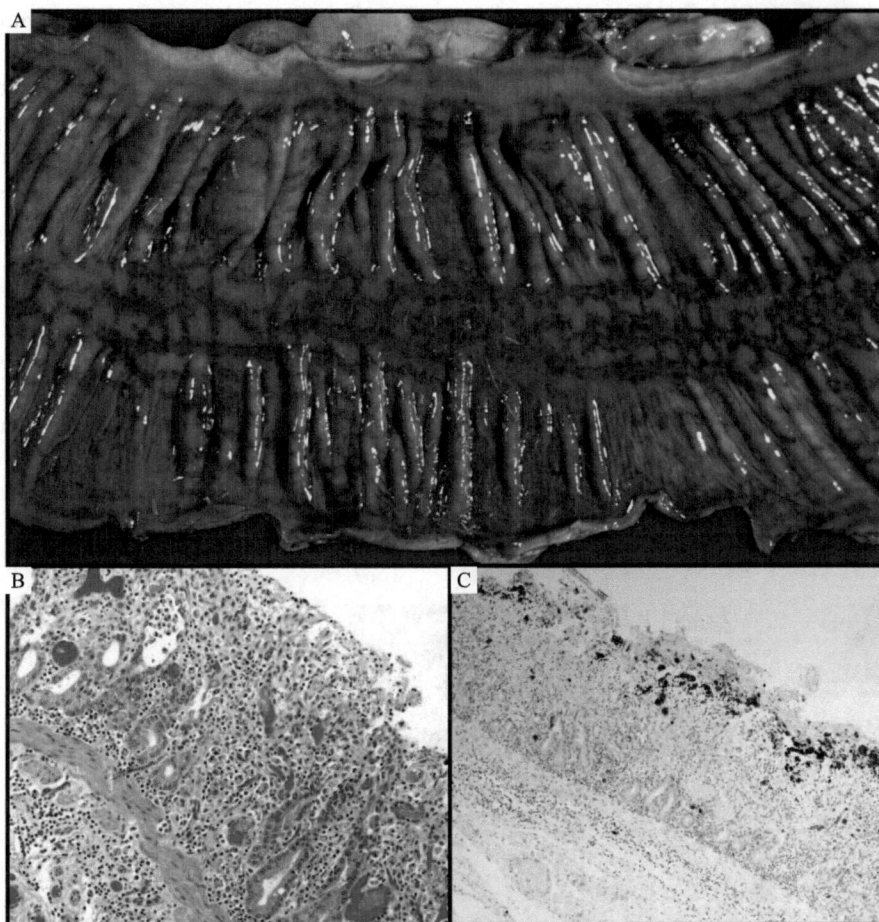

图 32-5 马冠状病毒相关肠炎

A. 中度坏死出血性马结肠炎；B. 混合炎性马空肠炎伴隐窝扩张和坏死（隐窝"脓肿"）以及微血管血栓；C. 马空肠坏死绒毛尖的弥漫性免疫反应。图像来源见英文。

Figure 32-5 Equine coronavirus-associated enteritis

A. Moderate necrohemorrhagic colitis, colon, horse; B. Mixed inflammatory enteritis with crypt ectasia and necrosis (crypt "abscesses") and microvascular thrombi, jejunum, horse; C. Diffuse immunoreactivity at the tips of necrotic villi, jejunum, horse. Source: Haake C, Cook S, Pusterla N, et al. 2020. Coronavirus infections in companion animals: virology, epidemiology, clinical and pathologic features. Viruses, 12 (9): 1023.

病毒性心肌炎和充血性心力衰竭，以及肠道疾病。但该病毒作为肠道疾病主要致病因素还不明确，其生物学特性需要进一步阐明。

with the virus exhibit viral myocarditis and congestive heart failure, and intestinal disease. However, this virus as the main pathogenic factor of intestinal disease is still unclear, and its biological characteristics need to be further clarified.

丙型冠状病毒属

禽传染性支气管炎病毒（AIBV）血

Gammacoronavirus (γ-CoV)

Avian infectious bronchitis viruses (AIBVs) have

清型众多，病毒通过消化道、呼吸道或接触传播。AIBV可感染所有年龄的鸡，但因其组织嗜性、血清亚型不同，引起的临床症状不一，包括呼吸道感染、肾炎、畸形卵和肠炎（**图32-6**）。感染禽的生产性能受到严重影响，发病率可达100%，但死亡率低。免疫组织化学、RT-PCR、实时RT-PCR等可用于检测，可结合序列分析进行分型。SPF鸡胚用于病毒分离，感染胚胎表现出特征性的侏儒症变化。中和试验和血凝抑制试验可进行血清型特异性检测。目前市场有多种减毒活疫苗和灭活疫苗可用。

many serotypes. The virus is transmitted through the digestive tract, respiratory tract or contact. AIBV can infect chickens at all ages. Diverseness of virus tissue tropism and its serotypes lead to different clinical symptoms in infected birds, causing respiratory infections, nephritis, malformed eggs and enteritis (**Figure 32-6**). The production performance of infected poultry is severely affected. The morbidity can reach 100%, but the mortality is low. Immunohistochemistry, RT-PCR, and real-time RT-PCR can be used for detection. Sequencing can be approached for subtyping. SPF chicken embryos are used for virus isolation. Infected embryos show characteristic dwarfism changes. Neutralization test and hemagglutination inhibition test can be performed for serotype-specific detection. A variety of live attenuated and inactivated vaccines are commercially available.

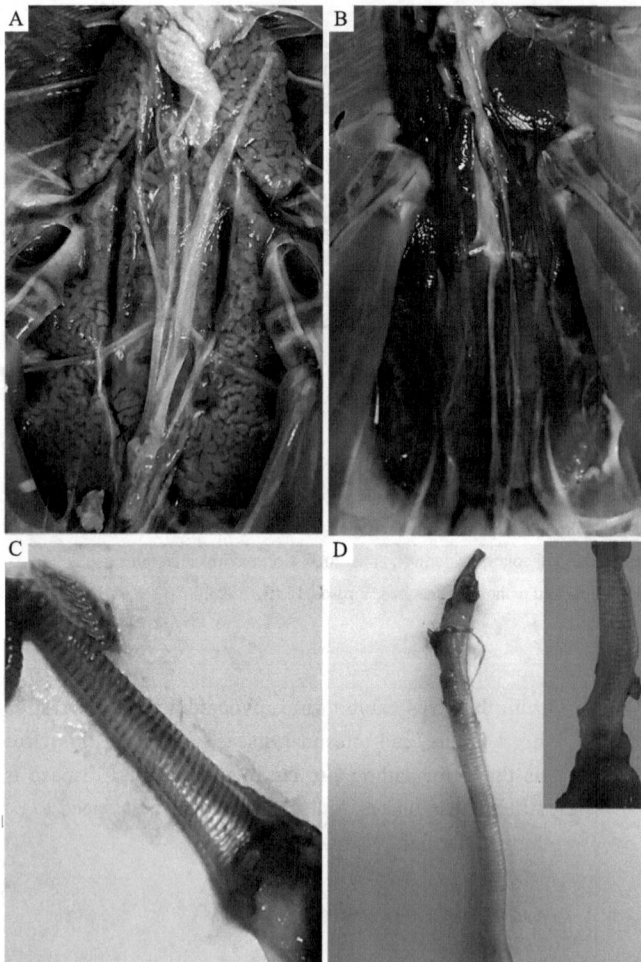

彩图

图32-6 禽传染性支气管炎病毒（AIBV）感染鸡的大体病变

A. 用AIBV CK/CH/JS/TAHY毒株感染鸡后，肾肿胀、苍白；B. 未感染病毒的鸡肾对照；C. 感染AIBV后，鸡气管出现充血和严重卡他性渗出物；D. 未感染病毒的鸡气管对照，内插图为气管黏膜表面。图像来源见英文。

Figure 32-6 Gross lesions of Avian Infectious bronchitis virus infected chickens

A. Swollen and pale kidney after challenging with strain AIBV CK/CH/JS/TAHY; B. Negative control kidney; C. Trachea illustrating hyperemia and serious catarrhal exudates after challenge with strain AIBV; D. Control trachea. Inset shows trachea's mucosal surface. Source: Ren G, Liu F, Huang M, et al. 2020. Pathogenicity of a QX-like avian infectious bronchitis virus isolated in China. Poult Sci, 99 (1): 111-118.

第2节 动脉炎病毒科
Section 2 *Arteriviridae*

动脉炎病毒科的特性是核衣壳立体对称，有囊膜紧贴并有蜂窝样的结构；基因组呈线状单链正链RNA，5′端有帽子结构，3′端有poly（A）。基因组编码核衣壳蛋白（N），主要囊膜蛋白（M和GP5）以及4种次要囊膜蛋白（GP2、GP3、GP4和E）。成员病毒在细胞质内复制，经出芽进入内质网，通过胞吐释放。成员病毒具遗传多样性，可引起从无到严重的临床症状或慢性持续感染，并导致严重经济损失。

猪繁殖与呼吸综合征病毒

病毒特性：猪繁殖与呼吸综合征病毒（PRRSV）有2个基因型：1型为欧洲型，2型为美洲型，猪和野猪易感。感染的动物分泌物和排泄物带毒，病毒可经垂直传播、水平传播等多种途径传播。原代猪肺泡巨噬细胞、非洲猴肾细胞（MARC-145）、永生化肺泡巨噬细胞系和ZMAC-1细胞用于病毒分离。PRRSV可被脂溶剂、普通消毒剂和清洁剂灭活。

致病性：PRRSV整个感染过程≥175 d。感染猪在若干周内抗体存在的同时出现病毒血症，已证实抗体可增强病毒感染。病毒感染单核细胞及巨噬细胞，造成免疫抑制。病毒可穿过胎盘感染仔猪，导致脐带出血性病变。感染猪表现为：厌食、发热、耳发绀（蓝耳病）、流涕；母猪子宫内膜炎，在怀孕110 d左右流产。

病毒受体：肺泡巨噬细胞上的CD163、唾液酸黏附素和硫酸乙酰肝素是PRRSV受体或黏附相关分子，在感染中发挥重要作用。

Arteritis viruses have the following characteristics: the nucleocapsid is stereo-symmetrical with a tightly attached envelope and honeycomb-like structure. Their genomes are linear positive single-stranded RNAs with a cap at the 5′ end and poly(A) at the 3′ end. The genome encodes nucleocapsid protein(N), major envelope proteins (M and GP5), and four minor envelope proteins(GP2, GP3, GP4 and E). These viruses replicate in the cytoplasm, enter the endoplasmic reticulum through budding and are released through exocytosis. The members are genetically diverse, causing from no to serious clinical symptoms or chronic persistent infection, leading to serious economic losses.

Porcine reproductive and respiratory syndrome virus

Virus characteristics: porcine reproductive and respiratory syndrome virus (PRRSV) has 2 genotypes: type 1 is a European type and type 2 is an American type. Pigs and boars are susceptible to the virus. The secretion and excreta of infected animals contain virus. PRRSV can be spread through multiple routes, such as vertical and horizontal transmission. Primary porcine alveolar macrophages, African monkey kidney cells (MARC-145), immortalized alveolar macrophage line, and ZMAC-1 cells are used for virus isolation. PRRSV can be inactivated by lipid solvents, common disinfectants and detergents.

Pathogenicity: the entire course of PRRSV infection is ≥ 175 d. Infected pigs develop viremia, with the antibodies being present for several weeks. It has been confirmed that antibodies against PRRSV can enhance viral infections. The virus infects monocytes and macrophages, leading to immune suppression. It can pass through the placenta and infect piglets, causing hemorrhagic changes in the umbilical cord. Infected pigs show anorexia, fever, cyanosis of the ears (blue ear disease), runny nose, as well as endometritis and abortion around 110 d post pregnancy.

Virus receptors: CD163, sialic acid adhesin and heparan sulfate on the surfaces of alveolar macrophages act as PRRSV receptors or attachment factors and play important roles in the infection.

实验室诊断：荧光抗体、免疫组织化学、RT-PCR用于检测病毒抗原或核酸，必要时配合ORF5基因序列测定确诊。病毒中和、ELISA和间接荧光技术均用于检测，有商品化的ELISA试剂盒可用。

防控：猪繁殖与呼吸综合征为WOAH通报疫病。无针对PRRSV的抗病毒治疗方法。商品化的PRRSV疫苗（包括活疫苗和灭活疫苗）有多种，灭活苗效果不佳，而弱毒苗可用于接种，但存在风险。

马动脉炎病毒

病毒特性：马动脉炎病毒（EAV）只有1个血清型，但有欧、美2个代表株。EAV感染主要为慢性持续感染。兔肾细胞（RK-13）最适合EAV分离，也可用Vero细胞和马肺细胞进行分离。该病毒可以被脂溶性溶剂、普通消毒剂和清洁剂灭活。

致病性：马、驴、斑马和骡容易感染。病毒最先在肺巨噬细胞中增殖，然后感染支气管淋巴结，通过血流传播全身，造成全身小动脉坏死、梗死，导致皮肤出现荨麻疹。该病毒通过接触和气雾传播，也能垂直传播；临床表现为高热41℃，白细胞减少，结膜炎，呼吸道和消化道黏膜卡他性炎症及四肢水肿。母马流产，公马暂时性不育、阴囊水肿。

实验室诊断：RT-PCR、免疫组织化学和原位杂交技术均可用于诊断，尤以RT-PCR为首选。WOAH的标准检测方法是通过RK-13细胞分离病毒。病毒中和试验是WOAH血清学诊断的标准，用于诊断或种马的血清学筛查。

防控：马动脉炎为WOAH通报疫病。无针对EAV的治疗方法，但有灭活苗或弱毒苗可用。

Laboratory diagnosis: fluorescent antibodies, immunohistochemistry, RT-PCR and sequencing are used to detect viral antigens or nucleic acids. These, if necessary, may be combined with ORF5 gene sequencing to confirm the diagnosis. Virus neutralization, ELISA and indirect fluorescence techniques are used for detection. Commercial ELISA kits are available.

Prevention and control: it is necessary to report PRRSV infection to WOAH. There are no antiviral treatments for PRRSV. There are many types of commercial PRRSV vaccines (including live and inactivated vaccines). However, the protective effect of inactivated vaccines is poor, while attenuated vaccines can be used, but may pose a risk of reversion.

Equine arteritis virus

Virus characteristics: equine arteritis virus (EAV) has only 1 serotype, but there are Europe and America representative strains. EAV infection is mainly chronic. Rabbit kidney cells (RK-13) are most suitable for EAV isolation. Vero cells and equine lung cells are alternatively used for isolation. EAV can be inactivated by lipid solvents, common disinfectants, detergents.

Pathogenicity: horses, donkeys, zebras and mules are susceptible to EAV. The virus first proliferates in lung macrophages, then infects bronchial lymph nodes, and spreads throughout the body through the bloodstream, causing necrosis and infarction in small arteries throughout the body and leading to urticaria on the skin. The virus is transmitted through contact and aerosol, and also vertically transmitted. Clinical manifestations are high fever of 41℃, leukopenia, conjunctivitis, catarrhal inflammation of the mucous membranes of the respiratory and digestive tracts and edema of the limbs. It also causes abortion in mare, and temporary infertility and scrotal edema in stallions.

Laboratory diagnosis: RT-PCR, immunohistochemistry and *in situ* hybridization techniques can be used for diagnosis. Especially, RT-PCR is commonly used. Virus isolation using RK-13 cells is the standard detection method by WOAH. Virus neutralization test is the standard of serological diagnosis and serological screening in stallions by WOAH.

Prevention and control: it is necessary to report EAV infection to WOAH. There are no treatments for EAV infection, but inactivated or attenuated vaccines are commercially available.

第3节 托巴套式病毒科
Section 3 *Tobaniviridae*

托巴套式病毒科的环曲病毒属有3个种：猪环曲病毒（PToV）、牛环曲病毒（BToV）和马环曲病毒（EToV）。其中，EToV是唯一在传代马肾细胞系分离出来的病毒，所有BToV毒株均具有致病性。该科病毒粒子具多种形式，包括圆形、肾形和圆环形颗粒，主要致病特点是破坏肠上皮细胞和肠绒毛的完整性，引起水样腹泻，幼龄动物尤甚，但多为急性自限性。

Torovirus in the family *Tobaniviridae* consists of 3 recognized species: *Porcine torovirus* (PToV), *Bovine torovirus* (BToV) and *Equine torovirus* (EToV). Among them, EToV is the only torovirus isolated from passaged equine kidney cell lines. All BToV strains are pathogenic. Virus particles of this family exhibit in different forms, including round, kidney-shaped and toroidal form. The main pathogenic characteristics of the members include destruction of the integrity of intestinal epithelial cells and intestinal villi, causing watery diarrhea, especially in young animals. The infections are mostly acute, but self-limiting.

思考题 Questions

1. 套式病毒目的成员有哪些？

2. 冠状病毒科中对猪具有重要致病作用的病毒有哪些？

3. 冠状病毒的形态结构与基因组的特征有哪些？

4. 猪繁殖与呼吸综合征病毒的致病机理、检测方法和防控措施有哪些？

5. 马动脉炎病毒的致病机理、检测方法和防控措施有哪些？

6. 动脉炎病毒的形态结构与基因组的特征有哪些？

7. 冠状病毒科中可导致人感染的病毒有哪些？

1. What are the members of *Nidovirales*?

2. What are the important pathogenic viruses in the family *Coronaviridae* that cause swine infections?

3. What are the morphological structure and genomic characteristics of coronaviruses?

4. What are the pathogenic mechanisms, diagnostic methods, and prevention and control measures of PRRSV?

5. What are the pathogenic mechanisms, detection methods, and prevention and control measures of EAV?

6. What are the morphological structure and genomic characteristics of arteriviruses?

7. What are the viruses causing human infection in the family *Coronaviridae*?

第33章 微RNA病毒目
Chapter 33 / *Picornavirales*

内容提要 微RNA病毒目包括微RNA病毒科、嵌杯病毒科和双顺反子病毒科。这些病毒具有共同的特征：包括单链正链RNA基因组，无囊膜，二十面体对称等。

Introduction *Picornavirales* includes *Picornaviridae*, *Caliciviridae* and *Dicistroviridae*. These viruses share common characteristics, including positive single-stranded RNA genomes, absence of envelope and icosahedral symmetry, *etc*.

第1节 微RNA病毒科
Section 1 *Picornaviridae*

微RNA病毒科包括63个属，其中大多数属的共同特征是其增殖快，环境稳定性显著，易感宿主间传播迅速，单链正链RNA，病毒在胞质内成熟，呈晶格排列，无囊膜。这些病毒感染可导致动物的高发病率和严重经济损失。

口蹄疫病毒

病毒特征： 口蹄疫病毒（FMDV）是口蹄疫病毒属的代表，有7个血清型：A、O、C、亚洲1和南非1、2和3型（SAT1、SAT2和SAT3）。各型之间无交叉保护反应。病毒蛋白有很好的免疫原性。该病毒抵抗力较强，耐低温。病毒对脂溶性有机溶剂不敏感，2%乙酸、2%氢氧化钠等可有效杀灭病毒。该病毒在乳仓鼠肾传代细胞（BHK-21细胞）繁殖良好；雏鸡适应毒或细胞适应毒能适应于鸡胚并失去对牛的致病力。可引起偶蹄类动物包括牛、水牛、猪、绵羊、山羊、鹿等动物的口蹄

Picornaviridae includes 63 genera. Most of the genera share common characteristics, including rapid replication, remarkable environmental stability and rapid transmission among susceptible hosts, as well as positive single-stranded RNA, maturation in the cytoplasm, lattice arrangement, and absence of envelope. Infection by these viruses can lead to high morbidity in animals and subsequently cause severe economic losses.

Foot and mouth disease virus

Viral characteristics: foot and mouth disease virus (FMDV) is a member of *Aphthovirus*, with 7 serotypes: A, O, C, Asia 1, and South Africa 1, 2 and 3 (SAT1, SAT2 and SAT3). There is no cross-protection among the different types. Viral proteins have good immunogenicity. This virus is resistant to the environment, low temperature, and lipid solvents, but can be effectively inactivated by 2% acetic acid or 2% sodium hydroxide. The virus replicates well in neonatal hamster kidney passage cells (BHK-21 cells). Chickling-or cell-adapted strains can grow in chicken embryo and lose their pathogenicity to cattle. It can cause foot and mouth disease (FMD) in cloven-hoofed animals, including cattle, buffalo, pigs, sheep, goats, deer and others. FMDV seriously

疫，严重危害动物和公共卫生。

致病性：口蹄疫病毒具有高度传染性，偶蹄兽最容易感染。成年动物感染后发病率高但死亡率低，而幼龄动物死亡率高，妊娠母牛感染后可流产。FMDV可通过消化道、呼吸道或接触传播，也可经气溶胶远距离传播。牛和羊主要经吸入含病毒的气溶胶而感染，而猪经消化道、呼吸道或接触而感染。病毒主要在动物的鼻咽上皮细胞复制（并可长期存在），然后导致病毒血症。感染动物发热、跛行（**图33-1**），舌头、脚、鼻和乳头出现特征性的水疱病变。在急性期之后，半数以上的感染动物转为无症状携带者，如牛可持续带毒长达3.5～5年，羊可带毒达半年。

affects animal and public health.

Pathogenicity: FMDV is highly contagious, and cloven-hoofed animals are the most susceptible hosts. Adult animals with the infection have high morbidity but low mortality, while infected young animals have high mortality. Pregnant cows infected with FMDV can result in abortion. FMDV can spread through the digestive tract, respiratory tract and contact, as well as over long distances by aerosols. Cattle and sheep are infected mainly by inhalation of aerosols containing virus particles, while pigs are infected through the digestive tract, respiratory tract or direct contact. The virus mainly replicates in the nasopharyngeal epithelial cells (and can exist for a long time), and then causes viremia. Clinical signs of the infected animals include fever, lameness (**Figure 33-1**) and characteristic blister lesions on the tongue, feet, nose and teats. After the acute phase, more than a half of the infected animals become asymptomatic

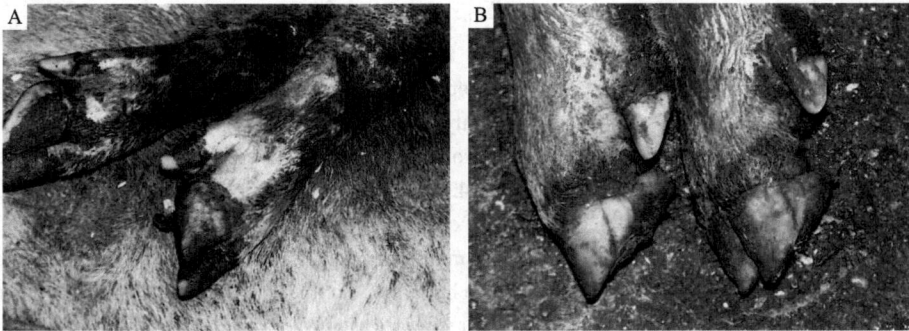

图33-1　猪口蹄疫
A. 蹄壳脱落，脚趾暴露；B. 蹄跟垫部严重病变。图像来源见英文。
Figure 33-1　FMD in pigs from the field
A. The horn is shed from a digit and the exposed surface beneath is observed; B. Severe lesions of the heel pad area. Source: Alexandersen S, Zhang Z, Donaldson AI, et al. 2003. The pathogenesis and diagnosis of foot-and-mouth disease. J Comp Pathol, 129 (1): 1-36.

实验室诊断：实时RT-PCR最为常用，可快速确诊口蹄疫病毒。通过BHK-21细胞或使用4～7日龄乳鼠分离病毒。用恢复期血清做中和试验、琼脂扩散试验、荧光抗体试验、夹心ELISA鉴定亚型，其中ELISA方法最为常用。临床上，口蹄疫应与水疱性口炎和猪水疱病鉴别。

防控：口蹄疫是WOAH通报疫病。无有效治疗药物，弱毒活疫苗因存在毒力返祖和影响净化问题已被淘汰。牧场可选择

carriers. Cattle can carry the virus for 3.5-5 years, and sheep can carry the virus for about half a year.

Laboratory diagnosis: real-time RT-PCR can be used for rapid diagnosis. The virus can be isolated in BHK-21 cells or by inoculating 4-7 days old suckling mice. Convalescent serum is used for neutralization test, agar diffusion test, fluorescent antibody test and sandwich ELISA to identify subtypes. ELISA is most commonly used. Clinically, FMD should be differentially diagnosed from vesicular stomatitis and swine vesicular disease.

Prevention and control: foot and mouth disease is WOAH notifiable disease. No effective therapeutic drugs are available. The live attenuated vaccines have been disused due to the problems of virulence reversion and interference

去除所有非结构病毒蛋白的灭活疫苗、合成肽疫苗免疫动物，既可区分野毒感染，也有利于口蹄疫净化。

猪水疱病病毒

病毒特性：该病毒属于肠道病毒属，被认为是人柯萨奇病毒B5血清型的猪变异株，只有1个血清型，但可分为4个抗原群，与口蹄疫病毒、水疱性口炎病毒无交叉反应。该病毒耐乙醚和氯仿，在pH 3.0时仍稳定，甲醛和高锰酸钾密闭熏蒸消毒24 h可有效杀灭病毒。病毒在仔猪原代肾细胞、IBRS-2、PK15等细胞上增殖并产生明显病变。猪是其唯一的自然宿主。

致病性：仅猪（包括野猪）感染，病毒主要经粪口途径、直接接触传播或者经皮肤的伤口传播。感染猪发热（41~42 ℃），在口腔黏膜、鼻镜、乳房、蹄冠部发生水疱，症状与口蹄疫类似（**图33-2**）。幼龄猪高发病率，但死亡率低。自然感染康复猪有坚强的抵抗力，可偶尔感染人。

实验室诊断：使用猪肾传代细胞或者2~3日龄的乳鼠或乳仓鼠分离病毒。如果初代分离阴性，可盲传2~3代。夹心ELISA可检测病毒抗原，RT-PCR可快速定性或定量检测病毒核酸，后者最为常

with eradication. The inactivated vaccines that remove all non-structural viral proteins or synthetic peptide vaccines can be used to immunize animals, which can be helpful to distinguish wild virus infections from vaccination, a beneficial feature for FMD eradication.

Swine vesicular disease virus

Virus characteristics: the virus belongs to the genus *Enterovirus*. It is considered to be a porcine variant of human Coxsackie virus B5 serotype. There is only 1 serotype, but it can be divided into 4 antigenic groups. It has no cross reactivity with foot and mouth disease virus and vesicular stomatitis virus. The virus is resistant to ether and chloroform and remains stable at pH 3.0. Closed fumigation with formaldehyde and potassium permanganate for 24 h can effectively kill the virus. The virus proliferates in primary piglet kidney cells, IBRS-2, PK15 and other cells and produces cytopathic effect. The pig is the only natural host.

Pathogenicity: only pigs (including wild boars) are susceptible to swine vesicular disease virus. The virus is mainly transmitted via the fecal-oral route, direct contact or wounds. Clinical signs of infected pigs include fever (41-42 °C), blisters on the oral mucosa, nose, breast, and hoof crown. The symptoms are similar to those of the foot and mouth disease (**Figure 33-2**). Young infected pigs have high morbidity but low mortality. Naturally infected and recovered pigs have strong resistance to reinfection. The virus can occasionally infect humans.

Laboratory diagnosis: porcine kidney passage cells or 2-3 days old suckling mice or suckling hamsters are used to isolate the virus. If the initial isolation is negative,

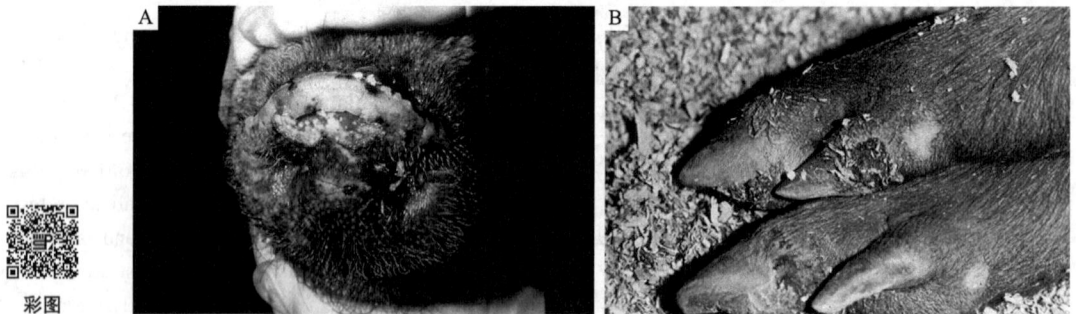

彩图

图33-2 猪水疱病的临床症状

A. 猪鼻背上有深溃疡；B. 猪蹄冠处有多处病变/溃疡。图像来源：艾奥瓦州立大学兽医学院。

Figure 33-2 Clinical signs of swine vesicular disease

A. There is a deep ulcer on the dorsum of the pig snout; B. There are multiple large erosions/ulcers of the coronary bands. Source: College of Veterinary Medicine, Iowa State University.

用。病毒中和试验也用于诊断。临床上应与口蹄疫病毒和最近出现的猪塞内卡病毒鉴别。

防控：该病是WOAH通报疫病，无有效治疗方法，但有商品化的乳鼠弱毒疫苗、组织培养弱毒疫苗、水疱皮和疱液灭活疫苗可用。在应急情况下，使用高免血清也可有效保护猪群。

禽脑脊髓炎病毒

病毒特性：禽脑脊髓炎病毒（AEV）是震颤病毒属的唯一成员，引起鸡流行性震颤。自然野毒株或胚适应株均可在敏感的雏鸡、鸡胚和鸡胚脑细胞、成纤维细胞、肾细胞等上生长。流行毒在细胞培养一般无细胞病变。抵抗力强，对甲醛熏蒸敏感。禽脑脊髓炎病毒主要感染鸡，并仅对1～21日龄鸡致病。

致病性：仅有一个血清型，但有两种不同的致病型：嗜肠型和鸡胚适应型。临床分离株多为嗜肠型，鸡胚适应型株具有高度嗜神经性。病毒通过粪口途径或垂直传播。潜伏期1～7 d，但1～2周龄雏鸡感染后就出现症状：病鸡眼睛无神，呆滞，腿无力或麻痹，跗关节着地或胫部行走；然后呈现病毒性脑炎病变，头颈震颤，受刺激后更为明显，存活鸡失明。成年鸡仅一过性产蛋下降。

实验室诊断：免疫荧光抗体技术和RT-PCR可用于诊断，尤以后者常用。接种易感的5～7日龄鸡胚卵黄囊可分离病毒，也可以使用易感鸡胚的脑组织细胞、肾细胞等分离病毒。临床上应与新城疫相鉴别。

防控：无有效治疗方法。育成鸡在8周到开产前4周点眼或者饮水接种一次活疫苗可以有效预防禽脑脊髓炎，也可使用灭活疫苗预防。

blind passage can be carried out for 2-3 generations. Sandwich ELISA can be used to detect viral antigens, and RT-PCR can quickly detect viral nucleic acid qualitatively or quantitatively. The latter is most commonly used. Virus neutralization test can also be used for diagnosis. Clinically, it should be differentially diagnosed from the foot and mouth disease virus, as well as the recent porcine Seneca virus.

Prevention and control: swine vesicular disease is WOAH notifiable disease. There are no effective treatments for this disease, but attenuated vaccines derived from suckling mouse or tissue culture, and inactivated vaccines made of blister skin or blister fluid are available. In case of emergency, using the high-titer immune sera has been reported to protect the pig herd effectively.

Avian encephalomyelitis virus

Virus characteristics: avian encephalomyelitis virus (AEV) is the only member of the genus *Tremovirus* and causes epidemic tremor in chickens. Natural wild or embryo-adapted virus strains can grow in susceptible chicklings, chicken embryos, and brain cells, fibroblasts, kidney cells, *etc.* of chicken embryo. The epidemic virus generally has no cytopathic effect in cell culture. It has strong resistance to environmental conditions, but is sensitive to formaldehyde fumigation. Avian encephalomyelitis virus mainly infects chickens and only causes clinical signs to chickens aged at 1-21 days.

Pathogenicity: there is only one serotype, but there are two different pathogenic types: enterophilic and chicken embryo adapted types. Field isolates are mostly enterophilic, and embryo adapted strains are highly neurophilic. The virus spreads via the fecal-oral route or vertically. The latent period of its infection is 1-7 days, while 1-2 weeks old chicks show symptoms after being infected with AEV. Clinical signs of sick chickens include dull eyes, sluggish legs, weakness or paralysis, tarsal joints on the ground or walking on the tibias. Then they show signs of viral encephalitis, such as head and neck tremor, which is more obvious upon stimulation. The survived chickens become blind. Adult chickens infected with this virus only show decreased egg production temporarily.

Laboratory diagnosis: immunofluorescence antibody technology and RT-PCR can be used for diagnosis, especially the latter that is more commonly used. The virus can be isolated by inoculating the yolk sac of susceptible 5-7 days old chicken embryos or using brain tissue and kidney cells of susceptible chicken embryos. Clinically, it should be distinguished from Newcastle disease.

Prevention and control: there are no effective treatments for this disease. Immunization with a live vaccine for the growing chickens from 8 weeks old to 4 weeks before the start of laying eggs can effectively prevent the disease. Inactivated vaccines can also be used for prevention.

鸭甲型肝炎病毒

病毒分类和特性：引起鸭肝炎的病毒有3种，分别是鸭甲型肝炎病毒（鸭肝炎病毒1型，DHAV）、鸭肝炎病毒2型和鸭肝炎病毒3型，其中1型为微RNA病毒科禽肝炎病毒属成员；2型和3型为星状病毒。

鸭甲型肝炎病毒无血凝性，可致死鸡胚和鹅胚并在其中增殖。鸡胚适应毒可在鸭胚成纤维细胞生长并有细胞病变。耐乙醚和氯仿，具有热稳定性。鸭甲型肝炎病毒是雏鸭高死亡率的重要原因之一。

致病性：自然病例仅见于雏鸭，28日龄以下雏鸭多发。DHAV对1周龄内雏鸭具有高致死率，几乎达100%。病死鸭见急性肝炎变化，肝肿大坏死、点状或者瘀斑样出血。肝脏炎性细胞渗出、胆管上皮细胞增生和脑炎病变，肾脏和脾脏肿大、充血；成鸭感染无症状（**图33-3**）。

实验室诊断：荧光抗体染色法和RT-PCR用于诊断。通过病料接种8～10日龄鸡胚尿囊腔，5～7 d后胚胎死亡，胚液发绿，胚体肝脏红黄色并有坏死灶可确诊。

防控：目前没有治疗方法。DHAV感染是WOAH通报疫病。使用弱毒疫苗免疫种鸭可使后代雏鸭获得良好免疫力。雏鸭发病后，被动抗体紧急接种有效。

Duck hepatitis A virus

Virus classification and characteristics: there are 3 kinds of viruses causing duck hepatitis, namely duck hepatitis A virus (DHAV) (duck hepatitis virus type 1), duck hepatitis virus type 2 and duck hepatitis virus type 3, of which type 1 is a member of the genus *Avihepatovirus* of the family *Picornaviridae*. Type 2 and 3 are astroviruses.

DHAV has no hemagglutination activity and can proliferate in and be lethal to chicken and goose embryos. Chicken embryo adapted virus strains can grow in duck embryo fibroblasts and have cytopathic effect. The virus is resistant to ether and chloroform and has thermal stability. DHAV is one of the important diseases that cause high mortality in ducklings.

Pathogenicity: natural infection is only found in ducklings. Ducklings under 28 days old are prone to be infected. DHAV infection is highly fatal for ducklings within 1 week old, with almost 100% mortality. Pathological changes associated with acute hepatitis include hepatomegaly and necrosis, punctate or ecchymosis-like bleeding, as well as hepatic inflammatory cell exudation, abnormal proliferation of bile duct epithelial cells, and encephalitis. Swelling and congestion of the kidney and spleen are also noted. Adult ducks infected by DHAV are usually asymptomatic (**Figure 33-3**).

Laboratory diagnosis: fluorescent antibody staining and RT-PCR are used for diagnosis. The diagnosis can also be made by inoculating the properly-treated clinical materials into the allantoic cavity of 8-10 days old chicken embryos. After 5-7 d, if the inoculated embryos die and there are greenish embryo fluid and red-yellow colored liver with necrotic foci, the diagnosis would be positive for duck hepatitis A virus infection.

Prevention and control: there are currently no treatments for this disease. DHAV infection is WOAH-listed disease. Immunization of breeding ducks with attenuated vaccines can provide good immunity to their offspring. Emergency vaccination with passive antibodies is also effective when ducklings are ill at the early stage.

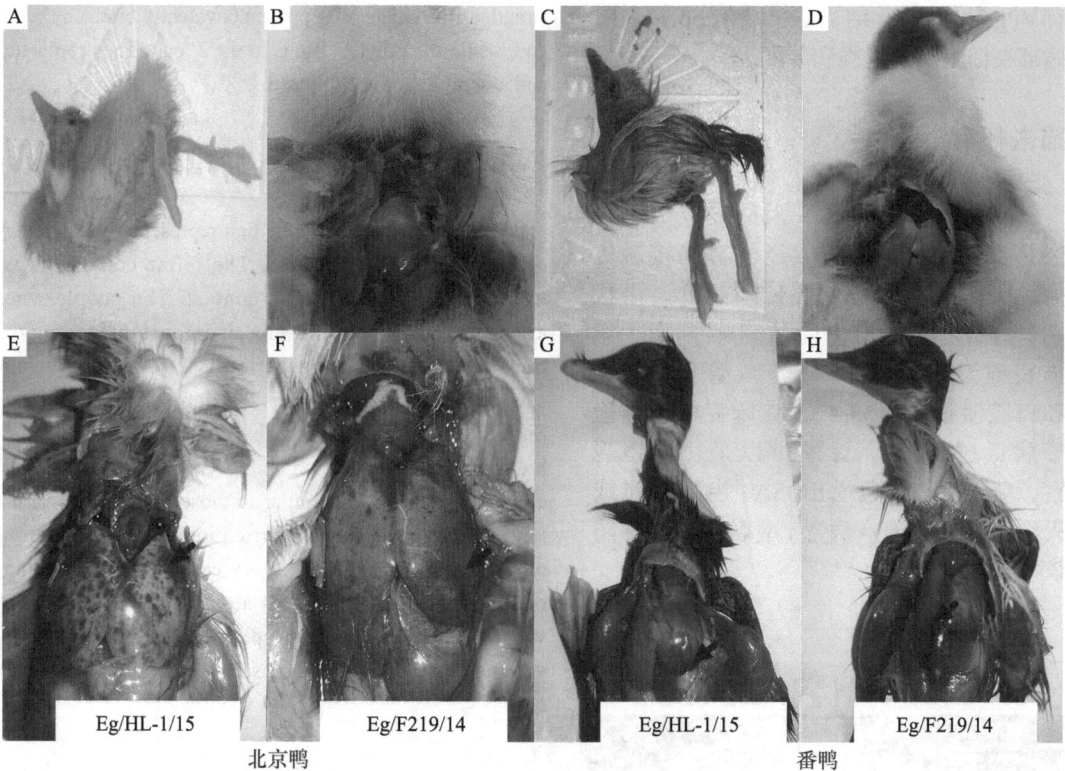

图33-3 实验感染鸭甲型肝炎病毒MK510860（Eg/HL-1/15）和MK510859（Eg/F219/14）的1日龄北京鸭和番鸭的临床症状和大体病变

A、C. 类似阿片过敏和痉挛性划水症状；B、D. 未感染病毒的北京鸭和番鸭对照组肝脏外观正常；E、F. 感染病毒的北京鸭肝脏表面有出血点；G、H. 感染病毒的番鸭肝脏严重充血。图像来源见英文。

Figure 33-3   Clinical signs and gross pathological lesions of 1 day old Pekin and Muscovy ducklings experimentally infected with duck hepatitis A virus MK510860 (Eg/HL-1/15), and MK510859 (Eg/F219/14)

A, C. Signs of opisthotonos and spasmodic kicking; B, D. Non-infected Pekin and Muscovy controls showed normal liver appearance; E, F. Hemorrhagic spots on the liver surface of Pekin ducklings; G, H. Muscovy livers showed severe congestion. Source: Hisham I, Ellakany HF, Selim AA, et al. 2020. Comparative pathogenicity of duck hepatitis A virus-1 isolates in experimentally infected Pekin and Muscovy ducklings. Front Vet Sci, 7: 234.

第2节 嵌杯病毒科
Section 2   *Caliciviridae*

　　嵌杯病毒科成员可以感染兔、猫、猪、牛、羊、人等。嵌杯病毒颗粒无囊膜，核衣壳二十面体对称。病毒衣壳由90个相同的拱形二聚体蛋白亚单位组成，从而在表面形成32个杯状凹陷，形成"嵌杯"样结构。基因组为线状单链正链RNA，其5′端

　　The members of the family *Caliciviridae* can infect rabbits, cats, pigs, cattle, sheep, humans, *etc*. Their virus particles don't have envelope and their nucleocapsids are of icosahedral symmetry. The virus capsids consist of 90 identical arched dimer protein subunits, which form 32 cup-like depressions on the surface, resembling an "inlaid-cup" in structure. Their genomes are positive single-

有VPg蛋白共价结合，3'端有poly（A）。病毒在细胞质内复制和装配。

猫嵌杯病毒

嵌杯病毒科的第一个代表成员是**猫嵌杯病毒**。猫嵌杯病毒经气雾和接触传播，买卖猫的人是重要传播媒介。潜伏期2~6 d，康复猫仍能从口咽排毒。病猫表现为结膜炎、鼻炎、气管炎、肺炎和口腔溃疡（**图33-4**），临床症状与猫疱疹病毒1型类似。诊断可以采用电镜观察，荧光抗体染色或ELISA。接种弱毒或灭活苗，如猫疱疹病毒1型联苗可以预防该病。

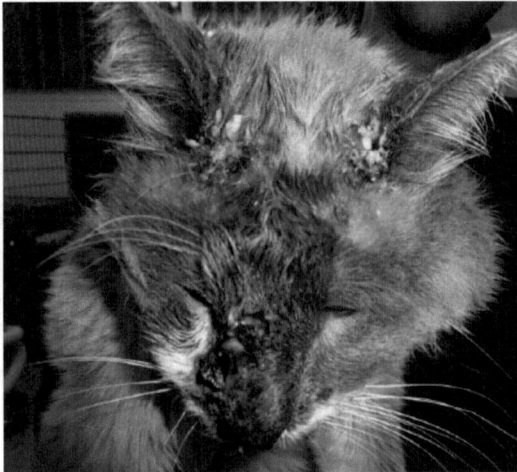

stranded RNA. The VPg protein covalently binds to the 5' end, while the poly(A) binds to the 3' end. They replicate and assemble in the cytoplasm.

Feline calicivirus

Feline calicivirus is the first representative member of the family *Caliciviridae*. The feline calicivirus is transmitted by aerosol and contact. The people who are engaged in cat sale are an important vector of transmission. The latent period is 2-6 d. Recovering cats can still discharge virus through their oropharynx. Sick cats are characterized by conjunctivitis, rhinitis, bronchitis, pneumonia, and oral ulcers (**Figure 33-4**). The clinical symptoms are similar to those of cat herpesvirus type 1. Diagnosis can be carried out by using electron microscopic observation, fluorescent antibody staining or ELISA. Vaccination with attenuated or inactivated vaccines, such as feline herpesvirus type 1 combined vaccine, can prevent this disease.

彩图

图33-4 感染猫嵌杯病毒的猫脸上的溃疡
图像来源见英文。

Figure 33-4 Ulcers on the face of a cat infected by feline calicivirus
Source: Deschamps JY, Topie E, Roux F. 2015. Nosocomial feline calicivirus-associated virulent systemic disease in a veterinary emergency and critical care unit in France. JFMS Open Rep, 1 (2): 2055116915621581.

兔出血症病毒

嵌杯病毒科的第二个代表成员是**兔出血症病毒**。兔出血症俗称兔瘟，病毒经粪口传播，2月龄以上兔易感。最急性者6~24 h内死亡，出现口鼻流血和神经症状，剖检可见血凝块充满全身血管，引发肝坏死和出血性败血症（**图33-5**）。病毒能凝集人红细胞，蚊蝇及肉食性鸟为传播

Rabbit hemorrhagic disease virus

Rabbit hemorrhagic disease virus is the second representative member of the family *Caliciviridae*. Rabbit hemorrhagic disease is commonly known as rabbit plague. The virus is transmitted by the fecal-oral route. 2 months old rabbits are susceptible. The most acutely sick rabbits die within 6-24 h, with nose and mouth bleeding and neurological symptoms. Blood clots filling the blood vessels of the whole body are visible after autopsy. It can cause liver

媒介。诊断可用人红细胞血凝试验，电镜观察和ELISA。兔出血症为WOAH通报疫病，接种灭活苗可以预防。

necrosis and hemorrhagic septicemia (**Figure 33-5**). The virus agglutinates human red blood cells. Mosquitoes, flies and carnivorous birds are vectors of transmission. Diagnosis can be carried out by using human erythrocyte hemagglutination test, electron microscopic observation, and ELISA. Rabbit hemorrhagic disease is a notifiable disease by WOAH and preventable by vaccination.

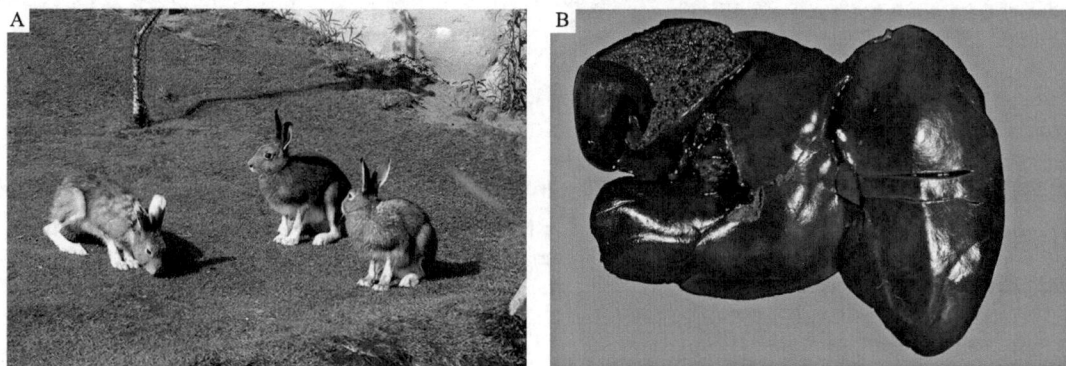

图33-5　感染兔出血症病毒的圈养野兔
A．动物园里的野兔；B．感染兔出血症病毒野兔的肝脏易碎，中度至重度充血。图像来源见英文。

Figure 33-5　Rabbit hemorrhagic disease virus infection in captive mountain hares (*Lepus timidus*)
A. Mountain hares in a zoo facility; B. Macroscopic picture of an infected liver. The liver of rabbit hemorrhagic disease virus infected mountain hares presented with friable consistency and moderate to severe congestion. Source: Buehler M, Jesse ST, Kueck H, et al. 2020. Lagovirus europeus GI.2 (rabbit hemorrhagic disease virus 2) infection in captive mountain hares (*Lepus timidus*) in Germany. BMC Vet Res, 16 (1): 166.

彩图

思考题　Questions

1. 微RNA病毒目的成员有哪些？

1. What are the members of the family *Picornavirales*?

2. 微RNA病毒科中具有重要致病作用的病毒有哪些？

2. What are the important pathogenic viruses in the family *Picornaviridae*?

3. 口蹄疫病毒的形态结构与基因组的特征有哪些？

3. What are the morphological structure and genomic characteristics of foot and mouth disease virus?

4. 口蹄疫病毒的致病机理、检测方法和防控措施有哪些？

4. What are the pathogenic mechanisms, diagnostic methods, and prevention and control measures of foot and mouth disease virus?

5. 猪水疱病病毒的致病性、检测方法和防控措施有哪些？

5. What are the pathogenicity, diagnostic methods, and prevention and control measures of swine vesicular disease virus?

6. 鸭甲型肝炎病毒的形态结构与基因组的特征有哪些？

6. What are the morphological structure and genomic characteristics of duck hepatitis A virus?

第34章 其他正链RNA病毒

Chapter 34 / Other positive-stranded RNA viruses

内容提要 其他正链RNA病毒成员包括戊肝病毒科、披膜病毒科、黄病毒科、星状病毒科、野田村病毒科等。

Introduction Members of other positive-stranded RNA viruses include *Hepeviridae*, *Togaviridae*, *Flaviviridae*, *Astroviridae*, *Nodaviridae*, etc.

第1节 戊肝病毒科
Section 1 *Hepeviridae*

戊型肝炎病毒是戊肝病毒科唯一的成员。该病毒在我国猪群抗体阳性率50%以上，戊肝病毒目前有5个基因型，其中基因1型和2型感染人，基因3型和4型感染人和动物，基因5型感染鸡。猪源毒株为基因3型和4型。因此，人食用了含有戊型肝炎病毒的猪肉、猪肉制品、瓜果蔬菜、地表水，以及鱼虾蟹等有感染的风险（**图34-1**）。

Hepatitis E virus (HEV) is the only member of the family *Hepeviridae*. The antibody positive rate of the virus in pig population in China is more than 50%. There are 5 genotypes in HEV. Genotype 1 and 2 infect humans, genotype 3 and 4 infect both humans and animals, and genotype 5 infects chicken. Pig-origin strains are genotype 3 and 4. Therefore, HEV contaminated pork products, fruits and vegetables, surface water, fish, shrimp and crab are risk factors (**Figure 34-1**).

第2节 披膜病毒科
Section 2 *Togaviridae*

马脑炎病毒为披膜病毒科成员，包括东部马脑炎病毒和西部马脑炎病毒。病毒颗粒为球形，有囊膜及细小的膜粒。核衣壳二十面体对称，内含线状正链单链RNA。马脑炎病毒为虫媒病毒，可引起关节炎、皮疹、脑炎和发热，鼻腔流出血

Equine encephalitis virus is the member of the family *Togaviridae* and can be divided into eastern equine encephalitis virus (EEEV) and western equine encephalitis virus (WEEV). The virus particles are spherical. The virion has an envelope and tiny spikes. The nucleocapsid is in icosahedral symmetry. It has linear positive single-stranded RNA. It is an arbovirus and causes arthritis,

沫。东部和西部马脑炎病毒为地方流行性人兽共患病毒（**图34-2**）。病毒的分离和鉴定必须在生物安全3级及以上实验室进行。病毒在细胞质复制，从细胞膜出芽成熟。

rash, encephalitis, fever, and foamy bleeding from nasal cavity. The EEEV and WEEV are enzootic (**Figure 34-2**). Isolation and identification of the viruses must be carried out in the laboratories of BSL-3 or BSL-4. The virus replicates in the cytoplasm and buds from the plasma membrane.

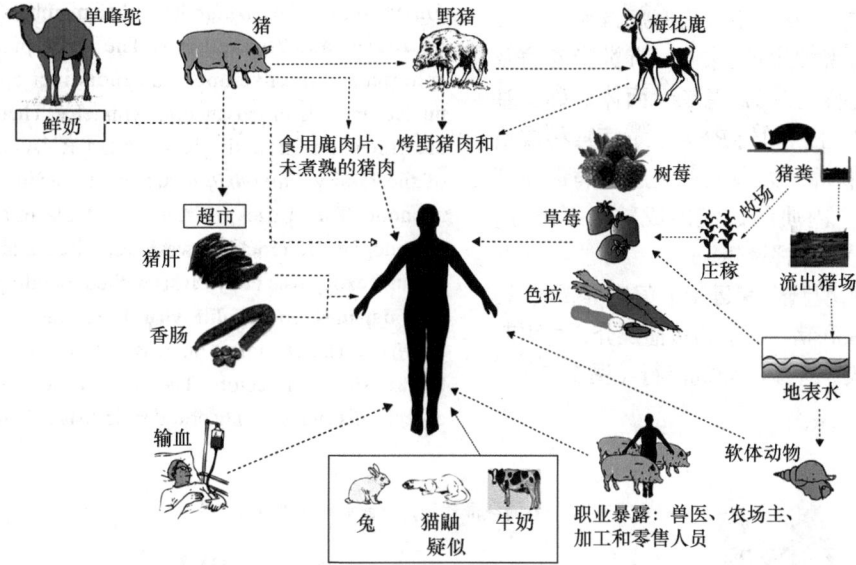

图34-1 戊型肝炎病毒的传播
图像来源见英文。

Figure 34-1 Zoonotic transmission of hepatitis E virus
Source: Khuroo MS, Khuroo MS, Khuroo NS. 2016. Transmission of hepatitis E virus in developing countries. Viruses, 8 (9): 253.

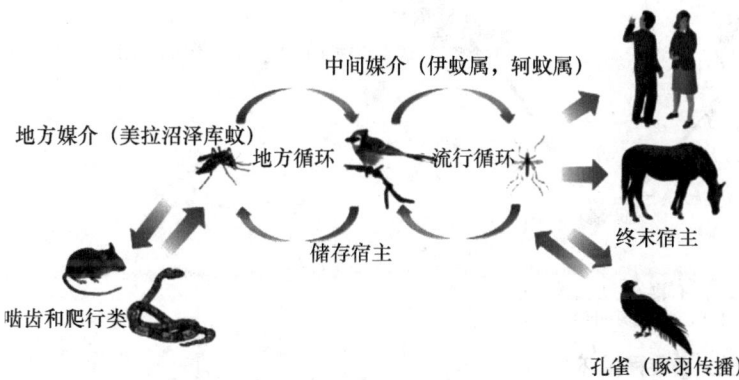

图34-2 东部马脑炎病毒的地方性人兽传播循环
雀形目鸟类和美拉沼泽库蚊为东部马脑炎病毒的储存库和放大宿主，维持病毒在不同动物间的传播循环。啮齿/有袋动物可能是南美洲东部马脑炎病毒的主要载体和宿主。雀形目鸟类因产生极高水平的病毒血症，故可感染人兽共患媒介和各种中间载体。人和马是终末宿主，因不能产生足够的病毒血症，故不能进行病毒传播。图像来源见英文。

Figure 34-2 Enzootic and epizootic/epidemic transmission cycles of eastern equine encephalitis virus (EEEV)
The enzootic EEEV transmission cycle is maintained between passerine birds as reservoir/amplification hosts and *Culiseta melanura*, as the main enzootic vector in swamp habitats. Rodents/marsupials may serve as principal enzootic vectors and reservoirs in South America. Passerine birds develop extremely high levels of viremia, enough to infect both enzootic vectors as well as a variety of bridge vectors. Humans and equids are dead-end hosts since they do not develop sufficient viremia to transmit the virus. Source: Go YY, Balasuriya UB, Lee CK. et al. 2014. Zoonotic encephalitides caused by arboviruses: transmission and epidemiology of alphaviruses and flaviviruses. Clin Exp Vaccine Res, 3 (1): 58-77.

第3节 黄 病 毒 科
Section 3 *Flaviviridae*

黄病毒属归属黄病毒科，该科的成员除了黄病毒之外，还包括瘟病毒属、丙肝病毒属等。病毒颗粒球形，有类脂囊膜及不明显的膜粒。核衣壳二十面体对称，基因组为线状正链单链RNA。黄病毒科成员多为虫媒病毒，大多人兽共患。病毒在细胞质复制，内质网贮泡内成熟，不出芽，通过胞吐或细胞裂解释放。

日本乙型脑炎病毒属于黄病毒属成员，可以感染人和猪。库蚊为传播媒介，可引起流产和脑炎，主要在亚洲流行（图34-3）。

The genus *Flavivirus* belongs to the family *Flaviviridae*. *Flaviviridae* has other members including *Pestivirus* and *Hepacivirus*. The virus particles are spherical with envelopes and indistinct spikes. The nucleocapsid is in icosahedral symmetry. Their genomes are linear positive single-stranded RNAs. Members of the family *Flaviviridae* are mostly arboviruses and zoonotic. The viruses replicate in cytoplasm and mature in endoplasmic reticulum vesicles. Viruses are released through exocytosis or cytolysis without budding.

Japanese encephalitis virus is a member of the genus *Flavivirus* and can infect humans and pigs. *Culex* spp. are the transmission vectors. The virus causes abortion and encephalitis and is mainly prevalent in Asia (**Figure 34-3**).

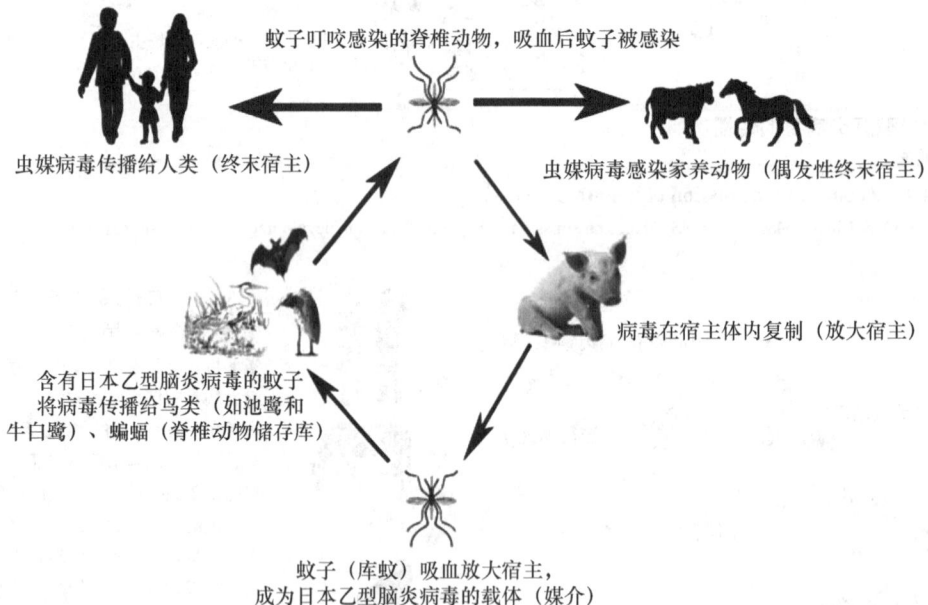

蚊子叮咬感染的脊椎动物，吸血后蚊子被感染

虫媒病毒传播给人类（终末宿主）

虫媒病毒感染家养动物（偶发性终末宿主）

病毒在宿主体内复制（放大宿主）

含有日本乙型脑炎病毒的蚊子将病毒传播给鸟类（如池鹭和牛白鹭）、蝙蝠（脊椎动物储存库）

蚊子（库蚊）吸血放大宿主，成为日本乙型脑炎病毒的载体（媒介）

图34-3 日本乙型脑炎病毒的地方性人兽传播循环
该病毒通过库蚊传播至脊椎动物宿主。猪为放大宿主，是传播周期中的关键环节。蝙蝠和鹭科水鸟是病毒的储存库。因病毒不能在感染的人群之间传播，故人为终末宿主。图像来源见英文。

Figure 34-3 Enzootic transmission cycle of Japanese encephalitis virus
The virus is transmitted to vertebrate hosts by mosquitoes belonging to the *Culex* spp. Pigs serve as amplification hosts and form a critical link in the transmission cycle. Ardeid water birds and bats serve as virus reservoirs. Humans are dead end hosts as the virus cannot be transmitted from an infected person to another. Source: Dutta K, Nazmi A, Basu A. 2011. Chemotherapy in Japanese encephalitis: are we there yet? Infect Disord Drug Targets, 11 (3): 300-314.

西尼罗病毒属于黄病毒属成员，感染人和马，库蚊和鸟为传播媒介，引起发热，主要在地中海地区、亚洲、非洲和美洲流行（图34-4）。

West Nile virus is a member of the genus *Flavivirus* and can infect humans and horses, causing fever. *Culex* spp. and birds are the transmission vectors. The disease is mainly prevalent in Mediterranean, Asia, Africa and America (**Figure 34-4**).

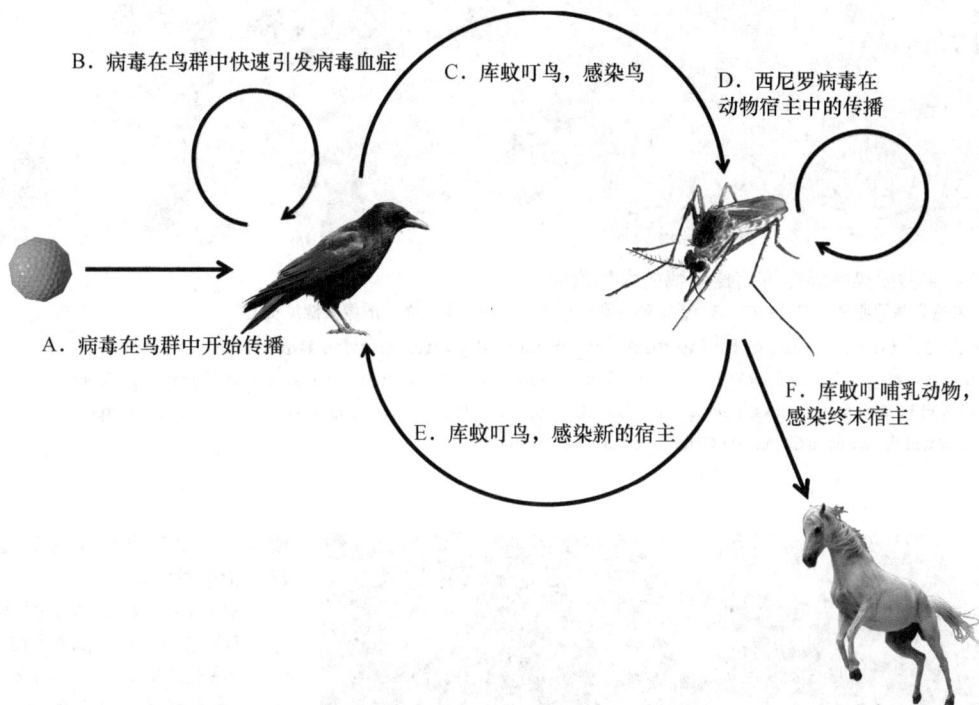

B. 病毒在鸟群中快速引发病毒血症
C. 库蚊叮鸟，感染鸟
D. 西尼罗病毒在动物宿主中的传播
A. 病毒在鸟群中开始传播
E. 库蚊叮鸟，感染新的宿主
F. 库蚊叮哺乳动物，感染终末宿主

图34-4 西尼罗病毒在动物宿主中的传播
A、B. 鸟类感染西尼罗病毒后病毒浓度增加；C、D. 鸟类将病毒传播至蚊子；E. 蚊子将病毒传播给鸟类，导致地方性传播；F. 蚊子或将病毒传递给人和马，即终末宿主。图像来源见英文。

Figure 34-4 Transmission cycle of West Nile virus through its animal hosts
A, B. Birds become infected with West Nile and viral titer increases; C, D. Birds transmit the virus to mosquitoes; E. Mosquitoes transmit the virus to birds, causing enzootic transmission; F. Mosquitoes bridge the virus to humans and horses, the common dead-end hosts. Source: Ahlers LRH, Goodman AG. 2018. The immune responses of the animal hosts of West Nile virus: a comparison of insects, birds, and mammals. Front Cell Infect Microbiol, 8: 96.

鸭坦布苏病毒属于黄病毒属成员，感染鸭和鹅，引起卵巢炎，主要在我国流行（图34-5）。

牛病毒性腹泻病毒（BVDV）是瘟病毒属的代表成员。该病毒有BVDV1和BVDV2两型，均可引起牛腹泻和黏膜病。其中BVDV1与猪瘟病毒有抗原交叉，可感染猪；而BVDV2与猪瘟病毒无抗原交叉。各种年龄的牛都易感，从口部直至肠道出现糜烂性或溃疡性病灶（图34-6）。

Duck Tembusu virus is a member of the genus *Flavivirus* and can infect ducks and geese, causing ovarian inflammation and being mainly prevalent in China (**Figure 34-5**).

Bovine viral diarrhea virus (BVDV) is the representative member of the genus *Pestivirus*. This virus has two types, BVDV1 and BVDV2, causing bovine diarrhea and mucosal disease. BVDV1 has cross-antigenicity with classical swine fever virus and infects pigs. BVDV2 doesn't have cross-antigenicity with classical swine fever virus. Cattle at all ages are susceptible. From the mouth to intestinal tract, there are

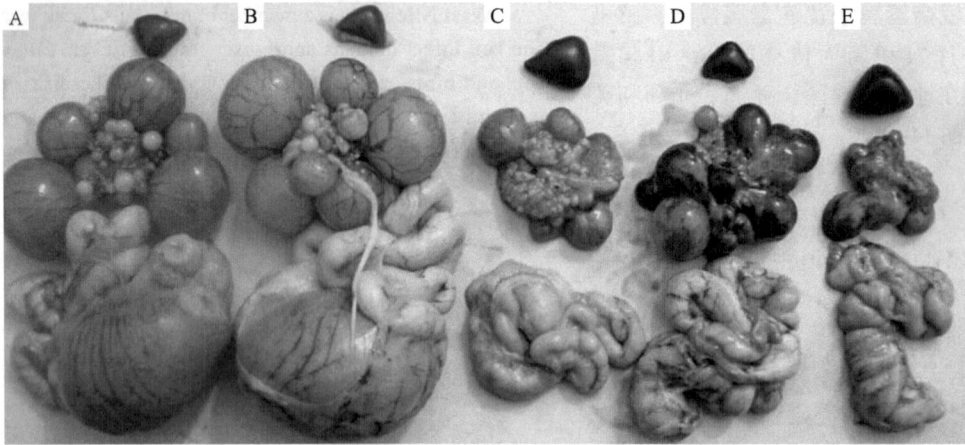

图34-5 实验感染鸭坦布苏病毒蛋鸭的大体病变

A、B. 未感染鸭的卵巢和脾脏；C~E. 感染鸭的卵巢出现退化和严重出血。图像来源见英文。

Figure 34-5 Gross lesions of laying ducks experimentally infected with Duck Tembusu virus

A, B. The ovaries and spleens of ducks from the non-infected group; C-E. Degeneration and severe hemorrhage in ovaries were observed in all infected ducks. Source: Liu P, Lu H, Li S, et al. 2013. Duck egg drop syndrome virus: an emerging Tembusu-related flavivirus in China. Sci China Life Sci, 56 (8): 701-710.

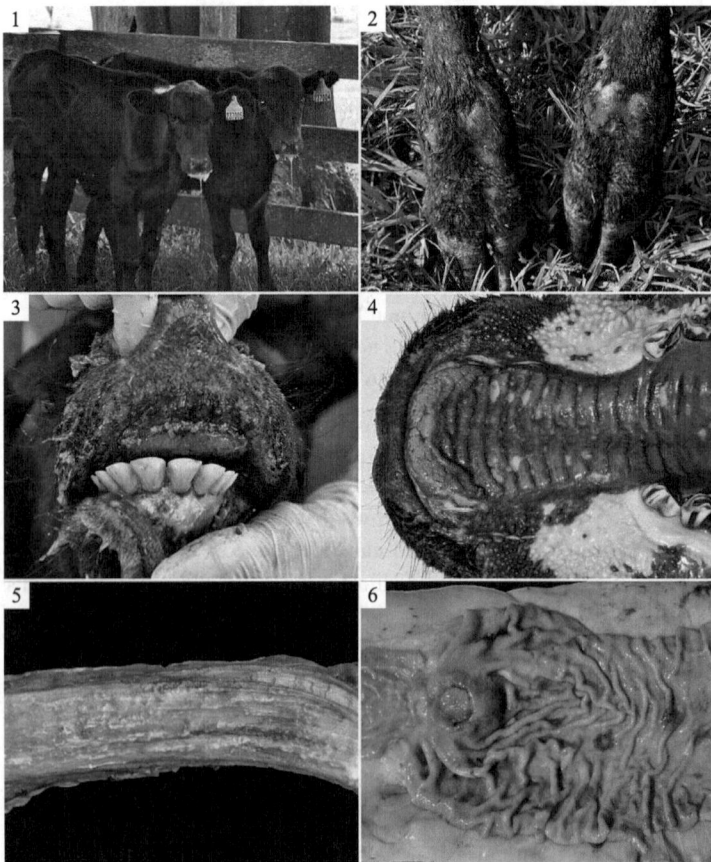

图34-6 牛病毒性腹泻病毒引起的犊牛黏膜病

1. 流涎和消瘦；2. 远端肢体；局部有广泛的皮肤溃疡；3. 口腔黏膜。多灶性溃疡至黏膜皮肤交界处；4. 硬腭。多灶性溃疡；5. 食管。纤维蛋白和坏死物覆盖的多灶性线状溃疡；6. 皱胃（幽门）黏膜。多灶性溃疡。图像来源见英文。

Figure 34-6 Mucosal disease caused by bovine viral diarrhea virus (BVDV) in calves

1. Severe sialorrhea and emaciation; 2. Distal limbs. There are focally extensive areas of cutaneous ulceration; 3. Oral mucosa. Multifocal to coalescing areas of ulceration extending to the mucocutaneous junction; 4. Hard palate. Multifocal areas of ulceration; 5. Esophagus. Multifocal severe linear ulcers covered by fibrin and necrotic material; 6. Abomasal (pyloric) mucosa. Multifocal ulcers. Source: Bianchi MV, Konradt G, de Souza SO, et al. 2017. Natural outbreak of BVDV-1d-induced mucosal disease lacking intestinal lesions. Vet Pathol, 54(2): 242-248.

病毒可通过胎盘感染胎儿，怀孕早期胎牛死亡，幸存犊牛终身感染，产生免疫耐受，不产生抗体但排毒。诊断可采用细胞培养或RT-PCR检测方法。该病为WOAH通报疫病，灭活苗效果不佳，接种弱毒苗可能诱发严重的黏膜病。

　　经典猪瘟病毒（CSFV）为瘟病毒属代表成员，引起经典猪瘟，或称猪霍乱。目前，美国、加拿大、澳大利亚已消灭此病。有基因1群和2群流行，我国猪群以2群感染为主。扁桃体是该病毒的主要定植组织，其他感染组织器官包括淋巴结、脾脏、肾脏等，病毒可在血管内皮细胞复制，导致皮肤出血性斑点、组织器官出血和脾梗死（**图34-7**）。病毒可在猪肾细胞

erosive or ulcerative lesions (**Figure 34-6**). The virus can infect fetuses through the placenta. The fetuses die at early stages of the pregnancy. The survivors have lifelong infection, becoming immune tolerance without antibodies, but they shed the viruses. Diagnosis can be performed by cell culture or RT-PCR method. This is WOAH notifiable disease. Inactivated vaccine is not effective. Severe mucosal disease can occur after administration of attenuated vaccines.

Classical swine fever virus (CSFV) is the representative member of the genus *Pestivirus*, causing classical swine fever (CSF), alternatively called hog cholera. At present, US, Canada and Australia have eliminated this disease. This virus has two genetic groups, group 1 and group 2. CSF in pigs in China is dominated by infections of group 2 virus strains. The virus has high preference to colonization in tonsils. Other affected

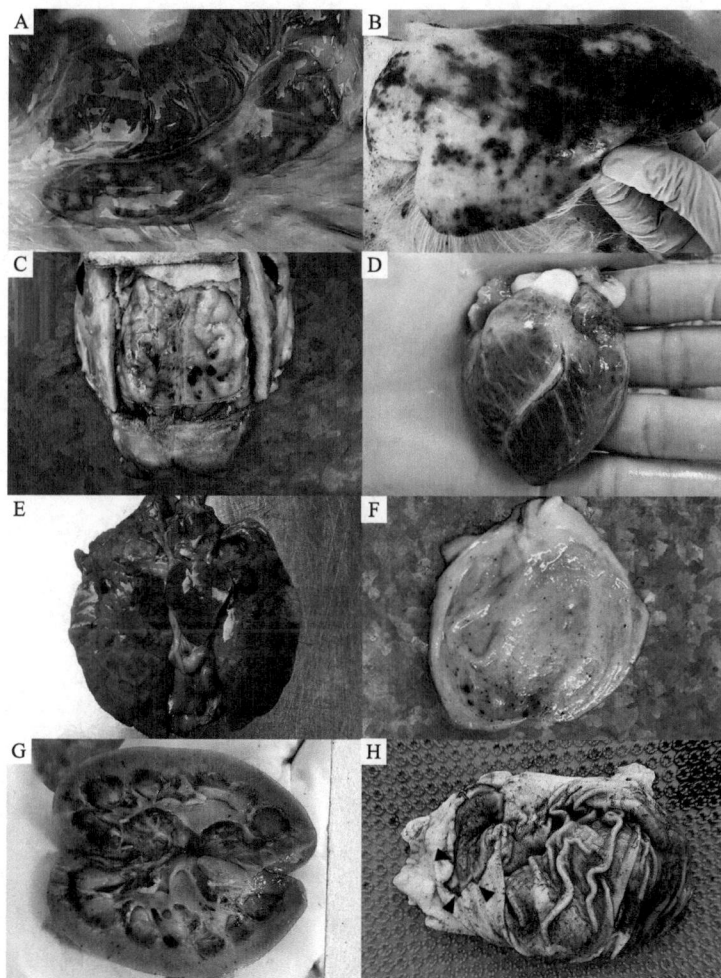

彩图

图34-7　经典猪瘟病毒感染猪的大体病变
A. 肠系膜淋巴窦出血；B. 耳垂红斑和充血；C. 脑膜出血；D. 心包出血；E. 胸膜淤血；F. 膀胱黏膜瘀点出血；G. 肾皮质和髓质瘀点出血；H. 胃溃疡（三角箭头）。图像来源见英文。

Figure 34-7　Gross lesions of classical swine fever virus infection in pigs
A. Sinus hemorrhages in mesenteric lymph node; B. Erythema and hyperaemia in ear lobe; C. Meningeal hemorrhages; D. Pericardial hemorrhages; E. Ecchymotic pleural hemorrhages; F. Petechial hemorrhages on urinary bladder mucosa; G. Petechial hemorrhages in renal cortex and medulla; H. Gastric ulcer (arrowhead) Source: Izzati UZ, Hoa NT, Lan NT, et al. 2021. Pathology of the outbreak of subgenotype 2.5 classical swine fever virus in northern Vietnam. Vet Med Sci, 7 (1): 164-174.

系（PK-15）或猪睾丸细胞系（ST）复制。诊断可用荧光抗体、免疫组化、RT-PCR检测等。该病为WOAH通报疫病，可接种C株猪瘟兔化弱毒疫苗预防。

organs include lymph nodes, spleen, kidneys, *etc*. The virus can replicate in the endothelial cells of the blood vessels, causing hemorrhagic spots or even ecchymosis of the skin, hemorrhage of the affected organs, and splenic infarction (**Figure 34-7**). The virus can replicate well in porcine kidney cell line (PK-15) or swine testis cell line (ST). Diagnosis can be made by immunofluorescent technique, immunohistochemistry and RT-PCR. This is WOAH notifiable disease. A lapinized classical swine fever virus C strain has long been used as an attenuated vaccine for effective prevention.

思考题 Questions

1. 简述经典猪瘟病毒的传播特点和检测手段。

2. 简述黄病毒科的分类和特点。

1. Please briefly describe the transmission characteristics and detection methods of classical swine fever virus.

2. Please briefly describe the classification and characteristics of the family *Flaviviridae*.